X-RAY DIAGNOSIS

X-RAY DIAGNOSIS

PETER ARMSTRONG
MB BS FRCR
Associate Professor of Radiology
University of Virginia
Charlottesville, Virginia, USA
formerly Consultant Radiologist
King's College Hospital, London

MARTIN L. WASTIE
MB BChir MRCP FRCR
Consultant Radiologist
University Hospital, Nottingham

with contributions on ultrasound by
ANTHONY J. BUSCHI MD
and
A. NORMAN A. G. BRENBRIDGE MD
University of Virginia Hospital
Charlottesville, Virginia, USA

Blackwell Scientific Publications
OXFORD LONDON EDINBURGH
BOSTON MELBOURNE

First published 1981

Printed in Great Britain at
The Alden Press, Oxford
Bound at Kemp Hall Bindery, Oxford

DISTRIBUTORS

USA
    Blackwell Mosby Book Distributors
    11830 Westline Industrial Drive
    St Louis, Missouri 63141

Canada
    Blackwell Mosby Book Distributors
    120 Melford Drive, Scarborough,
    Ontario M1B 2X4

Australia
    Blackwell Scientific Book
    Distributors
    214 Berkeley Street, Carlton
    Victoria 3053

British Library
Cataloguing in Publication Data

Armstrong, Peter
    X-ray diagnosis.
    1. Diagnosis, Radioscopic   2. X-rays
    I. Title   II. Wastie, Martin L.
    616.07'572   RC78

    ISBN 0–632–00173–9

Contents

Preface

The medical student often has great difficulty in interpreting x-ray films. Frequently the problem is deciding whether or not the examination is normal. When definite abnormalities are found the problems of diagnosis and differential diagnosis are at first overwhelming in their apparent complexity. A textbook can only partly help with these problems. What is needed is constant exposure to x-ray diagnosis in its clinical setting. In this book we have attempted to provide a framework on which to build this experience, using x-ray signs as our basic building blocks. The aim is to provide the answer in two common situations; 'Here is an x-ray film: what is wrong?', and 'I suspect this patient has such and such a disease: is there any evidence to support or refute this diagnosis?' In order to do this the chapters are, in general, divided into two sections. The first looks at the normal and discusses the abnormal signs which may be seen in disease; giving the differential diagnosis of these signs. The second section describes the x-ray findings in specific disorders.

We have not tried to be comprehensive in our coverage of techniques. Highly specialised examinations are described very briefly and several are omitted altogether. Ultrasound, nuclear medicine and computerised tomography frequently complement or even replace conventional radiology, but we have not dealt with these subjects in any depth, since this book is primarily designed as an introduction to the interpretation of conventional x-ray examinations.

One aspect of the presentation merits further discussion in order to avoid any misunderstanding, namely the very important consideration of how to integrate clinical information into x-ray film interpretation. There are many facets to this problem. The clinician inevitably approaches the films of his patient in a different way to the radiologist. The clinician has already seen the patient and decided to request the x-ray examination. He will necessarily view the film with predetermined questions. This is both an advantage and a disadvantage. The advantage is that he looks for, and therefore may find, subtle evidence of a disease he already suspects to be present. He also excludes many diseases from his initial list of possibilities. The disadvantage is that he may miss the unexpected and also that he may read into a film signs that are not truly present, because he is convinced on clinical grounds that such signs should be visible. The radiologist is, in this respect, in a different position. He can view each film twice; once before he knows the clinical details, in order to make an objective assessment, and once after he is given the clinical findings, at which time he can integrate his observations with the clinical information and can then check on any subtle findings that he may have missed on the initial examination of the film.

Though we have necessarily emphasised the objective approach in this book, we sincerely hope that a highly integrated approach will be followed by the reader in his day-to-day practice.

It is unfortunately beyond the scope of a small book, such as this one, to do full justice to two important subjects; the correlation of x-ray signs with pathology, and the role of radiology in clinical management, since this would involve including large sections on surgery, medicine, pathology, etc. Indeed, though we were often tempted to include sections on such subjects, we resisted. Consequently, this book cannot be read in isolation; it must be accompanied by the study of these other subjects.

ACKNOWLEDGEMENTS

It would not have been possible to write this book without the help of the many radiologists who have given ideas, valuable comments and inspiration. We would like to thank Dr J. W. Laws, Director of Radiology, King's College Hospital, London, for encouraging us in our teaching of radiology to medical students and for giving us so many ideas, many of which will be found in this book. We would also like to give particular thanks to the staff of the University and General Hospitals, Nottingham, particularly Dr E. J. Roebuck and Dr B. J. Preston for all their help and encouragement. Dr H. M. Saxton of Guy's Hospital, London, and Robert Auffenberg Jr, who at the time was a medical student at Washington University, St Louis, provided helpful detailed criticism of the manuscript.

Dr O. Olofsson and Dr O. Nylen of Falun, Sweden; Dr P. M. Dee and Dr T. E. Keats of the University of Virginia; Dr P. Davies and Dr D. H. Rose of Nottingham; Dr A. H. Chalmers of Bath; Dr K. C. Simpkins of Leeds, Dr I. H. Kerr of London; Professor T. Sherwood of Cambridge; Dr J. de Winter of Brighton; Dr J. Singh of Kuala Lumpur, Malaysia and Dr J. P. Pallan of Derby have all been most helpful in providing illustrations for the book.

Our particular thanks go to Mr M. G. Creasey of Nottingham for preparing photographic prints from the radiographs. This book would have been totally impossible without the endless typing and retyping undertaken by Miss Pauline Gamble of Nottingham and Mrs Patricia West and Mrs Shirley Yowell of the University of Virginia.

Finally, we would like to express our gratitude to Mr Peter Saugman of Blackwell Scientific Publications.

Peter Armstrong
Martin L. Wastie

1
Introduction

THE USE OF THE X-RAY DEPARTMENT

X-ray departments need to be well run and efficiently utilised in order to minimise radiation hazard and be cost-effective. Organising the department is in the hands of radiologists and radiographers, but the use to which it is put is largely up to the referring clinicians. Good communication between clinician and radiologist is vital. The x-ray staff need to know and understand the clinical problem in order to carry out the appropriate tests and to interpret the results in a meaningful way. The clinicians need to understand the strengths and limitations of the answers provided. Another important facet is sensible selection of investigations. There are two basic philosophies which take opposite paths to achieve a diagnosis. One approach is to request a battery of investigations, aimed vaguely in the direction of the patient's symptoms, hoping something will turn up. The other approach is 'trial and error'; decide one or two likely diagnoses and carry out the appropriate test to support or refute these possibilities. Each course has its proponents; we favour the selective approach since there is little doubt that the answers are usually obtained less expensively and with less distress to the patient. This approach depends on critical clinical evaluation; the more experienced the doctor, the more accurate he or she becomes.

Laying down precise guidelines for requesting radiological examinations is difficult because of a lack of uniformity in the management of patients and the variability in the information required. Certain generalisations do, however, seem possible.

1. An examination should only be requested when there is a reasonable chance that it will affect the management of the patient. There should be a question attached to every request, e.g. for a chest x-ray—what is the cause of this patient's haemoptysis? Apart from chest x-rays and questionably mammography, routine x-ray examinations in asymptomatic patients are to be avoided.

2. The time interval between follow-up examinations should be sensible, e.g. once pneumonia has been diagnosed, x-rays to assess progress can safely be left 7–10 days, unless clinical features suggest a complication. Some lesions do not require follow-up by x-ray at all, e.g. deformity of the duodenal cap due to benign ulceration does not change with treatment so there is no point in repeating the barium study just to see how 'the duodenal cap is getting on'.

3. Try to be specific about the localisation of problems. X-raying the clavicle, shoulder, humerus, elbow and forearm is ridiculous for a patient whose symptoms are clinically those of an abnormality in or immediately adjacent to the shoulder. Often a patient's symptoms can be localised to a portion of the gastrointestinal tract, the blunderbuss approach of indiscriminate barium meal, barium follow-through and barium enema in such patients is to be deplored. This approach is sometimes condoned by the argument that it shortens hospital stay. By all means, construct a reasonable programme of investigations but ask the radiologist to terminate the tests when the desired positive result is obtained.

4. Consider carefully which diagnostic imaging procedure will give the relevant information most easily, e.g. isotope bone scans should be the initial screening method in a search for asymptomatic bone metastases rather than skeletal survey. If positive, selected areas can be x-rayed for confirmation of the diagnosis where necessary. Isotope bone scans are, in general more sensitive than skeletal surveys for the detection of metastatic neoplasm.

X-RAYS

Production of x-rays (Fig. 1.1)

X-rays are produced when high-speed electrons rapidly decelerate. This is achieved by passing a very high voltage across two terminals placed in an evacuated tube.

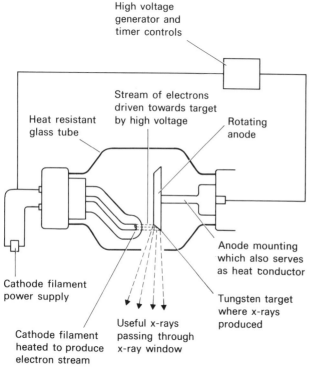

Fig. 1.1   Diagram of an x-ray tube.

One of the terminals, the cathode, is a tungsten alloy filament which is heated by a separate current. The other terminal, the anode, is a tungsten alloy target set in a disc of copper. The tungsten filament in the glass tube is heated to incandescence and it liberates free electrons. When a high voltage, usually in the range of 50 to 150 kV, is applied across the two terminals, the electrons are attracted towards the cathode at high speed. They hit the tungsten target on the anode and x-rays are produced.

Only a small portion of the x-ray beam escapes through an opening in the metal casing surrounding the tube, the remainder being absorbed by the casing.

Absorption of x-rays

X-rays are part of the electromagnetic spectrum of radiation. At the energies chosen for diagnostic radiology, some are absorbed by the tissues. An image of the resulting shadows can be recorded on film, or viewed on a fluorescent screen. There are four basic densities—gas, fat, other soft tissues and calcified structures. X-rays that pass through air are least absorbed and therefore cause the most blackening of the radiograph, whereas calcium absorbs the most and so the bones and other calcified structures appear virtually white. The soft tissues (except fat), e.g. the solid viscera, muscle, blood, bowel wall, etc., all have the same absorptive capacity and appear the same shade of gray on the radiograph. Fat absorbs slightly less x-rays and, therefore, appears a little blacker than the other soft tissues. The visibility of structures and disease depends on this differential absorption.

Photographic effect

X-rays cause blackening of the emulsion of a developed photographic film. In practice the effects of the x-ray beam are usually intensified by the use of fluorescent screens which emit light when exposed to x-rays. The photographic film is sandwiched between the two fluorescent screens in a special light tight cassette. When exposed to x-rays it is mainly the light emitted from the fluorescent screens that causes the blackening of the developed film.

RADIOGRAPHY

Projections in radiography

The projection is usually described by the path of the x-ray beam. Thus a posteroanterior (PA) view is where the beam passes from the back to the front, the standard projection for a routine chest film. An AP view is one taken from the front. The term 'frontal' refers to either PA or AP projection.

An x-ray film is a two-dimensional image. All the structures along the path of the beam will be projected on to the same portion of the film. Therefore it is often necessary to take at least two views to gain information about the third dimension.

Precise localisation of a shadow requires more than one view, usually at right-angles to one another, e.g. the PA and lateral chest film when examining pathology in the chest. Sometimes two views at right-angles are not appropriate and oblique views are substituted.

Horizontal ray films

The detection of air-fluid levels requires a projection using a horizontal x-ray beam, e.g. an erect film. Air-fluid levels are seen in several situations, e.g. in the bowel and in abscesses. The reason that air-fluid levels can only be seen on a film taken with a horizontal beam can best be understood by analogy with the fluid level in a glass of water. The only way to see the air-fluid level is to look at it from the side, i.e. in a horizontal direction (Fig. 1.2).

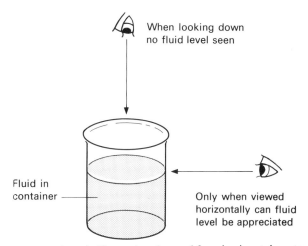

Fig. 1.2 Diagram illustrating the need for a horizontal ray to visualise an air-fluid level. Note that it is the direction of the ray that matters, not the position of the container. It makes no difference if the container is turned on its side since the level always remains horizontal.

Portable films

Films can be taken on patients in bed or in the operating theatres by using portable x-ray machines. Such machines have substantial limitations on the exposures they can achieve. This usually means longer exposure times and poorer quality films. Positioning of patients in bed is often inferior to that which can be achieved within the x-ray department. Radiation protection is another problem when portable equipment is used. Consequently, portable films should only be requested when the patients are unable to be moved to the x-ray department.

Magnification in radiography

All x-ray images shows some magnification, because the x-ray tube sends out a diverging beam of x-rays. The closer the object is to the film, the less the magnification (Fig. 1.3).

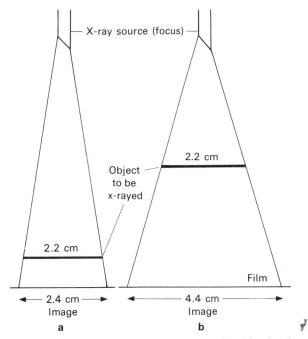

Fig. 1.3 Magnification in radiography. (a) The object is relatively close to the film so the magnification is slight; (b) the object is placed midway between the x-ray source and the film. The image in B is twice its actual size.

CONTRAST AGENTS

We saw earlier that structures are only visible on a radiograph because of differential absorption of the x-ray beam by the tissues of the body. Structures which are ordinarily invisible because they have the same density as the surrounding tissues may become visible when contrast material is introduced into them. Plain films, i.e. films without the addition of contrast agents, are usually taken prior to giving contrast in order to facilitate interpretation of the contrast examinations and to ensure that the details visible on the plain film are not obscured by the addition of contrast material.

Radiopaque contrast media contain atoms of high atomic number, which absorb x-rays and so appear white on the radiograph. Barium sulphate is used in the examination of the gastrointestinal tract. All the other opaque contrast media rely on iodine solutions, e.g. urographic and angiographic media (the same substances are used for both) and biliary contrast media.

Gas may be used as a contrast agent since it absorbs few x-rays and will appear black on the radiograph, e.g. air in pneumocephalography, or air in double contrast examinations of the colon.

SPECIAL TECHNIQUES

There are a large number of special techniques in radiology. Many of these are described elsewhere in this book in relation to specific organ systems or diseases.

Tomography

The aim of a tomogram (laminogram) is to blur out overlying structures, but to keep a selected plane of the body in sharp focus. This is achieved by moving the x-ray tube and film about an axis at the level of interest (Fig. 1.4). The thickness of the body in sharp focus (often referred to as the tomographic 'cut') can be varied by controlling the movement of the equipment. In chest

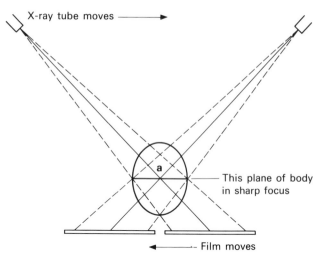

Fig. 1.4 Principle of tomography. The x-ray tube and film are fixed together by a bar. The axis of rotation of the bar is set at the desired level of (a) the tomographic section. The result is that the structures in the same plane as the axis of rotation are not blurred, whereas the image of structures above and below are blurred out.

work 'cuts' 1 cm thick are usually used. Sections of as little as 1 mm can be achieved with suitable apparatus, the use of such thin 'cuts' is virtually confined to the skull.

Angiography

An angiogram is an x-ray examination in which the blood vessels are opacified by an iodine-containing contrast medium (see p. 191 for example). Angiograms are broadly divided into arteriograms and venograms depending on the vessels injected.

There are two ways of injecting the contrast material for arteriography:

1. Directly through a needle. This method applies only to those vessels where needle puncture can be easily, safely and reliably accomplished, e.g. the abdominal aorta, the femoral artery in the groin, and the carotid artery in the neck.

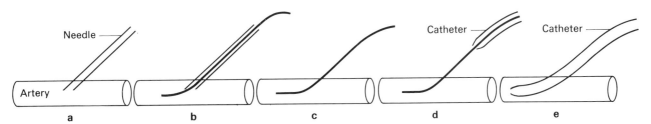

Fig. 1.5 'Seldinger technique' for catheterising blood vessels. The femoral or axillary arteries are the usual arteries utilised. (a) A needle is inserted through the skin into the artery; (b) a guide wire is passed through the needle into the lumen of the vessel; (c) the needle is withdrawn, leaving the guide wire in the lumen of the vessel; (d) A catheter is threaded over the guide wire and passed into the lumen of the vessel; (e) The guide wire is withdrawn leaving the catheter in position in the lumen of the vessel.

2. Through a catheter. The most widely used method of catheterising an artery is the 'Seldinger technique' illustrated in Fig. 1.5. At the end of the procedure the catheter is pulled out. A 5-minute compression of the puncture site with the fingers is enough to stop the bleeding in most patients. The advantage of the 'Seldinger technique' is that it is easy and quick to do, that the hole in the artery is no bigger than the catheter and that catheters of any length may be used.

With either method the problems are the same. The punctured artery may occasionally become blocked, or alternatively, it may be difficult to stop the bleeding once the catheter has been withdrawn. Either complication may require surgical intervention. There may be problems due to the contrast medium. Some of these are due to the fact that the contrast agents are hyperosmolar, which can be a problem since large quantities are sometimes given. The other complications due to contrast media are the same as those that occur with intravenous urography (see p. 177).

Some arteriograms are very painful and best done under general anaesthesia, while others, e.g. selective renal arteriography, can be done with local anaesthesia, sedation and analgesics.

Information obtained from arteriography

Arteriography enables the radiologist to demonstrate:

1. Disorders of the blood vessels—occlusions, stenoses, thrombi, aneurysm formation and angiomas.

2. Displacement or distortion of vessels by masses.
3. New vessels formed within tumours or walls of abscesses.
4. Arterial anatomy before surgery. This is only necessary where such information will influence the surgical procedure.

Subtraction films

It is possible to photographically eliminate the shadows that are present on the plain film from the films taken after contrast has been injected for the angiograms. The result is a film containing details of the opacified structures only (see Fig. 13.22, p. 330).

Lymphography

The lymphatic vessels and draining lymph nodes can be opacified wherever it is possible to directly cannulate a lymph vessel. In practice the commonest lymph vessels to be cannulated are those on the dorsum of the foot, though lymphograms via lymphatic vessels in the hand are occasionally performed. A blue-green dye is injected subcutaneously into one or more web spaces between the digits. The dye, which is absorbed into the lymph, enables the operator to identify the lymphatic vessels with certainty. A very fine needle on the end of polythene tubing can be inserted into a distended lymphatic vessel. Radio-opaque contrast medium—an oily one is usually chosen—can then be injected slowly over 1–3 hours. If the lymphatics on both feet are cannulated, the lymph vessels of the legs and the retroperitoneal nodes in the inguinal, iliac and para-aortic regions can be opacified. The pre-aortic and mesenteric nodes are not normally filled by this technique. Disease of the lymph vessels and lymph nodes can be documented. The pattern of lymph node replacement by disease processes is not specific enough to make a diagnosis; therefore, lymphography is used to assess the extent of disease, usually malignant neoplasm, the histological diagnosis of which has been, or will be, established by biopsy (Fig. 1.6). Oily contrast medium remains in the nodes for several

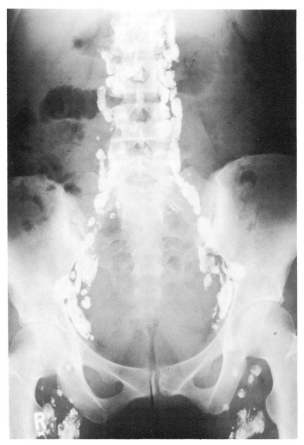

Fig. 1.6 Lymphogram showing normal inguinal, iliac and para-aortic nodes. This film was taken 24 hours after the injection of oily contrast medium into lymphatic vessels in the feet.

months making it possible to follow the progress of the opacified nodes with plain abdominal films.

Mammography

X-ray examinations of the breast utilises low kV x-rays in the range of 20–35 kV. At these energies there is a greater possibility of distinguishing fat from other soft tissue densities, and it is possible to identify the 'micro-calcifications' that are commonly seen in carcinoma of the breast. The technique of xeroradiography is often used in the place of conventional film systems.

Xeroradiography

Xeroradiography uses a different image-recording system from conventional radiography. The x-rays, instead of striking a photographic film, strike a charged selenium plate. Where the x-rays strike the plate, the charge is lost. A fine blue powder is then blown on to the plate. It adheres only to those portions of the plate which are still charged. The powder on the plate is then fused on to paper in a heat press. It is this image that is used for diagnosis. The selenium plate is cleaned and reused.

Wherever there is an edge between the charged and discharged areas of the plate, there is a dense accumulation of the powder. The edge enhancing effect is what distinguishes xeroradiography from conventional film imaging. It is exploited particularly in mammography (Fig. 1.7) and other soft tissue examinations.

ULTRASOUND

In diagnostic ultrasound examinations, very high frequency sound is directed into the body from a transducer placed in contact with the skin. As the sound travels through the body echoes are reflected by tissue interfaces and picked up by the transducer to be converted into an electrical signal.

Since air, bone and other heavily calcified materials, absorb nearly all the ultrasound beam, diagnostic ultrasound plays little part in the diagnosis of lung or bone

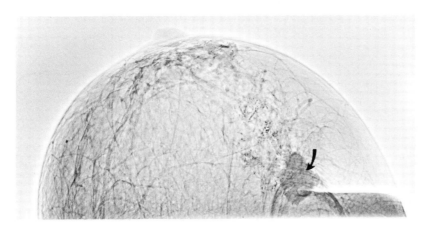

Fig. 1.7 A xeromammogram showing a mass in the breast (arrow) and microcalcifications extending irregularly from the mass towards the nipple. These microcalcifications are in carcinomatous tissue.

disease. Clearly, gas in the bowel will also interfere with the transmission of sound.

Fluid is a good conductor of sound. When fluid is bounded by a smooth wall, as in a cyst, or in organs such as the bladder, there is a large change in acoustic impedance resulting in a large echo. The fetus, within its amniotic sac surrounded by fluid and containing no air, can be well demonstrated by ultrasonic examination, but a solid tumour situated within a solid organ, e.g. a liver tumour may be more difficult to demonstrate since the acoustic impedance of tumour and liver tissue may be similar. Despite this limitation, ultrasound is used to define solid lesions in the liver and renal substances.

With pulsed ultrasound, very short pulses of sound lasting about a millionth of a second are transmitted from a transducer into the patient about 500 times/ second.

The ultrasound is produced by causing a special crystal to oscillate at a predetermined frequency. The same crystal 'listens' to the returning echoes, which are electronically amplified to be recorded as signals on a cathode ray oscilloscope. Photographic reproductions of the oscilloscope can then provide a permanent record. The time taken for each echo to return to the transducer is proportional to the distance travelled and it is, therefore, possible to measure the distance between interfaces. This is of great practical importance as, for example, in obstetrics where the measurement of the diameter of the fetal head has become the standard method of estimating fetal age.

Pulsed ultrasound has a number of uses:

1. To measure the distance between two interfaces, e.g. fetal measurement, the size of a cyst, the diameter of a cardiac chamber.
2. To build up an ultrasound picture of one or more slices of the body, particularly to elucidate structures filled or surrounded by fluid (Fig. 1.8).
3. To determine whether a mass is solid or cystic. Cysts or fluid-filled structures produce large echoes from their walls but no echoes from the fluid contained within them. Another very helpful feature is that more echoes than usual are received from the normal tissues behind the cyst; an effect known as 'acoustic enhancement'.
4. To plot the movement of structures. This application is particularly exploited in cardiac ultrasound where the movements of the valves and the walls of cardiac chambers can be demonstrated.

Continuous ultrasound utilises a somewhat different diagnostic principle to pulsed ultrasound. With continuous ultrasound the Doppler effect is exploited. Sound received from a mobile structure shows a variation in frequency with movement. Continuous ultrasound is used to detect the beating fetal heart and to demonstrate blood flowing in arteries and veins where the sound is reflected off the blood cells flowing in the vessels.

Ultrasound is becoming a speciality in its own right and is only very briefly discussed in this book. Examples

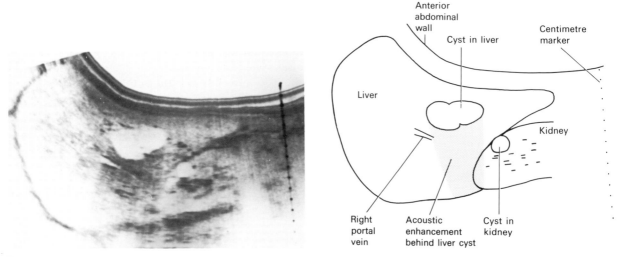

Fig. 1.8 Ultrasound examination of right upper quadrant of abdomen. The liver and right kidney are shown in this longitudinal scan. There is a bilocular cyst in the liver and a small cyst is demonstrated in the upper pole of the kidney.

of its use in obstetrics will be found on p. 223 and in the diagnosis of renal masses on p. 204.

At the energies and doses currently used in diagnostic ultrasound no harmful effects on living tissues have so far been demonstrated.

'Real time' scanning uses a method of observing moving images ultrasonically. A variety of 'real time' units are now available. These machines utilise one or more transducers. Image resolution varies; however, the better systems provide images that are almost as good as those obtained with static scans. The ability to image movement by 'real time' systems is exploited when either the fetus or the heart are examined. Another major application is that 'real time' ultrasound can rapidly survey large body areas, e.g. the upper abdomen. This can be likened to conventional x-ray fluoroscopy, with each image on the screen being a tomogram. Since the 'real time' system is a good survey method, it is often desirable to combine 'real time' and static imaging, using the 'real time' to define abnormal areas which can then be imaged with the high resolution static techniques.

RADIONUCLIDE IMAGING

The radioactive isotopes used in diagnostic imaging emit gamma rays as they decay. Gamma rays like x-rays are a form of electromagnetic radiation. Many naturally occurring radioactive isotopes, e.g. potassium-40, uranium-235 have half lives of hundreds of years. The radioisotopes used in medical diagnosis are artificially produced and most have short half lives, usually a few hours or days. To keep the radiation dose to the patient to a minimum the smallest possible dose of an isotope with a short half life should be used. Clearly the radio-pharmaceuticals should have no undesirable biological effects and should be rapidly excreted from the body following completion of the investigation.

The gamma rays emitted by the isotope are detected by a scintillation counter. This works on the principle that when gamma rays strike an activated sodium iodide crystal, light is produced. When this light hits the photo-cathode of a photomultiplier tube, electrons are released which are amplified to give an electrical pulse. This pulse is further amplified and analysed by an electrical processing unit and a recording is then made. The radiation from the isotope taken up by organs in the body can be recorded by a rectilinear scanner or a gamma camera.

The rectilinear scanner. In these machines the crystal detector mechanism is moved backwards and forwards in a grid-like pattern, over the area being scanned. The result is usually recorded photographically, as a print out of dots which vary in number with the intensity of radiation at different sites.

The gamma camera consists of a sodium iodide crystal coupled to a number of photomultiplier tubes arranged as a circular detector with a large field of view. This means that large areas such as both lungs or liver and spleen can be examined without having to move either the patient or the detector. The gamma camera is much more sensitive than the rectilinear scanner and takes much less time to build up a picture (Fig. 1.9).

Fig. 1.9 Liver scan. This gamma camera scan taken 30 minutes after injection of ^{99m}Tc sulphur colloid shows accumulation of isotope in a normal liver. Uptake is also present in the spleen (arrow).

The radioactive isotope most commonly used is technetium-99m. It is readily prepared, has a convenient half life of 6 hours and emits gamma radiation of a suitable energy for easy detection. Many other isotopes are available including iodine-131, gallium-67 and indium-113m.

Radionuclide imaging depends on the fact that certain radiopharmaceuticals concentrate selectively in different parts of the body. Some radiopharmaceuticals are in ionic form, e.g. ^{99m}TcO$_4^-$ in brain imaging, whereas others are used in colloid or particulate form. If a sulphur colloid is labelled with technetium-99m and injected intravenously it will be taken up by the reticuloendothelial system and can be used to visualise the liver and spleen. Larger particles are used in lung perfusion imaging; macroaggregates of albumin with a particle size of 10–75 μm when injected intravenously are trapped in the pulmonary capillaries. If the macroaggregates are labelled with technetium-99m then the blood flow to the lungs can be visualised.

COMPUTERISED TOMOGRAPHY

The recording system used in computerised tomography (CT) is quite different to that used for conventional x-ray films. Both systems use x-rays generated in much the same way, but CT uses a crystal or gas detector system which is more sensitive than film or fluorescent screen. The x-ray tube and dectector rotate around the patient so that a horizontal slice of tissue is visualised. By using a computer to reconstruct the image it is possible to build up a picture of a slice of the body, composed of small blocks of varying density. Compared to conventional film techniques, the range of densities recorded is increased from approximately 20 with conventional film to 2000 with CT. The resolution is such that not only can fat be distinguished from other soft tissues but gradations of density within the soft tissues can also be recognised, e.g. blood can be distinguished from muscle; brain tissue from CSF and CSF from blood. Even differences between certain soft tissue neoplasms and adjacent normal tissue can be identified. This revolutionary technique has completely altered the diagnostic approach to intracranial disease. The full impact on abdominal and thoracic diagnosis remains to be evaluated.

The principles for brain and body CT scanning are similar. The patient lies with the part to be examined within the gantry housing the x-ray tube and the array of

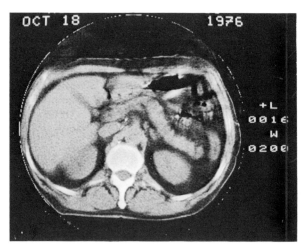
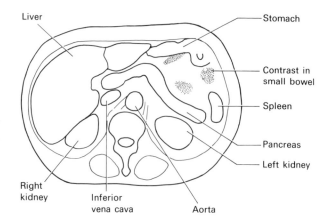

Fig. 1.10 CT scan. Normal transverse scan through the upper abdomen.

detectors which maintain a constant alignment with one another while travelling around the patient. In the latest equipment a fan-shaped beam of x-rays is emitted which rotates 360° around the patient. A computer analyses the information from the scan, by calculating the absorption coefficient of each of the small blocks of tissue 1·5 mm square and 3–13 mm thick in the path of the x-ray beam. In this way the picture of a horizontal slice of the body 3–13 mm thick can be constructed. This information is transferred to a cathode ray tube and can be recorded on magnetic tape or disc. It can then be displayed in numerical form for analysis of precise tissue absorption coefficients. Contrast agents can be utilised to enhance various aspects of tissue density. Only very low concentration of contrast agents are required since the recording system is so sensitive to minor changes in absorption coefficient.

The absorption coefficients are expressed on an arbitrary scale (Hounsfield units) with water density being zero, air density being −1000 units and dense bone +1000 units. The range in level of density to be displayed can be selected by controls on the computer. This selection can be made after the patient has left the department since all the information is stored. The range of densities to be visualized is known as the 'window width' and the mean level as the 'window height' (Fig. 1.10).

In summary, the major advantage of CT over conventional radiography is that it can distinguish differences between various soft tissues, e.g. in the brain it can distinguish CSF, blood, brain substance and can also distinguish diseased tissue from healthy tissue. Intravenous contrast may also be injected to improve the information but invasive techniques such as angiography and pneumoencephalography are not required.

Another advantage is that CT presents a horizontal tomographic section through the body, free of overlying blurred images.

RADIATION HAZARDS

X-rays and other ionising radiations are harmful. Natural radiation from the sun, radioactivity in the environment, and atmospheric radioactivity from nuclear bombs and other man-made sources contribute a genetic risk over which the individual doctor has no control. However, radiation for medical purposes, a far greater potential source of ionising radiations, is under the control of doctors. It is their responsibility to limit the use of x-rays for medical purposes to those situations where the benefit clearly outbalances the risks. Unnecessary radiation is to be deplored. This is prevented by good technique—limiting the size of the x-ray beam to the required areas, limiting the number of films to those that are necessary and keeping repeat examinations to a minimum. Just as important as these factors, all of which are really the province of those who work in the x-ray department, is avoiding unnecessary requests for x-ray examination, particularly those that involve high radiation exposure.

Radiation is particularly harmful to dividing cells. Genetically adverse mutations may occur following radiation of the gonads, resulting in congenital malformations and a genetic risk to the population. There is no threshold for the mutation rate, hence there is no such thing as a safe radiation dose.

Radiation to the developing fetus can have catastrophic effects. As well as the increased incidence of congenital malformations induced in the developing fetus it has been shown that the frequency with which leukaemia and other malignant neoplasms develop within the first 10 years of life is increased, probably by about 40% compared to the normal population, in children exposed to diagnostic x-rays while *in utero*. X-raying a fetus should, therefore, be kept to the absolute minimum and preferably avoided. Because radiation is especially dangerous to the fetus in its early stages of development the 10-day rule has been introduced in Britain to avoid the possibility of the pregnant patient being x-rayed before she is aware that she is pregnant. It has been recommended that all x-ray examinations involving appreciable radiation to the genital organs in women of child-bearing age should be taken in the 10 days following the first day of the menstrual period. This rule has to be waived in urgent cases.

2
The Respiratory System

The standard films of the chest comprise a posteroanterior view (PA) and a lateral view (Fig. 2.1). Both the PA and lateral films are exposed on full inspiration with the patient in the upright position. Films taken on expiration are difficult to interpret because in expiration the lung bases appear hazy and the heart shadow increases in size (Fig. 2.2).

Even though chest films are the commonest x-ray examinations performed they are also one of the most difficult plain film examinations for the student to

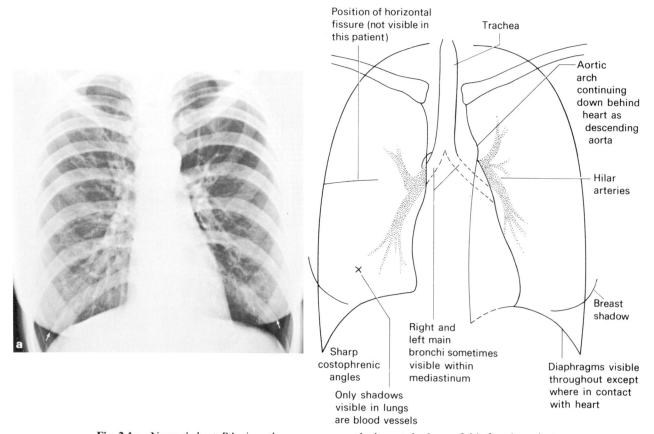

Fig. 2.1a Normal chest. PA view: the arrows are on the breast shadows of this female patient.

12

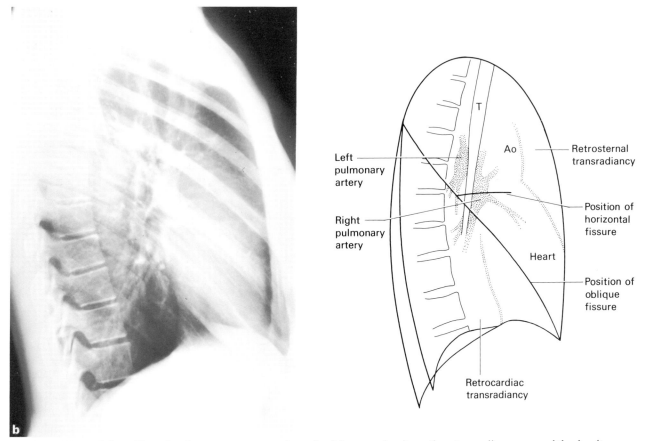

Fig. 2.1b Normal chest. Note that the upper retrosternal area is of the same density as the retrocardiac areas, and the density over the heart is the same as over the upper thoracic vertebrae. The vertebrae are more transradiant, i.e. blacker as one travels down the spine until the diaphragm is reached.

interpret. The trained radiologist often looks over a chest film in an apparently random fashion, and when he finds an abnormality his analysis is dictated by the possibilities that come to mind for that particular shadow. For example, if he sees a nodule in the lung he looks at the shape of the nodule, looks for other lung lesions and for evidence of spread of disease to the hilum, pleura or rib cage, and so on. This problem-orientated approach—the observer constantly asking himself questions, not only about the shadows he sees but also about the patient's clinical findings—is the quickest and most accurate way of achieving a diagnosis. However, this approach takes time to learn and in

the early stages a routine is necessary in order to avoid missing valuable radiological signs. The order in which one looks at the structures is unimportant; what matters is to follow a routine, otherwise important abnormalities will be overlooked. One way of examining the frontal and lateral chest films is presented below (Fig. 2.1).

Trace the diaphragms

The upper surfaces of the diaphragm should be clearly visible from one costophrenic angle to the other, except where the heart is in contact with the diaphragm. On a good inspiratory film the level of the right diaphragm is

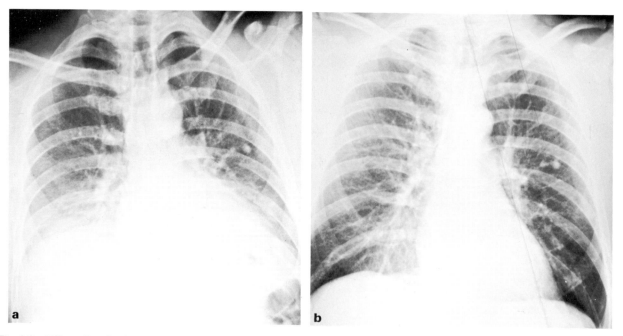

Fig. 2.2 Effect of expiration on chest film. (a) Expiration, (b) inspiration. On expiration the heart appears larger and the lung bases are hazy.

usually near the anterior end of the sixth rib, the right diaphragm being up to 2½ cm higher than the left.

Check the size and shape of the heart

See p. 85 for the details

Check the position of the heart and mediastinum

Normally, the trachea lies midway between the medial ends of the clavicles. The position of the heart is very variable; on average one-third lies to the right of the midline, but anything from one-half to one-fifth of the heart lying to the right of the midline is within the normal range.

Look at the mediastinum

The right superior mediastinal border is usually straight or slightly curved as it passes downward to merge with the right heart border. The left superior mediastinal border is ill defined above the aortic knuckle.

The outline of the mediastinum and heart should be clearly seen except where the heart lies in contact with the diaphragm.

In young children the normal thymus is clearly visualised. It may be very large and should not be mistaken for disease (Fig. 2.3).

Examine the hilar shadows

Normally, the hilar shadows are composed exclusively of pulmonary arteries and veins. Air in the major bronchi can be recognised but their walls are not usually visible. The hilar lymph nodes in the normal patient are too small to recognise as discrete shadows.

The left hilum is usually slightly higher in position than the right.

Examine the lungs

The only structures that can be identified within normal

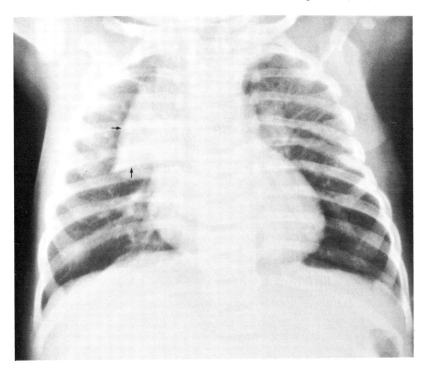

Fig. 2.3 Normal but prominent thymus in a child aged 3 months. The thymus shows the characteristic 'sail shape' projecting to the right of the mediastinum (arrows). This should not be confused with right upper lobe consolidation.

lungs are the blood vessels and the interlobar fissures. The fissures can only be seen if they lie tangential to the x-ray beam; they are after all composed of only two layers of pleura. Usually, only the horizontal fissure (minor fissure) is visible in the frontal projection, running from the right hilum to the sixth rib in the axilla. There is no equivalent to the horizontal fissure on the left. (In about 1% of people there is an extra fissure visible in the frontal view—the so-called azygos lobe fissure; Fig. 2.4.) The oblique fissures (major fissures) are only visible on the lateral view. The fissures form the boundaries of the lobes of the lungs so a knowledge of their position is essential for an appreciation of lobar anatomy (see Fig. 2.16).

Look for abnormal pulmonary opacities or translucencies. Do not mistake the pectoral muscles, breasts (Fig. 2.5) or plaits of hair for pulmonary shadows. Skin lumps or the nipples may mimic pulmonary nodules. The nipples are usually in the fifth anterior rib space. In general, if one nipple is visible the other will also be seen.

A good method of finding subtle shadows on the frontal film is to compare one lung with the other, zone by zone. Detecting ill-defined shadows on the lateral view can be difficult. It is helpful to bear in mind that in most people the density of the retrosternal space is very similar to that of the retrocardiac density, and that the density over the heart is similar to that over the upper thoracic spine. Another helpful feature is that as the eye travels down the thoracic vertebral bodies as far as the diaphragm, each body should appear more lucent than the one above.

Check the integrity of the ribs, clavicles and spine

In females check that both *breasts* are present. Following mastectomy the breast shadow cannot be defined. The reduction in the soft tissue bulk leads to an increased transradiancy of that side of the chest, which should not be confused with pulmonary disease.

Assess the technical quality of the film

This is necessary since incorrect exposure or faulty centring or projection may either hide or mimic disease.

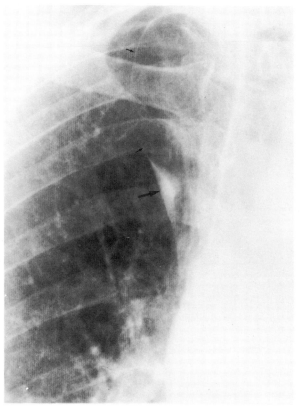

Fig. 2.4 The azygos lobe fissure. During intrauterine development the azygos vein migrates from the chest wall to lie within the mediastinum. In patients with an azygos 'lobe' the vein has migrated through the lung drawing the pleura with it. The azygos vein (large arrow) is seen at the medial end of the fissure (small arrows) instead of in its normal position in the tracheobronchial angle. This variant is of no clinical significance.

The correctly exposed routine PA chest film is one in which the ribs and spine behind the heart can just be identified. Unless one can see through the heart lower lobe lesions may be completely missed. A straight film is one where the medial ends of the clavicles are equidistant from the pedicles of the thoracic vertebrae.

Extra views

Oblique views. Films taken with the patient turned to one or other side, are useful for demonstrating the chest wall and occasionally for showing intrathoracic shadows to better advantage.

Lordotic view (Fig. 2.6). A lordotic view is an AP film taken with the patient leaning backwards. In this projection the clavicles and first ribs are projected above the lungs and so do not obscure the apices. Lateral views are of little value for demonstrating apical lesions because the lung apices are poorly seen through the shoulders.

Lateral decubitus views are not as the name would suggest, lateral views; they are frontal projections taken with the patient lying on one or other side using a horizontal x-ray beam. Their purpose is to demonstrate free pleural fluid which will collect along the dependent chest wall (see Fig. 2.42b).

Expiration films. A frontal film may be deliberately exposed on expiration in order to demonstrate diaphragmatic movement or the ability of the lung to deflate. A pneumothorax may be more obvious on an expiration than an inspiration film.

Penetrated film

By using a more penetrating x-ray beam it may be possible to see more details of cardiac and mediastinal structures. Some centres use high-kV films routinely, others use low-kV techniques as a routine and may require penetrated films in selected cases.

Tomography

Tomography is an extremely valuable tool in chest radiology. The principles are discussed on p. 4.
    The main indications for chest tomography are:

1. To show the edge of an intrathoracic mass, e.g. a pulmonary nodule, and to detect any calcification within it.
2. To show the presence of cavitation in a pulmonary mass or an area of consolidation, and to show the state of the adjacent lung.

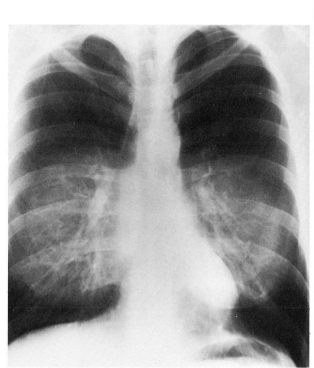 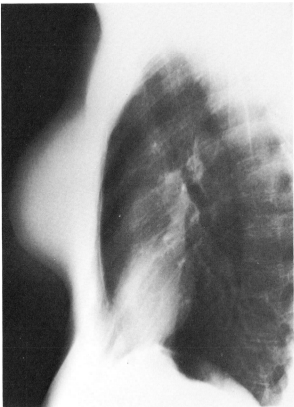

Fig. 2.5   Breast shadows. Do not confuse the breast shadows for pulmonary consolidation. In this instance the patient has had mammary implants.

3. To provide an accurate localisation of a mass or an area of consolidation.

4. To clarify the presence of rib destruction.

5. To analyse hilar shadows.

6. To clarify an indefinite pulmonary lesion (Fig. 2.7). It should, however, be realised that tomography will rarely demonstrate an abnormality in areas that are clearly normal on conventional films.

Fluoroscopy

The image at fluoroscopy is poor compared to that which can be achieved with x-ray film. It is limited to the following uses:

1. For observing movement of the diaphragm.

2. As an aid in selecting the optimum oblique view in problem cases.

3. In suspected inhalation of a foreign body to demonstrate obstructive emphysema.

Bronchography

Bronchography involves introducing an iodised oil into the bronchial tree. There are various ways of instilling the contrast agent; via a catheter introduced into the trachea through the nose or mouth, or alternatively following puncture of the cricothyroid membrane.

The major indication is for the assessment of bronchiectasis (see Fig. 2.80). Bronchograms are rarely used to demonstrate other abnormalities of the bronchi. The

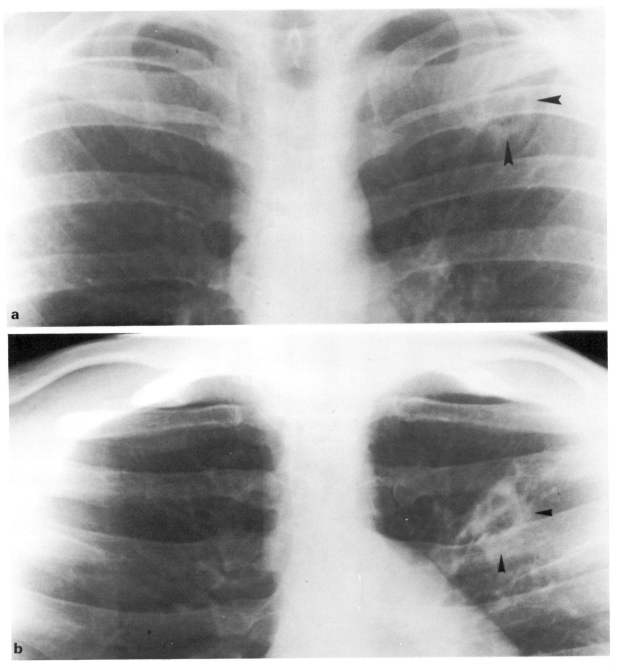

Fig. 2.6   The lordotic view of the lung apices. A tuberculous cavity is present in the left upper lobe (arrows). (a) On the conventional PA projection it is difficult to demonstrate because it is partly hidden by the overlying ribs and clavicle; (b) on the lordotic view the lesion is projected clear of the bone.

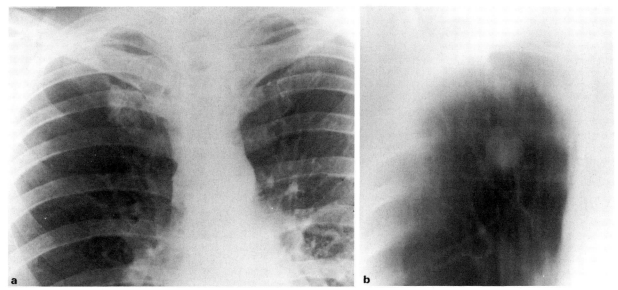

Fig. 2.7 The use of tomography to show an indefinite pulmonary lesion. (a) The PA film shows a 2.5 cm lobulated mass in the right upper lobe that is partly hidden by rib shadows; (b) on tomography this mass is clearly shown.

major bronchi are much better seen under direct vision. Bronchography rarely provides useful information about the nature of peripheral pulmonary masses, since regardless of aetiology they all displace and obstruct the smaller airways.

Pulmonary angiography

The pulmonary arteries and veins can be demonstrated by taking serial films following the rapid injection of angiographic contrast medium into the pulmonary artery through a catheter. The catheterisation is carried out under fluoroscopic control with continuous electro-cardiographic and pressure monitoring by an operator skilled in cardiac catheterisation. It carries a small but definite risk to the patient.

Its major use is in the demonstration of pulmonary emboli. Occasionally, it is needed to demonstrate congenital vascular anomalies.

Radionuclide lung scanning

There are two different kinds of lung scans that can be obtained using the gamma camera, namely perfusion and ventilation scans. For perfusion scans macroaggregates of albumin with an average particle size of 30 μm labelled with an isotope such as technetium-99m are injected intravenously. These particles become trapped in the pulmonary capillaries and their pattern when imaged reflects blood flow (Fig. 2.8).

A ventilation scan visualises air flow and distribution in the lungs. For this the patient inhales a radioactive gas such as xenon-133 (Fig. 2.9).

The major indication for perfusion scans is in the assessment of blood flow to the lungs, particularly in order to diagnose pulmonary embolism. In pulmonary embolism one or more filling defects are seen on the perfusion scan, but since many other lung abnormalities, e.g. pneumonia, emphysema and pleural effusion also give rise to perfusion defects, when a defect is seen, the perfusion scan may be combined with a ventilation scan. By comparing the ventilation and perfusion scans it is possible to diagnose pulmonary embolus by demonstrating parts of the lung that are ventilated but not perfused.

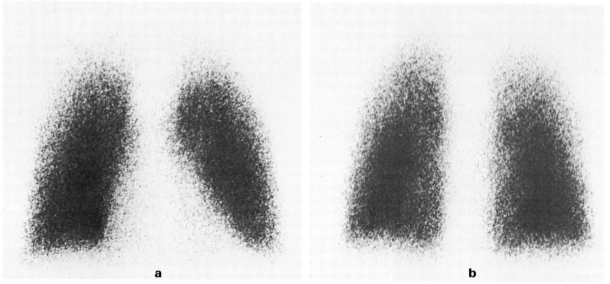

Fig. 2.8 Normal radionuclide perfusion scan using ^{99m}Tc-labelled macroaggregates of albumin. (a) Anterior view; (b) posterior view.

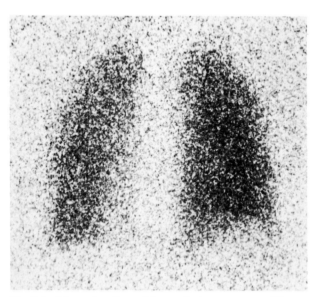

Fig. 2.9 Normal radionuclide ventilation scan using ^{133}Xe; posterior scan.

DISEASES OF THE CHEST WITH A NORMAL CHEST FILM

Life-threatening respiratory disease may exist in patients with a normal chest x-ray. Sometimes it is only possible to detect abnormality by comparison with previous or later examinations, e.g. minor elevation of one diaphragm due to pulmonary embolism, or early pulmonary shadows due to fibrosing alveolitis. Respiratory disease with a normal chest x-ray can be conveniently divided into:

1. Obstructive airways disease often produces no recognisable abnormality. Asthma and acute bronchiolitis may produce overinflation of the lung, a sign that may be difficult to be certain about, but in many cases the chest film is normal. Emphysema, when severe, gives rise to the signs described on p. 67 but when moderate, the chest x-ray may be normal or very nearly so. Uncomplicated acute or chronic bronchitis does not produce any radiological signs, so if a patient with chronic bronchitis has an abnormal film he either has some other disease or he has developed a complication, e.g.

pneumonia or cor pulmonale. A small proportion of patients with productive cough due to bronchiectasis show no plain film abnormality.

2. Small lesions. It is impossible to see solitary lung masses or consolidations of less than 6 mm in diameter and it is very rare to see them if they are less than 1 cm in size. Even lesions as large as 2–3 cm may be very difficult to identify on routine films if they are hidden behind overlapping rib and clavicle shadows or behind the heart.

Endobronchial lesions, such as carcinoma, cannot be diagnosed on routine films unless they cause collapse/consolidation or considerable obstructive emphysema.

3. Infections. Most patients with acute bacterial pneumonia present with recognisable consolidation, but in other infections, particularly with viruses, the consolidation may only develop after the onset of symptoms. Patients with active tuberculosis, including miliary tuberculosis, may initially have a normal chest film.

4. Diffuse pulmonary infiltrations, particularly pulmonary fibrosis of many different aetiologies, may be responsible for breathlessness with substantial alteration in lung function tests before any clear-cut radiological abnormalities are evident.

5. Pleural abnormality. Dry pleurisy will not produce any radiological findings, and even 300 ml of pleural fluid may be impossible to recognise on PA and lateral chest films.

THE ABNORMAL FILM

When faced with an abnormal chest film the first questions to ask oneself are 'where is the abnormality?' and 'how extensive is it?' Only then does one move to the question 'what is it?' Clearly, the differential diagnosis for pulmonary lesions is quite different from that for mediastinal, pleural or chest wall disease. The first step is to examine both the frontal and lateral films; usually the location of any lesion will then be obvious. If the shadow is surrounded on all sides by aerated lung it must arise within the lung, but when a lesion is in contact with the pleura or mediastinum it may be difficult to decide the site from which it originated.

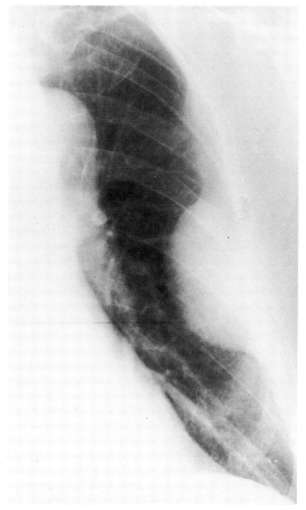

Fig. 2.10 Extrapleural mass (soft tissue mass of multiple myeloma arising in a rib). The mass has a smooth convex border with a wide base on the chest wall. This shape is quite different from a peripherally located pulmonary mass, an example of which can be seen in Fig. 2.27.

If the shadow has a broad base with smooth convex borders projecting into the lung and a well-defined outline it is likely to be pleural, extrapleural or mediastinal in origin (Fig. 2.10).

Particular care is needed when evaluating shadows in the hilar region on the frontal view. It is always essential to check on the lateral film whether the shadow is *at* the

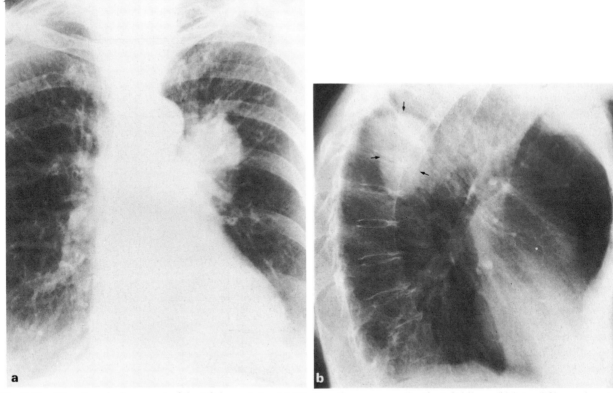

Fig. 2.11 Mass in apical segment of the left lower lobe. (a) PA film: the mass overlies the left hilum; (b) lateral film: unless one studies the lateral film, it would be easy to misdiagnose this lesion as a hilar mass rather than as a peripherally located intrapulmonary mass (arrows). (The diagnosis in this instance was primary carcinoma of the bronchus.)

hilum or just projected over it (Fig. 2.11) since the causes and subsequent investigation of hilar enlargement are different to those of pulmonary shadows.

The silhouette sign (Fig. 2.12)

The silhouette sign is an invaluable sign for localising disease in the chest. The information on the chest film is largely dependent on the contrast between the radiolucent air in the lungs compared with the opacity of the heart, blood vessels, mediastinum and diaphragm. An intrathoracic lesion touching a border of the heart, aorta or diaphragm will obliterate that border on the chest x-ray. This sign was named 'the silhouette sign' by Felson and has two important applications:

1. It often makes it possible to localise a shadow by observing which borders are lost, e.g. loss of the heart border must mean that the shadow lies in the anterior half of the chest. Alternatively, loss of part of the diaphragm outline indicates disease of the pleura or lower lobes.

2. It makes it possible on occasion to diagnose disorders such as pulmonary consolidation even when one is uncertain as to the presence of an opacity. It is a surprising fact that a wedge or lens-shaped opacity may be very difficult to see because of the way the shadow fades out at its margins, but if such a lesion is in contact with the mediastinum or diaphragm it causes loss of their normally sharp boundaries.

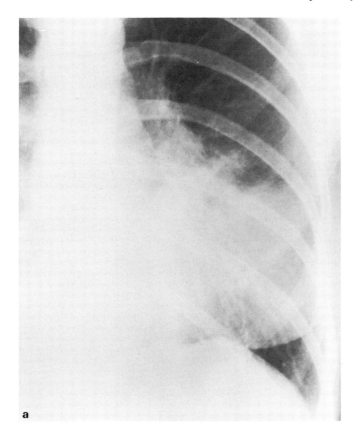
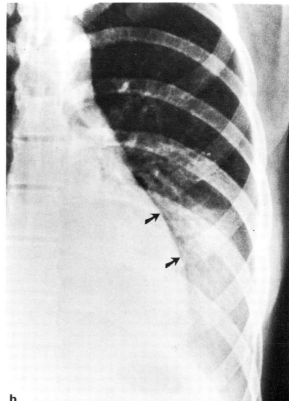

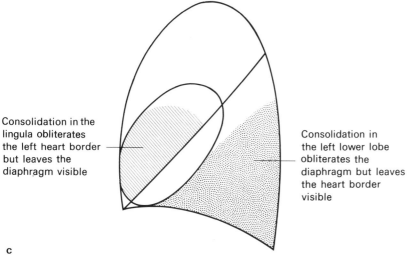

Consolidation in the lingula obliterates the left heart border but leaves the diaphragm visible

Consolidation in the left lower lobe obliterates the diaphragm but leaves the heart border visible

Fig. 2.12 'The silhouette sign'. (a) The left heart border is invisible because it is in contact with the adjacent lingular consolidation. The cardiac outline disappears because of the intimate contact with the pulmonary pathology; (b) the consolidation lies in the left lower lobe and obliterates the left diaphragm, but the heart border can still be seen arrows because it is in contact with the normally aerated left upper lobe. These relationships are easily appreciated from the lateral projection (c).

THE LUNGS

When looking at a chest film it is of practical help to try and place any abnormal intrapulmonary shadows into one or more of the following broad categories:

1. Air-space filling.
   (a) Pulmonary oedema.
   (b) Pulmonary consolidation ('alveolar infiltrates') with or without atelactasis.
2. Spherical shadows.
3. Line shadows.
4. Widespread small shadows.

The presence of cavitation or calcification should be noted.

Air-space filling

'Air-space filling' means the replacement of air in the alveoli by fluid, or rarely, by other materials. The fluid can be either a *transudate* (pulmonary oedema) or an *exudate*. The causes of an alveolar exudate include infection, infarction, pulmonary contusion, haemorrhage, collagen disease and allergy.

The signs of 'air-space filling' are:

1. A shadow with ill-defined borders (Fig. 2.13) except where the disease process is in contact with a fissure, in which case the shadow has a well-defined edge.
2. An air bronchogram (Fig. 2.14). Normally, it is not possible to identify air in the bronchi within the lung

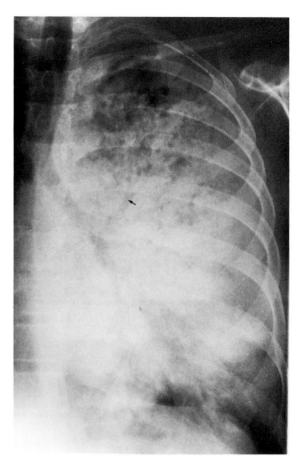

Fig. 2.14 The air bronchogram. An extensive air bronchogram is seen in this patient with pneumonia of both lobes of the left lung. The arrow points to some bronchi that are particularly well seen.

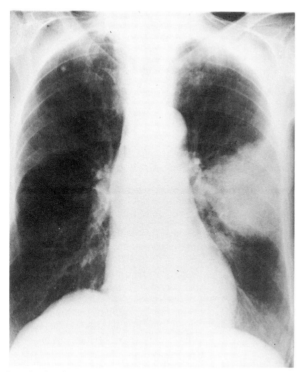

Fig. 2.13 A pulmonary infiltrate. In this case the homogeneous shadow in the left lung is due to a pulmonary infarct.

substance because the walls of the bronchi are too thin and they are surrounded by air in the alveoli, but if the alveoli are filled with fluid the air in the bronchi contrasts with the fluid in the lung.

3. The silhouette sign, namely loss of visualisation of the adjacent mediastinal or diaphragm outline (see p. 22 for explanation of this sign).

Pulmonary oedema

There are two forms of pulmonary oedema; interstitial and alveolar. Since the oedema fluid, initially, collects in the interstitial tissues of the lungs, all patients with alveolar oedema also have interstitial oedema. Alveolar oedema is always acute. The commonest causes are acute left ventricular failure, renal failure and over-transfusion.

Alveolar oedema (Fig. 2.15) is almost always bila-

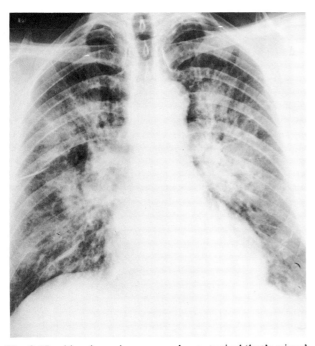

Fig. 2.15 Alveolar pulmonary oedema: typical 'bat's wings' pattern. The pulmonary shadows are bilateral (greater on the left than the right) and maximal in the perihilar region, fading towards the lung periphery.

teral, involving all the lobes. The shadowing is maximal close to the hila and fades out peripherally, leaving a relatively clear zone around the edges of the lobes. This pattern of oedema is sometimes called the 'butterfly pattern' or the 'bat's wing pattern'.

An identical appearance to pulmonary oedema may also be seen with pulmonary exudates due to aspiration of gastric contents, inhalation of noxious gases and in some cases of pneumonia, particularly in patients with reduced immunity.

One helpful feature in distinguishing pulmonary oedema from widespread exudates is the rapidity with which oedema appears and the speed with which it disappears on treatment. Substantial changes in the severity of air-space filling in a 24-hour period are virtually diagnostic of pulmonary oedema.

Consolidation (alveolar infiltrates) and collapse

Consolidation and collapse often coexist. It is, however, convenient to consider these two phenomena separately. Pure consolidation shows no loss of volume.

Consolidation of a whole lobe is virtually diagnostic of bacterial pneumonia. The recognition of lobar consolidation requires an appreciation of the radiological anatomy of the lobes (Fig. 2.16). Lobar consolidation produces an opaque lobe, except for air in the bronchi (air bronchograms). Since the consolidated lobe is airless the fissure between it and the normal lung does not appear as a line, but is seen as a clear-cut border to the opacity. Because of the silhouette sign the boundary between the affected lung and the adjacent heart, mediastinum and diaphragm will be invisible. The appearance of consolidation of the right lower lobe is seen in Fig. 2.17.

Patchy consolidation i.e. one or more patches of ill-defined shadowing up to the size of one or more segments (Fig. 2.18) is usually due to either infection, infarction or, less commonly, allergy. There is no reliable way of telling from the films which of these possibilities is the case. In most instances the clinical and laboratory findings point to one of these alternatives.

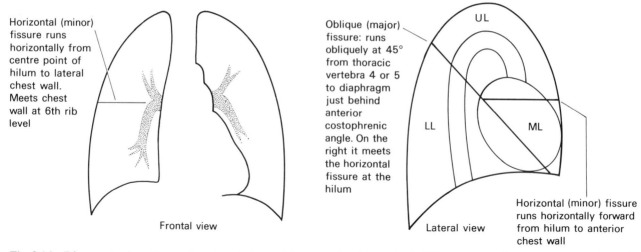

Horizontal (minor) fissure runs horizontally from centre point of hilum to lateral chest wall. Meets chest wall at 6th rib level

Oblique (major) fissure: runs obliquely at 45° from thoracic vertebra 4 or 5 to diaphragm just behind anterior costophrenic angle. On the right it meets the horizontal fissure at the hilum

Horizontal (minor) fissure runs horizontally forward from hilum to anterior chest wall

UL

LL

ML

Frontal view

Lateral view

Fig. 2.16 Diagram to show the position of the lobes and fissures. The oblique (major) fissure runs a similar course through each lung. On the left it separates the upper and lower lobes (LL). On the right there is an extra fissure—the horizontal (minor) fissure separating the upper lobe (UL) and the middle lobe (ML).

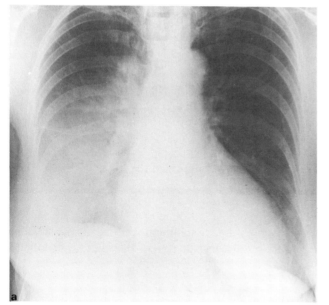

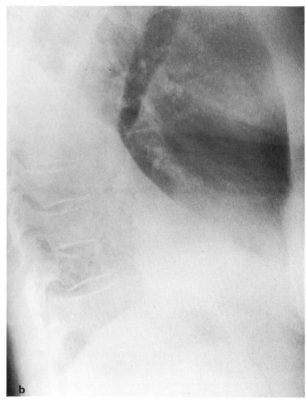

Fig. 2.17 Consolidation of the right lower lobe. The shadow corresponds to the position of the right lower lobe. The heart border and medial half of the right diaphragm can still be identified, whereas the lateral half of the diaphragm is invisible. On the lateral view (b) the oblique fissure forms a well-defined anterior boundary and the right diaphragm is ill defined. Only the left diaphragm is seen clearly.

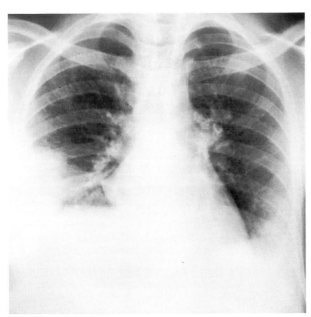

Fig. 2.18 (*left*) Patchy consolidation in bronchopneumonia. There are areas of consolidation (infiltrate) in both lower lobes laterally. (Calcified lymph nodes are, incidentally, present at the left hilum.)

When spherical in shape consolidation may be difficult to distinguish from a lung tumour, but usually serial films show a change over a short interval if the shadow is due to consolidation, whereas no change will be apparent if it is due to a tumour.

Cavitation (abscess formation) within the consolidated areas in the lung may occur with many bacterial infections (Fig. 2.19), but the organisms that are particularly liable to produce cavitation are staphylococci,

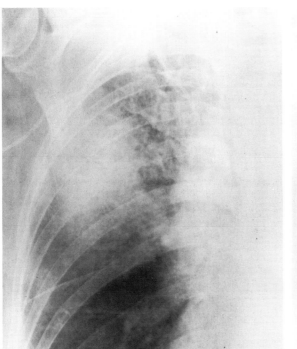

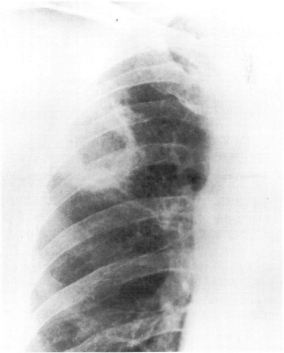

Fig. 2.19 Cavitation in staphylococcal pneumonia. Above left shows an area of consolidation which 7 days later (right) shows central translucency due to the development of cavitation.

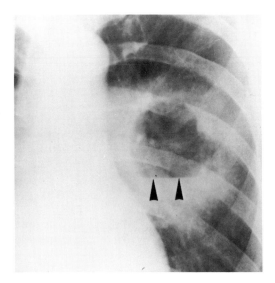

Fig. 2.20 Fluid level (arrows) in a lung abscess in the superior segment of the left lower lobe.

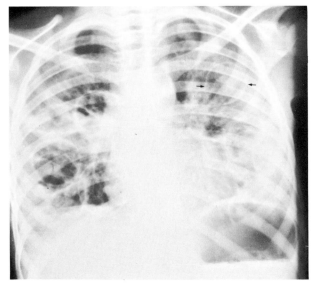

Fig. 2.21 Pneumatocele formation in staphylococcal pneumonia. The pneumatoceles are seen as large air-containing ring shadows; there are several in the right mid and lower zones. One in the left upper zone is arrowed.

klebsiella, *Mycobacterium tuberculosis* and anaerobic bacteria. Cavitation or abscess formation is only recognisable once the abscess communicates with the bronchial tree allowing the liquid centre of the abscess to be coughed up and replaced by air. The air is then seen as a transradiancy within the opacity of the consolidation, this may only be visible on tomography. In the erect position an air-fluid level will be visible if enough of the fluid contents of the abscess cavity remains (Fig. 2.20).

Cavitation is occasionally seen in other forms of pulmonary consolidation, e.g. infarction and Wegener's granulomatosis.

Pneumatoceles (Fig. 2.21) are thin-walled air cysts which are associated with a number of consolidative processes; staphylococcal pneumonia and pulmonary contusion are the commonest. The air cyst is believed to be the result of an air leak into the lung from a bronchiole. Pneumatoceles may increase or decrease in size very rapidly or, alternatively, they may persist for many years after the original consolidation has resolved, but eventually almost all will disappear.

Collapse

Collapse (loss of volume of a lung or lobe) may be due to any of the following:

—bronchial obstruction
—pneumothorax or pleural effusion
—fibrosis of a lobe, usually following tuberculosis
—bronchiectasis
—pulmonary embolus

Collapse due to bronchial obstruction occurs because no air can get into the lung to replace the air absorbed from the alveoli. The commoner causes are:

(a) Bronchial lesions
    —usually primary carcinoma
    —rarely endobronchial tuberculosis.
    —rarely bronchial adenoma
(b) Intraluminal occlusion
    —mucous plugging, particularly in postoperative or asthmatic patients or in pneumonia
    —by foreign body

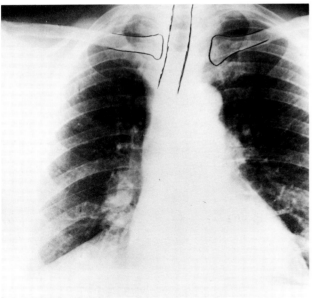

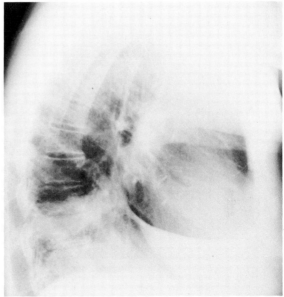

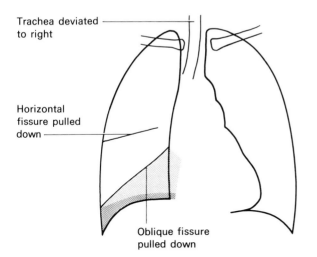

Trachea deviated to right

Horizontal fissure pulled down

Oblique fissure pulled down

Position of oblique fissure

Dense shadow due to overlapping of opaque right lower lobe on heart

Fig. 2.22 Collapse of the right lower lobe (the apical segment is relatively well aerated).

(c) Invasion or compression by an adjacent mass
 —invasion by adjacent malignant tumour
 —compression by enlarged lymph nodes

The signs of collapse are:

1. The shadow of the collapsed lobe or lobes.
2. The silhouette sign.
3. Displacement of structures to take up the space normally occupied by the collapsed lobe.

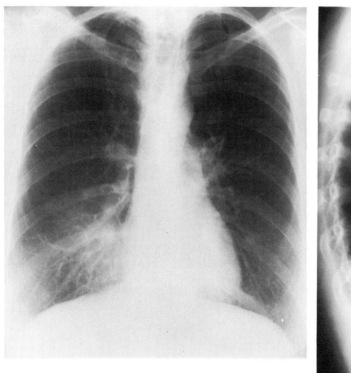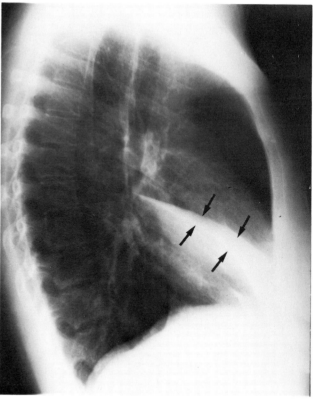

Fig. 2.23 Collapse of the middle lobe. The collapsed lobe is most obvious on the lateral view (arrows). Note the silhouette sign obliterating the lower right heart border.

Consolidation almost invariably accompanies obstructive lobar collapse, so the resulting shadow is usually obvious. Occasionally, the loss of volume is very severe and the lobe becomes so shrunken that unless it is precisely tangential to the x-ray beam the shadow may be difficult to see. The silhouette sign can be very useful in this situation, since the mediastinal and diaphragmatic borders will be ill defined adjacent to the collapsed lobe.

The silhouette sign also helps in deciding which lobe is collapsed. Collapse of the anteriorly located lobes (the upper and middle) will obliterate portions of the mediastinal and heart outlines, whereas collapse of the lower lobes obscures the outline of the adjacent diaphragm and descending aorta.

When a lobe collapses other structures move to take up the space. The unobstructed lobe(s) on the side of the collapse expands (compensatory emphysema) and this is reflected by displacement of fissures and the movement of the hilum towards the collapsed lobe. The fissure in this condition is seen as a boundary to an airless lobe *not as a line* between aerated lobes. The mediastinum and diaphragm may move towards the collapsed lobe. Since lobar collapse is such an important diagnosis and one that is often misinterpreted it is worth devoting time to study the appearance of collapse of each of the lobes (Figs. 2.22–2.26).

In collapse of the whole of one lung the entire hemithorax is opaque and there is substantial mediastinal and tracheal shift.

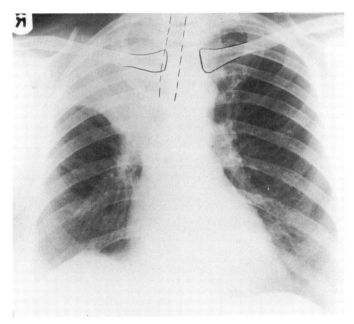

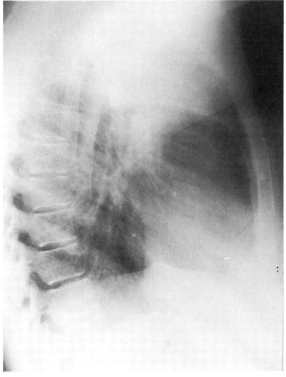

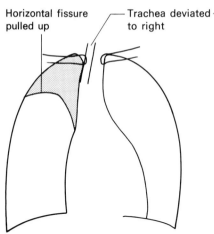

Horizontal fissure pulled up — Trachea deviated to right

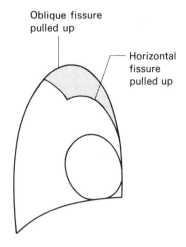

Oblique fissure pulled up

Horizontal fissure pulled up

Collapse in association with pleural abnormality. The presence of air or fluid in the pleural cavity will allow the lung to collapse. In pneumothorax the diagnosis is obvious but if there is a large pleural effusion with underlying pulmonary collapse it may be difficult to diagnose the collapse, and even if this is recognised it can be difficult to tell whether the collapse is due to pleural fluid or whether both the collapse and the effusion are due to the same process, e.g. carcinoma of the bronchus.

Fig. 2.24 Collapse of the right upper lobe.

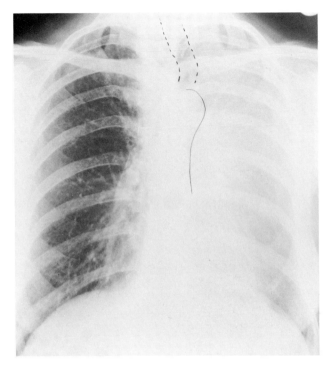

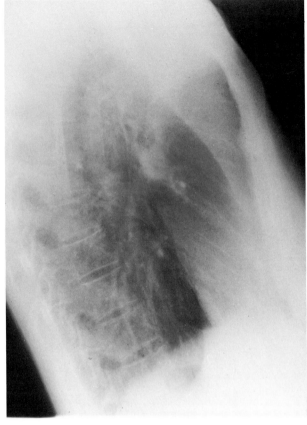

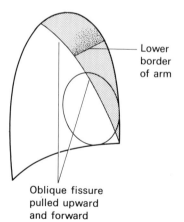

Fig. 2.25 Collapse of the left upper lobe. Note that the lower border of the collapsed lobe is ill defined on the PA view and that the upper two-thirds of the left mediastinal and heart borders are invisible, but that the aortic knuckle and descending aorta are identifiable (the visible portions of the aorta have been drawn in for greater clarity).

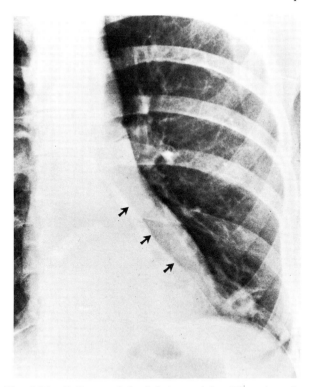

Fig. 2.26 Collapse of the left lower lobe. The triangular shadow of the collapsed lobe is seen through the heart. Its lateral border is formed by the displaced oblique fissure (arrows).

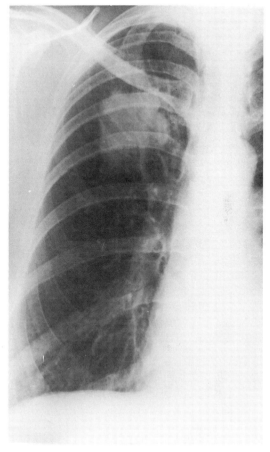

Fig. 2.27 Solitary spherical shadow. The large size and the lobulated outline are both strongly in favour of the diagnosis of primary carcinoma of the lung.

Collapse due to lobar fibrosis or bronchiectasis. In both these states a lobe may be reduced in volume, sometimes very severely, but usually the lobe remains aerated. The fissure in this situation will be seen as a line since there is air on both sides of it.

Collapse due to pulmonary embolus. With collapse due to pulmonary embolus the involved lobe usually shows a combination of patchy consolidation and loss of volume. The loss of volume is rarely very great.

Spherical shadows (lung mass)

Multiple well-defined spherical shadows in the lungs are virtually diagnostic of metastases. Occasionally, such a pattern is seen with pyaemic abscesses.

The diagnosis of a solitary spherical shadow in the lung, often referred to as a 'coin lesion' or a 'nodule' (Fig. 2.27), is a common problem.

The usual causes are

—primary carcinoma
—benign tumour of the lung, hamartoma being the commonest
—infective granuloma, tuberculosis being the commonest in the United Kingdom, fungus granulomas being commonest in the U.S.A.
—metastasis
—lung abscess

Frequently the patients have no symptoms, the shadow being noted on a routine chest film. When a 'coin lesion' is discovered in a patient over 40 years of age with haemoptysis or weight loss bronchial carcinoma becomes the major consideration. Primary carcinoma is unlikely particularly in a patient under 30 years of age.

The list of possible diagnoses given above contains lesions requiring totally different forms of management. Hamartomas and inactive tuberculomas are best left alone, whereas carcinoma, active tuberculosis and lung abscess require treatment. Careful observation of the features to be discussed below may help in making the diagnosis:

Comparison with previous films

Being able to assess the rate of growth of a spherical lesion in the lung is of great help in making the diagnosis. Lack of change over a period of a year or more is a strong pointer to either a benign tumour or an inactive granuloma.

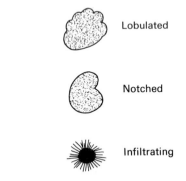

Lobulated

Notched

Infiltrating

Fig. 2.28   Outline of primary carcinoma of the lung.

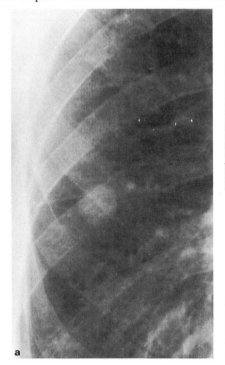

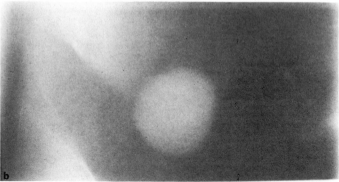

Fig. 2.29 Smooth round nodule without calcification or cavitation. (a) plain film; (b) tomogram. In this instance the lesion proved to be a hamartoma. The differential diagnosis includes tuberculoma (or fungal granuloma) and metastasis. Primary carcinoma is unlikely in a spherical, completely smooth small nodule.

The shape of the shadow

Primary carcinomas nearly always show a lobulated, notched or infiltrating outline (Fig. 2.28). Even if only one small portion of the lesion has an ill-defined edge the diagnosis of primary carcinoma should be seriously considered. If the shadow is perfectly spherical and the edge very well defined it is likely to be a hamartoma, a tuberculoma or a metastasis (Fig. 2.29).

Calcification (Fig. 2.30)

Substantial calcification virtually rules out the diagnosis of a malignant lesion. It is a common finding in hamartomas and tuberculomas. In hamartomas the calcifica-

tion is often of the 'popcorn' type. Tomography is of great value in detecting small calcifications and confirming that the calcification is within the lesion, not just projected over it.

Cavitation (Figs. 2.31 & 2.32)

If the centre of the mass undergoes necrosis and is coughed up air is seen in the mass. Such air is often accompanied by fluid, in which case an air-fluid level will be visible on erect films. The air, which may be difficult to appreciate without tomography, is seen as a translucency within the mass.

The presence of a cavity always indicates a significant lesion. It is very common in lung abscess, relatively common in primary carcinomas and occasionally seen with metastases. It does not occur in benign tumours or inactive tuberculomas.

The distinction between cavitating neoplasms and lung abscesses can be very difficult and sometimes impossible, particularly if the walls are smooth. If, how-

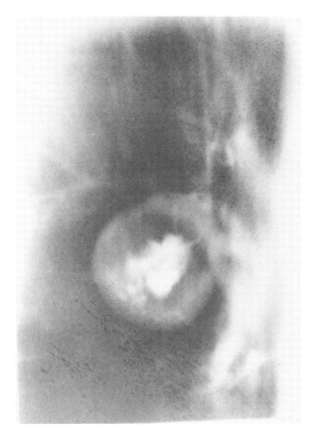

Fig. 2.30 Calcification in a pulmonary hamartoma (tomogram). The central flocculant calcification is typical of that seen in hamartomas.

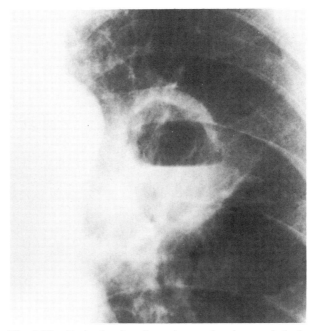

Fig. 2.31 Cavitation in a lung abscess showing a relatively thin, smooth wall and an air-fluid level.

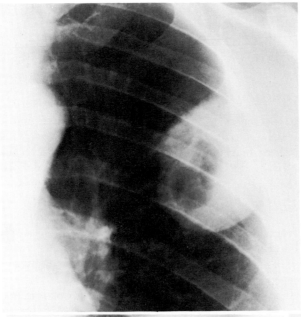

Fig. 2.32 (*left*) Cavitation in a primary carcinoma of the lung. The cavity is eccentric and its internal margins are irregular.

ever, either the inner or outer walls are irregular the diagnosis of carcinoma is highly likely.

An important point to bear in mind is that an infective lung abscess may be secondary to obstruction of a central bronchus, e.g. from carcinoma or foreign body.

Size

A solitary lesion over 6 cm in diameter which does not contain calcium is nearly always either a primary carcinoma or a lung abscess.

Involvement of adjacent ribs

Destruction of the adjacent ribs is virtually diagnostic of invasion by carcinoma. Tumours of the lung apex are particularly liable to invade the chest wall and adjacent bones (Pancoast's tumour) (Fig. 2.94).

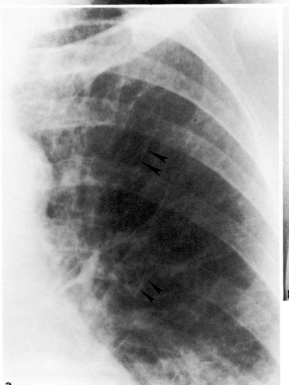

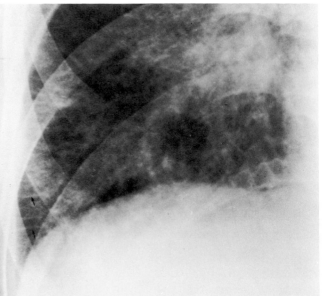

Fig. 2.33 Septal lines. (a) Deep septal lines (Kerley A lines); (b) Kerley B lines. Note that the septal lines (arrows) in these two patients with pulmonary oedema are thinner than blood vessels and that the B lines are seen in the outer cm of lung where the blood vessels are invisible or very difficult to identify.

The appearance of the adjacent lung

There are often small shadows or calcifications in the lung adjacent to a tuberculoma or other infective granulomas. The presence of old tuberculous disease in the surrounding lung does not, however, exclude the diagnosis of carcinoma, since carcinomas are known to develop in lung previously infected by tuberculosis.

Other lesions

Always check the rest of the film extra carefully after you have noticed a lung mass. Finding a second lesion may alter the diagnostic possibilities. Finding a metastasis or a small pleural effusion may completely alter the management of the patient.

Line shadows

The term 'line shadow' refers to a line, usually no thicker than an ordinary pencil line, traversing the lung. All line shadows except fissures are abnormal. Septal lines are by far the most important. They should never be ignored since they always indicate significant disease.

Septal lines

The pulmonary septa are connective tissue planes containing lymph vessels. They are normally invisible. It is only when they become thickened that they can be seen on the chest film. There are two types of septal lines:

1. Kerley A lines which radiate towards the hila in the mid and upper zones. These lines are much thinner than the adjacent blood vessels and do not reach the lung edge (Fig. 2.33a).

2. Kerley B lines which are horizontal, never more than 2 cm in length and are seen best at the periphery of the lung bases. In contrast to the blood vessels they often reach to the edge of the lung (Fig. 2.33b).

There are two important causes of septal lines

—pulmonary oedema
—lymphangitis carcinomatosa (p. 80)

Pleuropulmonary scars

These scars from previous infection or infarction are a common cause of line shadows. They usually reach the pleura and are often associated with visible pleural thickening. Such scars are of no significance to the patient.

Emphysematous bullae

Bullae are often bounded and traversed by very thin line shadows, the bullae have no normal vessels within them and this makes the interpretation easy (Fig. 2.34).

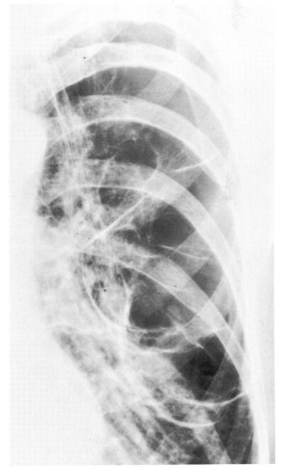

Fig. 2.34 Bullae. The bullae (blebs) are the air spaces devoid of blood vessels. The walls of the bullae are seen as line shadows.

The pleural edge in a pneumothorax

The pleural edge is seen as a line approximately parallel with the chest wall. No lung vessels will be seen beyond the pleural line. Once the line is spotted the diagnosis is rarely in doubt (Fig. 2.49).

Widespread small shadows

Nodular and reticular shadows

Chest films with widespread small (2—5 mm) pulmonary shadows often present a diagnostic problem. With few exceptions it is only possible to give a differential diagnosis when faced with such a film. A final diagnosis can rarely be made without an intimate knowledge of the patient's symptoms and signs.

Many descriptive terms have been applied to these shadows, the commonest being 'mottling', honeycomb', 'fine nodular', 'reticular' and 'reticulonodular' shadows. In this book we will use two basic terms: 'nodular', to signify discrete small round shadows and 'reticular' to describe a net-like pattern of small lines. Often there are both nodular and reticular elements and this is called 'reticulonodular' (Figs 2.35–2.37).

All these patterns are due to very small lesions in the lung, no more than 1 or 2 mm in size. Individual lesions of this size are invisible on a chest film. That these very small lesions are seen at all is explained by the phenomenon of superimposition; when myriads of tiny lesions are present in the lungs it is inevitable that many will lie in line with one another. It follows that when very small non-calcified shadows are visible the lung must be diffu-

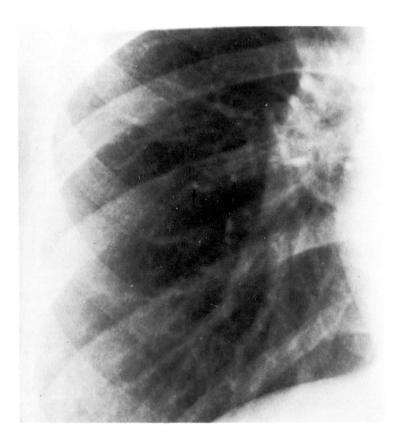

Fig. 2.35 Normal lung. Note that the lung vessels are the only identifiable shadows in those areas which are free of overlying ribs but no vessels are visible in the outer 2 cm of the lung. Vessels seen end-on appear as small nodules (arrows).

sely involved by disease. It is worth noting that the size of the multiple small shadows seen on the x-ray film gives no clue to the size of the responsible lesions, except to predict that they are small; nor can the shape of the lung shadows be reliably used to predict the shape of the lesions seen at pathology.

How to decide whether or not multiple small pulmonary shadows are present

Often the greatest problem is to decide whether widespread abnormal shadowing is present at all, since normal blood vessels can also appear as nodules and interconnecting lines. To be confident involves looking carefully at many hundreds of normal films to establish the normal pattern. Look particularly at the areas between the ribs where the lungs are completely free of overlying shadows. Note that in the normal (Fig. 2.35):

1. Vessels seen end-on appear as small nodules but these nodules are no bigger than the vessels seen in the immediate vicinity and their number corresponds to the expected number of vessels in that area.

2. The linear pattern is a branching system which connects up in an orderly way with the larger vessels more centrally.

3. There are no visible vessels in the outer 1–2 cm of the lung.

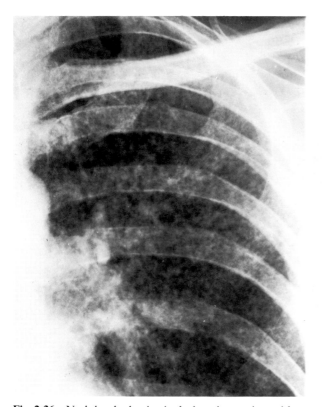

Fig. 2.36 Nodular shadowing in the lung in a patient with miliary tuberculosis.

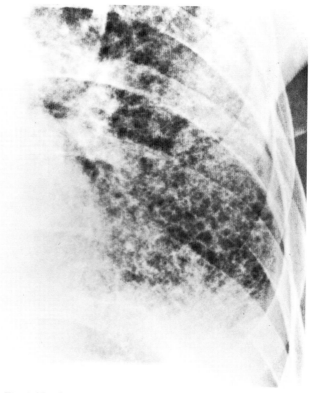

Fig. 2.37 Reticulonodular shadowing in the lung in a patient with fibrosing alveolitis.

Table 2.1   Commoner causes of nodular and reticular shadowing

Diagnosis	*X-ray pattern*	*Distribution of shadows*	*Other features which may be seen*
Miliary tuberculosis	Small nodules of uniform size	Uniform	± mediastinal/hilar lymph nodes One or more patches of consolidation
Sarcoidosis	(a) Fine nodular	Uniform	Hilar and paratracheal lymph nodes
	(b) Reticulonodular	Often predominant in mid and upper zones	Hilar and paratracheal lymph nodes
Coal miners' pneumoconiosis	Nodular	Predominant in upper zones	Progressive massive fibrosis in the complicated form of the disease Emphysema
Asbestosis	Fine reticulonodular	Predominant in lower zones	Pleural thickening and/or calcification
Fibrosing alveolitis	Reticulonodular	Often predominant in lower zones, but may show a variety of patterns	Diaphragms often high and indistinct
Lymphangitis carcinomatosa	Reticulonodular	No predominant pattern	Septal lines Bronchial wall thickening Hilar adenopathy Other signs of carcinoma
Pulmonary oedema	Ill-defined nodules	Often central predominance with clear zone at periphery of lobes	Cardiac enlargement Left atrial enlargement Septal lines

Another important sign is that the abnormal shadows obscure the adjacent vessels and the borders of the mediastinum and diaphragms may be less sharp than normal.

Having decided that abnormal shadowing is present, and determined its pattern, the next step is to decide whether or nor the shadows are maximal in one or more zones of the lung, or whether they are uniformly distributed. Then other abnormalities on the film should be looked for.

Having made all these observations one is in a position to provide the differential diagnostic list. Rather than discuss all 130 possibilities (more are described every year) we have tabulated the more important and common ones (Table 2.1).

Multiple ring shadows of 1 cm or larger are diagnostic of bronchiectasis (Fig. 2.38). The shadows represent dilated thick-walled bronchi. If they contain air and fluid they will show fluid levels.

Widespread pulmonary calcification may occur following pulmonary infection with tuberculosis, histoplasmosis or chickenpox (Fig. 2.39). Tuberculosis is the com-

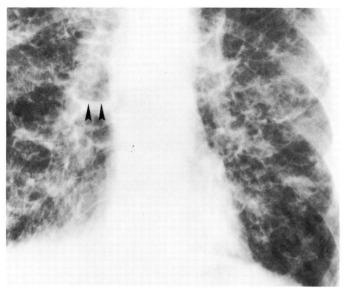

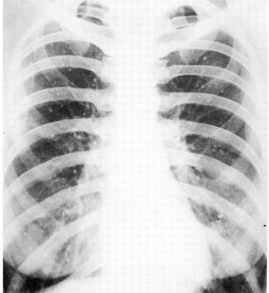

Fig. 2.38 Bronchiectasis causing large ring shadows. Each ring shadow represents a dilated bronchus. A fluid level in one of the dilated bronchi is arrowed.

Fig. 2.39 Calcified chicken-pox pneumonia. The small round well-defined scattered calcifications are the late result of adult chicken pox pneumonia. A similar appearance is seen in calcified histoplasmosis.

monest cause of widespread pulmonary calcification in most parts of the world, but in parts of the U.S.A. histoplasmosis is a more common cause.

Small calcified nodules may form in the lungs, particularly in the lower zones, of patients with long-standing rheumatic mitral valve disease. They are believed to be the result of pulmonary haemorrhage.

Increased transradiancy of the lungs

1. Generalised increased transradiancy of the lungs is one of the signs of emphysema. The other signs outlined on p. 67 must be present for the diagnosis to be made.
2. When one hemithorax appears more transradiant than normal the following should be considered:

(a) Compensatory emphysema. Occurs when a lobe or lung is collapsed or has been excised and the remaining lung expands to fill the space.

(b) Pneumothorax. The diagnosis of a pneumothorax depends on visualising the lung edge with air peripheral to it, and checking that the space one believes to be a pneumothorax does not contain any vessels (see p. 48).

(c) Reduction in the chest wall soft tissues, e.g. mastectomy.

(d) Overinflation due to central obstruction (Fig. 2.40). Most obstructing lesions in a major bronchus lead to lobar collapse. Occasionally, particularly with an inhaled foreign body, a check valve mechanism occurs leading to air-trapping and overinflation. The foreign body usually lodges in a major bronchus; the right because of its more vertical orientation being commoner than the left. The affected lung becomes abnormally transradiant and the heart is displaced to the opposite side. These signs are particularly obvious on expiration since air is trapped when the patient, usually a child, breathes out. Air trapping is best appreciated at fluoroscopy when the fixed position of the diaphragm is noted

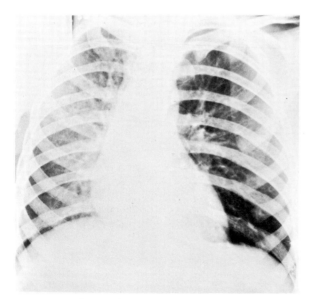

Fig. 2.40 Inhaled foreign body causing check valve obstruction of the left main bronchus. Note the increased transradiancy of the left lung, and the slight displacement of the heart to the right.

and the mediastinum can be seen to swing away from the obstructed side on expiration.

(e) Massive pulmonary embolism to one or other pulmonary artery may lead to unilateral reduction in the size of all the blood vessels beyond the hilum.

(f) McLeod's syndrome, which is also known as the Swyer–James syndrome, is due to pneumonia in childhood when the growing lung is damaged by infection. The affected lung, when seen in adults, is reduced in volume and shows reduction in blood flow with air trapping on expiration.

PLEURAL PATHOLOGY

Pleural effusion

The radiological appearances of fluid in the pleural cavity are the same regardless of whether the fluid is a transudate, an exudate, pus or blood. Because of its opacity a large effusion may hide abnormality in the

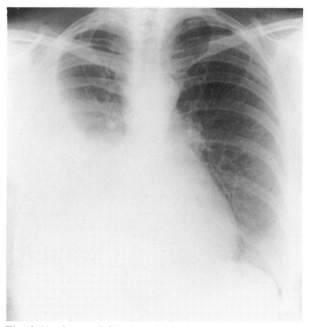

Fig. 2.41 Large right pleural effusion. The shadow of the pleural fluid is entirely homogeneous and lies outside the lung edge. The fluid appears higher laterally than medially, a point that can be useful in differentiating pleural fluid from pulmonary shadows. In this instance the trachea and heart are in normal position but no pulmonary disease was present.

underlying lung, which may be revealed only if lateral decubitus views or pleural aspiration is undertaken.

Free pleural fluid (Fig. 2.41)

Free fluid collects in the most dependent portion of the pleural cavity and always fills in the costophrenic angles. Free pleural effusions assume two basic shapes; they are usually seen in combination with one another:

1. One is analogous to what happens when a large balloon is pressed down into a bucket of water. The water is forced around the balloon up the sides of the bucket. Similarly, with a pleural effusion, some of the fluid 'runs up' all sides of the lung and appears as a concave edge. It also runs into the fissures, particularly into the lower end of the oblique fissures. Very large effusions run over the top of the lung.

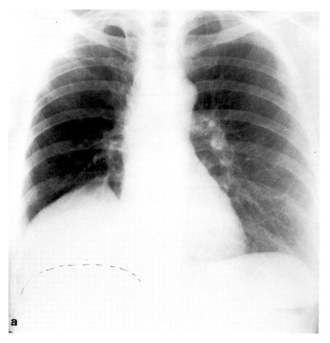

Fig. 2.42 Large right subpulmonary effusion. Almost all the fluid is between the lung and the diaphragm. The right diaphragm cannot be seen. (a) Its estimated position has been pencilled in (the patient has had right mastectomy); (b) a lateral decubitus view of the same patient, the pleural fluid moves to lie between the lateral chest wall and the lung edge (arrows).

The smooth edge between the lung and the fluid can be recognised on an adequately penetrated film, providing that the underlying lung is aerated. This smooth edge should always be looked for: it is diagnostic of pleural pathology.

2. Sometimes little or no fluid is seen running up the chest wall. The fluid is then known as a 'subpulmonary effusion' (Fig. 2.42). The upper border of the fluid is much the same shape as the normal diaphragm, and since the true diaphragm shadow is obscured by the fluid it may be very difficult or even impossible to tell from the standard erect film if any fluid is present at all.

If there is doubt about the presence or amount of the fluid a frontal film taken with the patient lying on one side (a lateral decubitus view) can be of help. The fluid, if free to move, will then lie along the dependent lateral chest wall. This technique is particularly valuable when the effusion is largely or wholly subpulmonary.

Whenever pleural fluid comes into contact with the diaphragm, heart or mediastinum, the normal boundary is invisible—another application of the silhouette sign.

Since a pleural effusion occupies space in the thorax compression collapse of the underlying lung is inevitable, the compressed lung being otherwise normal. Alternatively, both the pleural effusion and the pulmonary collapse may be due to the same primary process, e.g. carcinoma of the bronchus. Distinguishing these two forms of pulmonary collapse can be difficult. The position of the trachea is the most useful sign, if the trachea is displaced to the side of the effusion there must be substantial collapse of the lung (Fig. 2.44b). If the trachea is central it is usually very difficult to decide whether the underlying collapse is purely due to compression or whether pulmonary disease is also present. Very large effusions with little underlying pulmonary collapse displace the mediastinum and trachea to the opposite side (Fig. 2.44a).

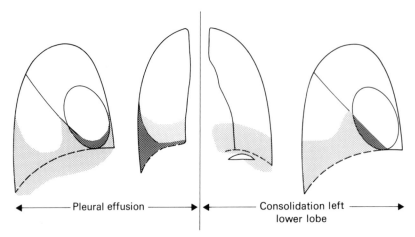

Fig. 2.43 Distinguishing pleural effusion from pulmonary consolidation or collapse

Effusion	*Consolidation/collapse*
Fluid is higher laterally than medially	Shape conforms to that of lobe(s) or segment(s)
Homogenous shadow outside lung. Lung bounded by a sharp concave boundary	No homogeneous shadow outside the lung edge
No air bronchogram	Air bronchogram may be present
Where effusion runs into a fissure both sides of the fissure are visible	Where the consolidation abuts against a fissure, the fissure is seen as an edge, therefore only one side of the fissure is visible
Trachea and mediastinum cental or pushed to opposite side (Fig. 2.44a)	Trachea and mediastinum central or pulled towards side of shadow (Fig. 2.44b)

Distinguishing pleural effusion from pulmonary consolidation or collapse requires careful observation. The table below Fig. 2.43 describes the points to look for.

Loculated pleural fluid (Fig. 2.45)

Loculated effusions occur when the free flow of fluid within the pleural cavity is prevented by pleural adhesions. Although loculation occurs in all types of effusion

Fig. 2.45 (*facing page*)   Interlobar effusion. (a) PA view: this patient with a right pleural effusion also has a shadow in the right upper zone; (b) the lateral view shows this has a characteristic lentiform shape due to an interlobar effusion in the upper part of the oblique fissure (arrow).

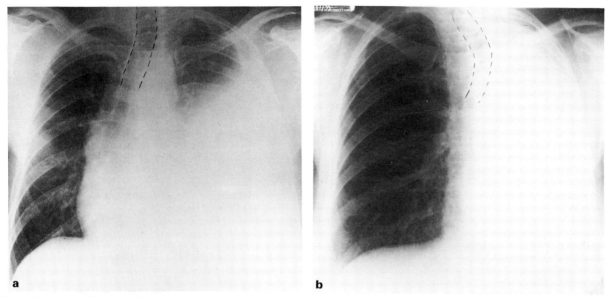

Fig. 2.44 (a) Mediastinal displacement due to a large pleural effusion; (b) collapse of the left lung showing tracheal and mediastinal displacement.

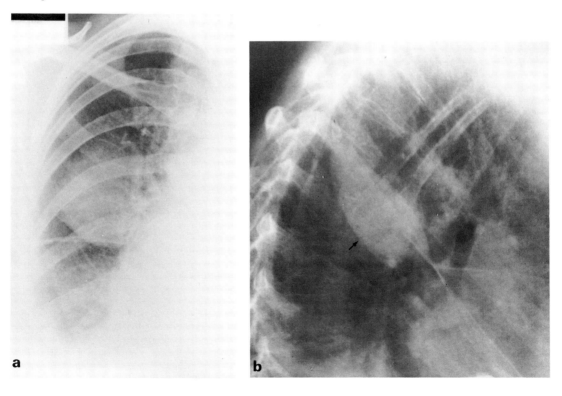

it is a particular feature of empyemas. Such loculations may either be at the periphery of the lung or within the fissures between the lobes. Loculated interlobar effusions may closely simulate lung tumours, particularly in the PA view, but the distinction can be made by noting that on the lateral film the effusion is lens-shaped, lying within the oblique or horizontal fissure. It is not surprising that the lateral view provides the answer since this is the one that demonstrates the fissures best.

Causes of pleural effusion

There are many causes for pleural effusion. In some cases the cause is visible on the chest film:

1. *Infection.* Pleural effusions due to pneumonia are on the whole small, and the pneumonia is usually the dominant feature on the chest film. Large loculated effusions in association with pneumonia often indicate empyema formation. In many cases of tuberculosis the effusion is the only abnormality and the effusion may be large.

   Subphrenic abscess nearly always produces a pleural effusion and commonly causes associated pulmonary consolidation.

2. *Malignant neoplasm.* If the effusion is due to bronchogenic carcinoma other signs of tumour are usually evident. Effusions occur with pleural metastases, but it is unusual to see the pleural deposits themselves. Such effusions are often large.

3. *Cardiac failure.* In acute left ventricular failure bilateral effusions are seen as small stripes of fluid in the costophrenic angles and fissures. In the longstanding cases the effusions are larger. They are usually bilateral often larger on the right than the left. Other evidence of cardiac failure, such as alteration in the size or shape of the heart, pulmonary oedema or the signs of pulmonary venous hypertension, are usually present.

4. *Pulmonary infarction* may cause pleural effusion. Such effusions are usually small and accompanied by a shadow of the pulmonary infarct.

5. *Collagen diseases.* Pleural effusions, either unilateral or bilateral, are relatively common in collagen diseases. They may be the only abnormal features on a chest film,

but other features of collagen disease may also be present (see p. 64).

6. *Nephrotic syndrome, renal failure, ascites and Meig's syndrome* are all associated with pleural effusions, the cause of which cannot be determined from the chest film.

Pleural thickening (pleural fibrosis) (Fig. 2.46)

Following resolution of a pleural effusion, particularly following pleural infection or haemorrhage, pleural thickening may occur. The appearances are similar to pleural fluid but always smaller than the original sha-

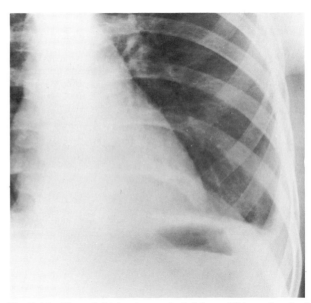

Fig. 2.46 Pleural thickening at the left base. This patient had been treated for a tuberculous pleural effusion which had resolved leaving pleural thickening, which obliterated the left costophrenic angle.

dow, often much smaller. The costophrenic angle remains obliterated. It is sometimes impossible to distinguish fluid from thickening on conventional projections, especially if comparison with previous films is not possible. The problem can be resolved by a lateral decubitus view, where free fluid will move to lie along the lateral chest wall, whereas pleural thickening is unaltered in appearance.

Localised plaques of pleural thickening along the lateral chest wall commonly indicates asbestos exposure.

Pleural tumours (Fig. 2.47)

Pleural tumours produce smooth lobulated masses based on the pleura. Malignant pleural tumours, both primary and secondary, frequently cause pleural effusions and the tumour itself may be partly or wholly hidden by the shadow of the fluid. The commonest pleural tumours are metastatic carcinoma. Primary pleural tumours (benign or malignant mesotheliomas) are relatively uncommon. Since many malignant mesotheliomas are secondary to asbestos exposure the other features of asbestosis (pulmonary fibrosis, pleural thickening and calcification) may be seen.

Pleural calcification (Fig. 2.48)

Irregular plaques of calcium may be seen with or without accompanying pleural thickening. When unilateral

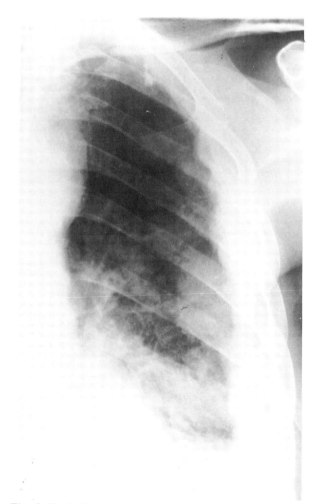

Fig. 2.47 Lobulated pleural thickening due to malignant neoplasm. The tumour in this instance was a mesothelioma of the pleura.

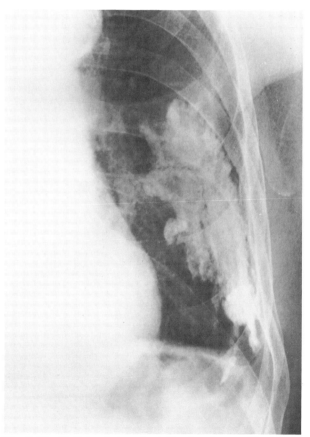

Fig. 2.48 Unilateral pleural calcification due to old tuberculous empyema.

it is likely to be due to either an old empyema, usually tuberculous, or an old haemothorax. Bilateral pleural calcification is often due to asbestosis but it is occasionally seen with old tuberculous disease and old haemothorax. Sometimes no cause can be determined.

Pneumothorax (Fig. 2.49)

The diagnosis of pneumothorax depends on recognising:

1. The line of pleura forming the lung edge separated from the chest wall, mediastinum or diaphragm by air.
2. The absence of vessel shadows outside this line.

Lack of vessel shadows alone is insufficient evidence on which to make the diagnosis, since there may be few, or no visible vessels in emphysematous bullae. One might have expected that the collapsed lung beneath the pneumothorax would be more opaque than normal, but in fact, unless the pneumothorax is very large, there is no appreciable increase in the density of the lung.

The detection of a small pneumothorax can be very difficult. The cortex of the normal ribs take a similar course to the line of the pleural edge, so the abnormality may not strike the casual observer. Sometimes a pneumothorax is more obvious on a film taken in expiration.

Once the presence of a pneumothorax has been noted the next step is to decide whether or not it is under tension. This depends on detecting mediastinal shift.

It is worth noting that since the underlying lung collapses with increased pressure in the pleural space tension pneumothoraces are usually large.

Causes of pneumothorax

Pneumothorax is associated with many underlying disorders of the lung including

—emphysema
—trauma
—certain forms of pulmonary fibrosis
—tuberculosis
—rarely metastases

The majority, however, occur in young people with no

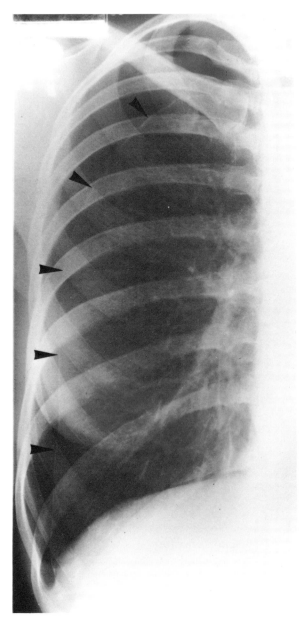

Fig. 2.49 Pneumothorax. The pleural edge is arrowed. The diagnosis of pneumothorax requires the identification of this edge and a clear space beyond it.

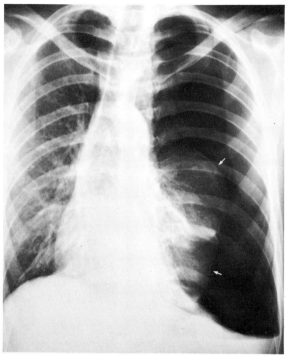

Fig. 2.50 Tension pneumothorax. The left diaphragm is depressed and the mediastinum is shifted to the right. The left lung (arrows) is substantially collapsed.

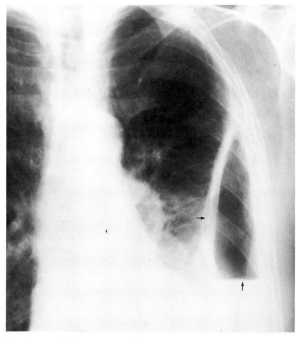

Fig. 2.51 Hydropneumothorax. There is a loculated hydropneumothorax. The vertical arrows point to the air-fluid level. The horizontal arrow points to the thickened pleura over the left lung.

recognisable lung disease. These patients have small blebs or bullae at the periphery of their lungs which burst.

Hydropneumothorax, haemopneumothorax and pyopneumothorax

In most cases of pneumothorax, whatever the cause, some fluid is present in the pleural cavity. In spontaneous pneumothorax the amount is usually small.

Fluid in the pleural cavity, whether it be a pleural effusion, blood or pus, assumes a different shape in the presence of a pneumothorax. The diagnostic feature is the air-fluid level (Fig. 2.51).

THE MEDIASTINUM

The mediastinum is one continuous space between the

sternum at the front, the spine and ribs at the back and the lungs on the two sides. The mediastinum is divided into anterior, middle and posterior divisions for descriptive purposes (Fig. 2.52). It should be realised, however, that masses often cross from one compartment to the other. Local or generalised widening can be due to many different pathological processes. These are usually classified according to their position in the mediastinum (Fig. 2.53); so if a mediastinal mass is identified on the frontal view the next step must be to attempt to localise it in the lateral view. This may be quite easy to do, but anterior mediastinal masses are sometimes difficult to visualise in the lateral view (Fig. 2.54). Therefore, if no obvious abnormality is seen in the lateral film, the anterior mediastinum should be carefully reviewed. In most people a transradiant area, known as the retrosternal space, can be identified behind the sternum in front of the ascending aorta. If this space is uniformly opaque it

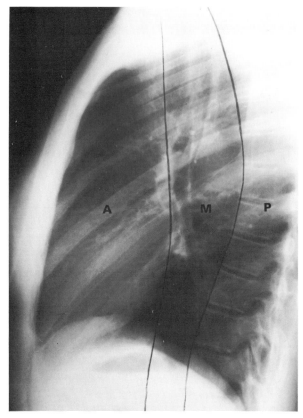

Fig. 2.52 The anterior (A), middle (M) and posterior (P) compartments of the mediastinum. The divisions are arbitrary and do not correspond to those used by anatomists. The anterior mediastinum refers to the structures anterior to a line forming the anterior boundary of the trachea and the posterior surface of the heart. The posterior mediastinum refers to structures posterior to a line forming the anterior boundary of the vertebral bodies.

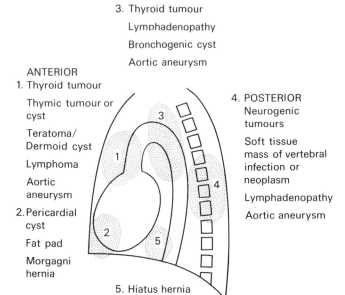

MIDDLE
3. Thyroid tumour
   Lymphadenopathy
   Bronchogenic cyst
   Aortic aneurysm

ANTERIOR
1. Thyroid tumour
   Thymic tumour or cyst
   Teratoma/ Dermoid cyst
   Lymphoma
   Aortic aneurysm
2. Pericardial cyst
   Fat pad
   Morgagni hernia

4. POSTERIOR
   Neurogenic tumours
   Soft tissue mass of vertebral infection or neoplasm
   Lymphadenopathy
   Aortic aneurysm

5. Hiatus hernia

Fig. 2.53 The causes of mediastinal masses divided according to location. Note that both lymphadenopathy and aortic aneurysms may occur in any of the three major compartments.

is probable that there is anterior mediastinal pathology. It should be realised, however, that this space is sometimes opaque in obese but otherwise normal subjects.

The various types of mediastinal masses are listed in Fig. 2.53. It is worth noting the following:

1. Lymphadenopathy is the commonest cause of a mediastinal mass. It may occur in any of the three compartments and it is often possible to diagnose enlarged lymph nodes from their lobulated outlines.

2. Neurogenic tumours are by far the commonest posterior mediastinal lesions. They frequently cause pressure erosions of the adjacent ribs and thoracic spine.
3. Certain tumours, such as dermoid cysts and thymomas, are confined to the anterior mediastinum.
4. Calcification occurs in many conditions but never in malignant neoplastic lymphadenopathy. Occasionally, the calcification is characteristic in appearance, e.g. in aneurysms of the aorta (Fig. 2.55).
5. A mediastinal mass due to a hiatus hernia is usually easy to diagnose on plain films because it often contains air and may have a fluid level; best seen on the lateral view (Fig. 2.56). A film taken after a mouthful of barium has been swallowed will easily confirm or exclude the diagnosis of hiatus hernia.
6. Masses in the right cardiophrenic angle anteriorly are virtually never of clinical significance. They are nearly all either large fat pads, benign pericardial cysts or hernias through the foramen of Morgagni (Fig. 2.57).

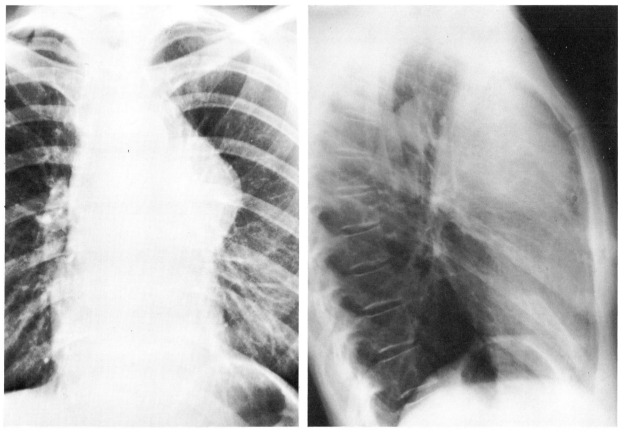

Fig. 2.54 Anterior mediastinal mass. There is a large mass situated anteriorly in the mediastinum projecting to the left side which was due to a mass of lymph nodes involved by malignant lymphoma. Diagnosing the anterior location of the mass depends on noting the density of the retrosternal areas. This area should normally have the same density as the retrocardiac area.

7. A mass in the superior mediastinum that extends into the neck and displaces the trachea is likely to be a goitre.

Barium swallow examination is of great value in assessing middle mediastinal masses, because they usually cause extrinsic compression of the oesophagus. The position and shape of the compression helps in localising and assessing the size of the lesion. Hiatus hernia is readily confirmed by barium swallow.

Pneumomediastinum

Air in the mediastinum indicates a tear in the oeso-phagus or an air leak from the bronchi. These may be spontaneous or follow trauma, namely: external injury, damage by a foreign body, endoscopy or surgery.

Spontaneous leakage from the bronchial tree is most commonly seen in patients with asthma where the air tracks through the interstitial tissues of the lung into the mediastinum following rupture of a small airway.

The air is seen as fine streaks of transradiancy within the mediastinum, often extending upward into the neck and supraclavicular fossae (Fig. 2.58).

Traumatic rupture of the aorta causes mediastinal widening. This is discussed in more detail on p. 75.

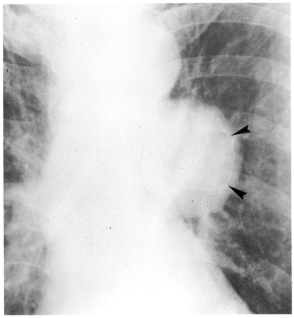

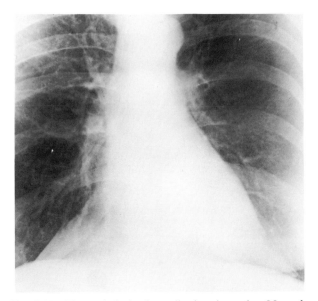

Fig. 2.55 Calcification in an aneurysm arising from the descending aorta. The arrows point to the distinctive curvilinear calcification within the mass, which is in intimate contact with the aorta.

Fig. 2.57 Fat pads in both cardiophrenic angles. Note the loss of clarity of the adjacent cardiac outline—an example of the silhouette sign. The anterior location was confirmed on the lateral view.

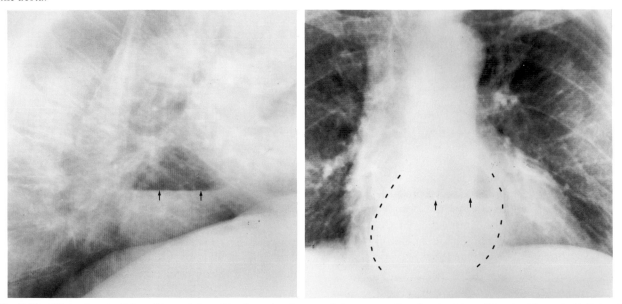

Fig. 2.56 Hiatus hernia. PA and lateral chest film show the characteristic retrocardiac density (dotted lines) with an air-fluid level (arrows).

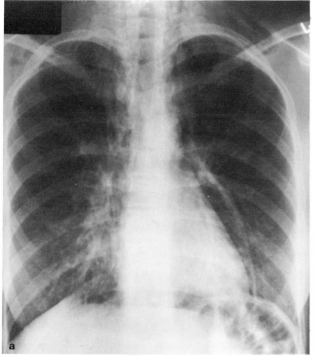

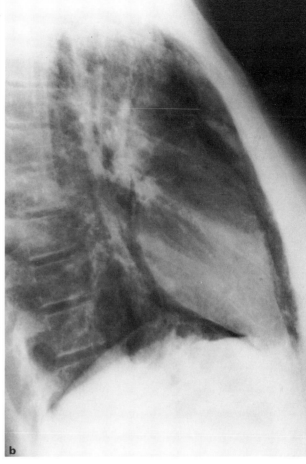

Fig. 2.58 Pneumomediastinum. (a) PA view shows the air in the mediastinum extending up into the neck; (b) lateral view shows how the heart is clearly outlined by the air.

HILAR ENLARGEMENT

Hilar enlargement presents two main diagnostic problems:

1. Is the enlarged hilum due entirely to large blood vessels, or is a mass present?
2. If a hilar mass is present, what is its nature?

How to decide whether a hilar mass is present

The normal hilar shadows are composed of the pulmonary arteries and veins. The hilar lymph nodes and bronchi cannot be identified as separate shadows. The main lower lobe arteries are the thickness of an adult's little finger (9–16 mm) with parallel walls except where they give off branches. Any lobulation of the hilum, any local expansion or any increase in the density compared with the opposite side are indications of a mass (Fig. 2.59).

Enlargement of the pulmonary arteries can usually be recognised by appreciating the branching nature of the shadow and the fact that vascular enlargement is usually bilateral and often accompanied by enlargement of the main pulmonary artery and heart (Fig. 2.60). The causes are discussed on p. 92.

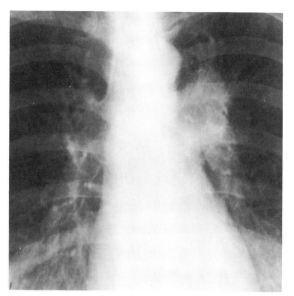

Fig. 2.59 Lobulated mass at left hilum due to enlarged lymph nodes. The right hilum is normal. The lymphadenopathy in this case was due to metastases from a bronchial carcinoma in the left lower lobe.

Hilar masses

Hilar masses are nearly always due to either lymph node enlargement or carcinoma of the bronchus. It is often possible, particularly with the help of tomography, to distinguish between these two alternatives.

Lymph node enlargement

Usually more than one lymph node is enlarged, so the hilum appears lobulated in outline (Figs. 2.59 & 2.61). The adjacent bronchi on tomography are normal or very slightly narrowed.

Unilateral enlargement of hilar lymph nodes may be due to:

1. Metastases from carcinoma of the bronchus, in which case the primary tumour is often visible; metastases from other sites are rare.
2. Malignant lymphoma.

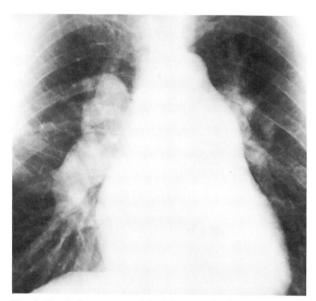

Fig. 2.60 Enlargement of the hilar arteries in a patient with severe pulmonary hypertension. Note that the heart and the main pulmonary artery are also enlarged and that the hilar shadows branch in the manner expected of arteries.

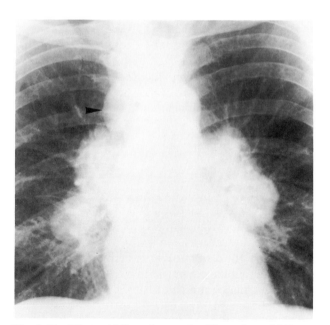

Fig. 2.61 Bilateral hilar adenopathy. The enlarged hila are clearly lobular in outline and there is also enlargement of the right paratracheal nodes (arrow). The diagnosis in this patient was malignant lymphoma.

3. Infections, particularly tuberculosis and histoplasmosis, in endemic areas. Hilar adenopathy is rarely seen in acute bacterial infections. Tuberculosis is the commonest cause of unilateral hilar adenopathy in children. It forms part of the primary complex.

Bilateral enlargement of hilar nodes (Fig. 2.61) occurs in:

1. Sarcoidosis. This is far and away the commonest cause. The diagnosis is almost certain if the patient is asymptomatic, or has either erythema nodosum or iridocyclitis. Simultaneous enlargement of the right paratracheal nodes is common. Lung changes are sometimes visible (see p. 61).
2. Tuberculosis. The African and Asian races show this form of the disease. It is rare to see bilateral hilar adenopathy due to tuberculosis in Caucasians.
3. Malignant lymphoma.
4. Glandular fever and fungus diseases are rare causes of bilateral hilar adenopathy.

Neoplasm

Primary carcinoma of the bronchus frequently presents as a hilar mass. If narrowing of the adjacent bronchus is seen on tomography or if lobar collapse/consolidation is present, the diagnosis of carcinoma is virtually certain. Lymphadenopathy due to metastatic spread from an adjacent primary carcinoma may also be visible.

THE DIAPHRAGM

The position of the diaphragm may reflect disease. Both diaphragms may be pushed up by abdominal distention or they may be high as a result of lung disease.

Unilateral elevation of the diaphragm occurs with loss of volume of the ipsilateral lung, or it may be due to an abdominal mass or a subphrenic abscess. In each of these situations the cause of the elevated diaphragm should be visible, or at least suspected, from the chest or abdominal film. Marked elevation of one diaphragm with no other visible abnormality suggests either eventration or paralysis. It should always be borne in mind that subpulmonary effusion may mimic elevation of the diaphragm (see p. 43).

Paralysis of the diaphragm results from disorders of the phrenic nerves, e.g. invasion by carcinoma of the bronchus. The signs are elevation of one hemidiaphragm (Fig. 2.91) which on fluoroscopy shows paradoxical movement, i.e. it moves upward on inspiration.

Eventration of the diaphragm is a congenital condition in which the diaphragm lacks muscle and becomes a thin membranous sheet. Except in the neonatal period it is almost always an incidental finding and does not cause symptoms. When the whole of one diaphragm is involved, almost invariably the left, that diaphragm is markedly elevated. On fluoroscopy, the diaphragm may remain fixed during inspiration and expiration, but when more severely involved it moves paradoxically and cannot be distinguished from a paralysed diaphragm. The eventration may only involve part of one diaphragm, resulting in a smooth 'hump' (Fig. 2.62).

Rupture of the diaphragm is discussed on p. 75.

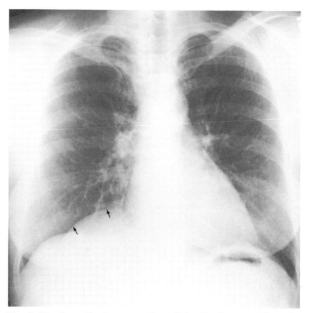

Fig. 2.62 Localised eventration of the diaphragm. There is a smooth localised elevation of the medial half of the right diaphragm (arrow) typical of a localised eventration. On the lateral view the eventration involved the anterior half of the diaphragm.

THE CHEST WALL

The chest wall should be examined for evidence of soft tissue swelling or rib abnormality. Rib abnormalities may be difficult to spot because of their curvature which results in foreshortening of some part of the rib on every view. If a rib abnormality is suspected oblique views should be obtained, this applies particularly to fractures.

Soft tissue swelling occurs with a number of rib lesions; fractures, infections and neoplasms (Fig. 2.63).

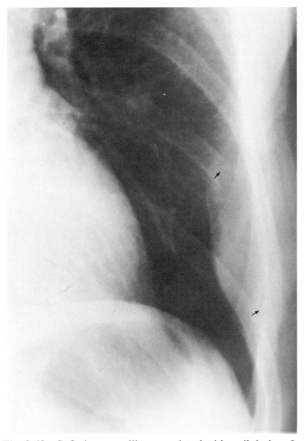

Fig. 2.63 Soft tissue swelling associated with a rib lesion. In this patient with myeloma in a rib, the soft tissue swelling is more obvious than the rib destruction. The bone between the two arrows has been destroyed. This important sign could be easily overlooked unless special attention is paid to identifying rib destruction in the region of soft tissue swelling.

The soft tissue swelling may be more obvious than the rib lesion. If an opacity suggesting soft tissue swelling is seen arising from the chest wall it is vital to obtain a good view of the underlying ribs using oblique views or tomography where necessary.

Clearly, many diseases of bone such as Paget's disease, myeloma and metastases, involve the ribs or sternum in much the same way as they do elsewhere in the skeleton.

Rib notching refers to small or well-defined notches with corticated margins on the inferior surfaces of the ribs. The commonest cause is coarctation of the aorta (Fig. 3.25). Included among the many differential diagnoses is neurofibromatosis.

Congenital abnormalities of the ribs are common but rarely of clinical significance. It is important not to mistake bifid ribs, or fused ribs for lung shadows.

PNEUMONIA AND LUNG ABSCESS

Bacterial pneumonia

The common feature of all pneumonias is a cellular exudate within the alveoli. The damage to the pulmonary parenchyma varies; with pneumococcal pneumonia, for example, complete resolution usually occurs, whereas with certain other infections, notably staphylococcal, klebsiella, anaerobic bacteria and tuberculosis, lung destruction with cavitation is common

Pneumonia may be secondary to obstruction of a major bronchus, carcinoma being the common cause. A central obstruction should always be considered in any patient presenting with consolidation of one lobe or in two lobes supplied by a common bronchus, e.g. the right middle and lower lobes.

The basic radiological features of pneumonia are one or more areas of consolidation, sometimes accompanied by loss of volume of the affected lobe. When the consolidation involves the whole of one lobe it is called *lobar pneumonia*. The common infecting organism in lobar pneumonia is *Streptococcus pneumoniae*. In pneumococcal pneumonia there is dense consolidation of one

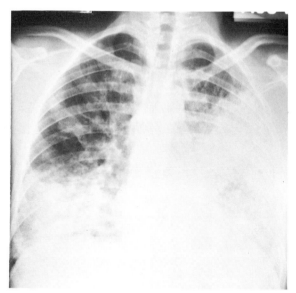

Fig. 2.64 Bronchopneumonia. There is widespread patchy consolidation involving particularly both lobes of the left lung and the right lower lobe.

lobe, usually without loss of volume. There may be an associated pleural effusion, but, if present, it is usually small.

When the consolidation is patchy, involving one or more lobes, it is commonly referred to as bronchopneumonia (Fig. 2.64). The common organisms are *Staphylococcus aureus*, various Gram-negative bacilli and anaerobic organisms. When widespread the radiological differentiation from pulmonary oedema, or pulmonary infarction can be difficult. Consolidation in pneumonia and infarction often abuts the fissures whereas pulmonary oedema often leaves a clear zone at the periphery of the lobes (see Fig. 2.15).

The differentiation from pulmonary emboli depends on the clinical findings since the radiological features of pulmonary infarcts and pneumonia are often identical.

Viral pneumonia

Viral pneumonia and pneumonia due to *Mycoplasma pneumoniae* often produce widespread ill-defined conso-

lidation and loss of clarity of the vascular markings, which on occasion may resemble pulmonary oedema. Alternatively, only a localised area of consolidation may be seen. Pleural effusions are rare. The radiological abnormality may persist for many weeks after clinical recovery.

Lung abscess

A lung abscess is a localised suppurative lesion of the lung parenchyma. The signs are described on p. 27. The commoner predisposing factors are:

1. Aspiration of food or vomit; these abscesses are usually in the apical segments of the lower lobes or in the posterior segments of the upper lobes.
2. Infection beyond an obstructing lesion in the bronchus.
3. Infected emboli, particularly in drug addicts.

PULMONARY TUBERCULOSIS

Pulmonary tuberculosis is usually divided into primary and postprimary forms. Primary tuberculosis is the result of the first infection with the bacillus *Mycobacterium tuberculosis* and usually occurs in childhood. Postprimary tuberculosis, the usual form in adults, is believed to be a reinfection, the patient having developed relative immunity following the primary infection. This is not, however, the whole explanation of the difference between the two forms of disease, and it is likely that the maturation of tissues plays a part; the older the patient, the less the tendency towards lymph node enlargement.

Primary tuberculosis

In primary tuberculosis an area of consolidation, known as the Ghon focus, develops in the periphery of the lung—usually in the mid or upper zones. Usually, the pulmonary shadow is small, but it may occasionally involve most of the lobe. Sometimes the pulmonary consolidation is so small that it is nearly invisible. The

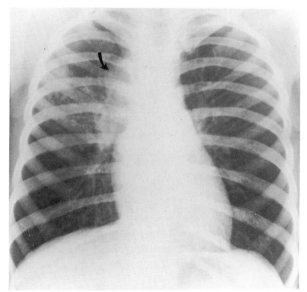

Fig. 2.65 Tuberculosis—the primary complex. This young child shows ill-defined consolidation in the right lung together with enlargement of the draining lymph nodes (arrow).

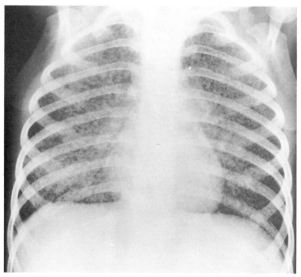

Fig. 2.66 Miliary tuberculosis. The innumerable small nodular shadows uniformly distributed throughout the lungs in this young child are typical of miliary tuberculosis. In this instance no primary focus of infection is visible.

consolidation is often accompanied by visibly enlarged hilar or mediastinal lymph nodes (Fig. 2.65). This combination of pulmonary consolidation and lymphadenopathy is known as the *primary complex*. The clinical features of the primary complex vary. The majority have few symptoms and pass unrecognised. The remainder have fever, cough and malaise and occasionally erythema nodosum.

In most cases, whether treated or not, the primary complex heals and often calcifies. A calcified primary complex remains visible throughout life.

Spread of infection may occur through:
1. The bronchial tree leading to tuberculous bronchopneumonia which radiologically appears as patchy or lobar consolidation. It often involves more than one lobe, may be bilateral and frequently cavitates.
2. The blood stream resulting in miliary tuberculosis.

In *miliary tuberculosis* (Fig. 2.66) there are innumerable small nodules visible in the lungs. They are all of much the same size and fairly evenly distributed. Usually, the nodules are well defined but in severe cases they become relatively confluent so that the individual nodules are difficult to appreciate. The primary complex may be visible and a pleural effusion may be present. Early in the course of the illness the chest film may be normal.

Primary tuberculosis may present with a pleural effusion. Occasionally the primary complex is also visible, but often the effusion is an isolated abnormality.

Postprimary tuberculosis

Postprimary tuberculosis usually presents as cough, haemoptysis, weight loss, night sweats or malaise. Occasionally, the disease is discovered on a routine chest film. Radiologically postprimary tuberculosis is usually confined to the lung in the upper posterior portions of the chest, namely the apical and posterior segments of the upper lobes and the apical segments of the lower lobes. The initial lesions are multiple small areas of consolidation (Fig. 2.67), often bilateral. If the infection progresses the consolidations enlarge and frequently cavitate. Cavities are seen as rounded air spaces (translucencies) completely surrounded by pulmonary

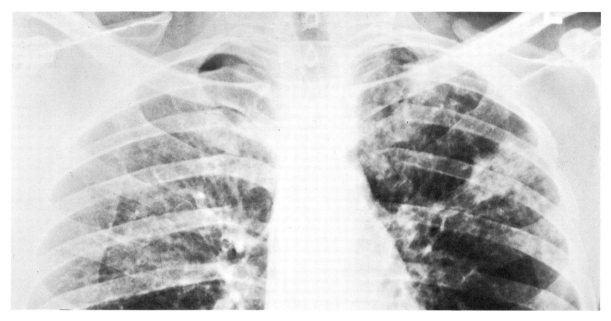

Fig. 2.67   Postprimary tuberculosis. There are ill-defined consolidations scattered in both upper lobes; their size and distribution should suggest the diagnosis of tuberculosis.

shadowing (Fig. 2.68). Tomography is often required before one can be certain of the diagnosis of cavitation. Even with tomography it may occasionally be difficult to decide whether cavities are present.

The infection may undergo partial or complete healing at any stage. Healing occurs by fibrosis and often calcification (Fig. 2.69), but both fibrosis and calcification may be seen in the presence of continuing activity.

Occasionally, the disease takes the form of lower or middle lobe bronchopneumonia.

As with the primary form, postprimary tuberculosis may spread to give widespread bronchopneumonia or miliary tuberculosis.

Pleural effusions are frequent. When they resolve they often leave permanent pleural thickening which may calcify.

Tuberculoma refers to a tuberculous granuloma in the form of a spherical mass; usually less than 3 cm in diameter. The edge is usually sharply defined and these lesions are often partly calcified. Tomography may be needed to demonstrate the calcium. Most tuberculomas are inactive but viable tubercle bacilli may be present even in the calcified lesions.

Mycetoma (Fig. 2.70). The fungus *Aspergillus fumigatus* may colonise old tuberculous cavities as a ball of fungus (mycetoma) lying free within the cavity. Since the fungus ball usually occupies only a portion of the available space a rim of air is seen between the mycetoma and the wall of the cavity. The shape and position of this rim varies with the position of the patient. Cavities containing mycetomas are usually surrounded by other evidence of old tuberculous infection; fibrosis and calcification of the adjacent lung.

Is the disease active?

An important role of radiology is to try and determine whether pulmonary tuberculosis is active or inactive. This can be very difficult and is sometimes impossible. It

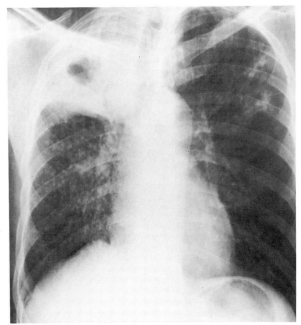

Fig. 2.68 Cavitation in tuberculosis. The right upper lobe is consolidated and contains a large central cavity. Patchy consolidation due to tuberculous bronchopneumonia which has spread via the bronchi is seen in the right mid and lower zones and in the left upper zone.

is important to realise that there is no way of excluding activity unless serial films are available. Valuable diagnostic signs of activity are:

1. Development of new lesions on serial films.
2. The demonstration of cavities.

Lack of change over a long period of time is useful evidence against activity, but the change in serial chest films may be very subtle, even with active disease. It should be remembered that the presence of calcification does not exclude activity.

Many routine chest films in asymptomatic patients show evidence of tuberculosis. In a few, the diagnosis of active disease will be readily apparent by the presence of cavities or by comparison with previous films. In the remainder it can be a considerable problem to decide which patients to investigate further and which to accept as having old inactive infection. The better defined the

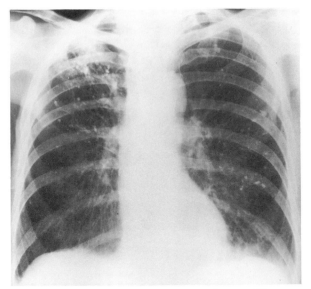

Fig. 2.69 Old calcified tuberculous disease. There are numerous foci of calcification in both lungs. The right upper lobe shows extensive fibrosis and bullae. There was no evidence in this patient that active infection was present. However, given this film in isolation, active disease could not be excluded.

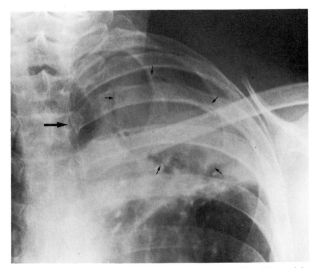

Fig. 2.70 Mycetoma. There is a mycetoma outlined by a thin rim of air (arrows) in a tuberculous cavity in a fibrotic left upper lobe. Also note the calcifications in the adjacent lung, and the way the trachea is pulled towards the contracted left upper lobe (large arrow).

shadows and the greater the calcification, the less the likelihood of activity. The presence of ill-defined shadows, even if partially calcified, is suggestive of active disease. However, the decision is often largely based on the clinical findings and the results of sputum examination for tubercle bacilli.

Fungus and parasitic diseases

When fungi are inhaled they may produce lung infection. The radiological appearances vary with the particular fungus, but two broad divisions can be made:

1. Infection of the otherwise normal patient by such organisms as histoplasmosis, coccidioidomycosis and blastomycosis (Fig. 2.71). These organisms, which are found chiefly on the continent of America and do not occur in the United Kingdom, produce lesions in the lung that are very similar and often identical to tubercu-

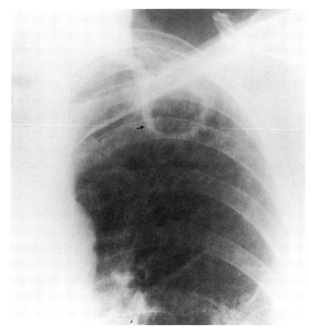

Fig. 2.71 Fungus infection. The cavity (arrow) in this patient from the southeast part of the U.S.A. was due to North American blastomycosis. Note the similarity to tuberculosis. Other fungi, e.g. histoplasmosis can give an identical appearance.

losis. Cavitation is a particular feature. Healing by fibrosis and calcification is frequent.

2. Infection in the compromised host. Due to reduced immune mechanisms fungi such as *Candida albicans* and *Aspergillus fumigatus* as as well as several of the fungi native to America may cause widespread pneumonias. It is not possible to predict the infecting organism from the chest film, indeed it is usually not possible to distinguish fungal infection from infection with bacteria, viruses and parasites, particularly *Pneumocystis carinii* in these patients.

Aspergillus fumigatus is a fungus that affects the lung in two other unusual ways. It may colonise a pre-existing cavity forming a fungus ball or mycetoma (see p. 59). Allergic bronchopulmonary aspergillosis is discussed on p. 66.

Hydatid disease. Pulmonary infection with *Echinococcus granulosus* may result in cysts in the lung or pleural cavity. These cysts may be solitary or multiple and are seen as spherical shadows with very well-defined borders.

SARCOIDOSIS

Sarcoidosis is characterised pathologically by non-caseating granulomas in many organs including lung, liver, spleen, lymph nodes, skin and bone. The aetiology is obscure and the diagnosis depends on a correlation of the clinical, pathological and radiological manifestations. The Kveim test is positive in half the patients and the Mantoux test is usually negative. Biopsy of lymph nodes, liver, lung or Kveim test sites may be necessary as part of the diagnostic investigation.

The radiological manifestations are largely confined to the chest (the bone lesions are described on p. 257). Nephrocalcinosis is another rare finding.

The chest x-ray features are:

1. Hilar and paratracheal lymphadenopathy. When hilar lymphadenopathy is present it is always bilateral (Fig. 2.72). Half the cases also show paratracheal nodal

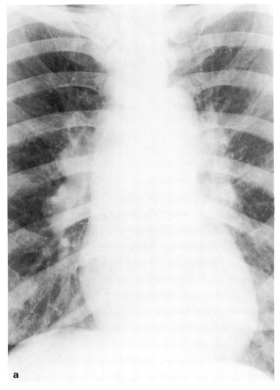

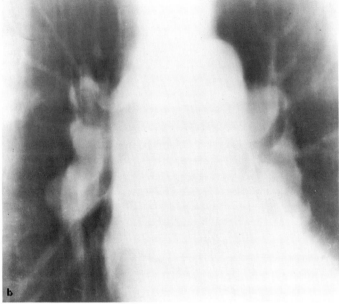

Fig. 2.72 Bilateral hilar lymphadenopathy in sarcoidosis. (a) Plain film; (b) Tomogram. The lobular outline to the hila is characteristic of lymph node enlargement. In this case the paratracheal nodes are not visibly enlarged. This patient had no symptoms or signs, the abnormality being discovered on a routine chest film.

enlargement, more obvious on the right than the left. The nodes may calcify after many years.
2. Reticulonodular shadowing in the lung. The pattern varies from uniform small nodular shadows, which may clear on steroid therapy, to the coarse reticular shadows maximal in the mid and upper zones, which represent gross pulmonary fibrosis (Fig. 2.73). At this stage the pulmonary disease is often irreversible.

The majority of patients with sarcoidosis of the chest have lymphadenopathy only, which clears without treatment and the disease does not progress to pulmonary involvement. Such patients usually do not have chest symptoms. Some present with iridocyclitis, some with erythema nodosum and fever. A few have in addition polyarthritis. Many are discovered on routine chest x-ray and have no symptoms or signs.

Approximately 10% develop lung involvement, some of these patients still have visibly enlarged lymph nodes

at this stage. Often the lymph nodes get smaller and may return to normal, even though the lung changes persist.

DIFFUSE LUNG FIBROSIS (FIBROSING ALVEOLITIS)

The known causes of diffuse pulmonary fibrosis include the following: allergic alveolitis, collagen diseases (including rheumatoid arthritis), pneumoconiosis and sarcoidosis, but a substantial proportion of cases are idiopathic in origin.

Idiopathic fibrosing alveolitis

In this disease there is thickening of the alveolar walls with fibrosis and desquamation. As the disease progresses the alveolar walls break down and rounded air

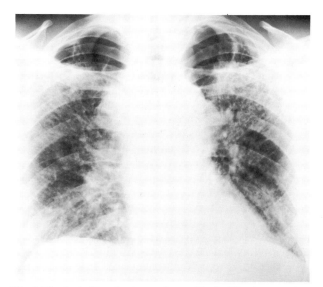

Fig. 2.73 Late fibrotic stage of sarcoidosis. The dense reticulonodular shadowing radiates outwards from the hila, maximally in the mid and upper zones. Enlarged lymph nodes are still visible at the hila and in the right paratracheal region. Many patients with this degree of pulmonary fibrosis, due to sarcoidosis, will not have visibly enlarged lymph nodes.

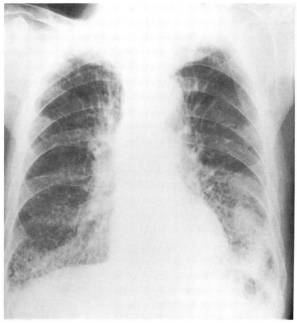

Fig. 2.74 Idiopathic fibrosing alveolitis. In this example there is reticulonodular shadowing with basal predominance. Scleroderma gives a similar picture. Notice that the diaphragms are indistinct because of the changes in the adjacent lungs.

spaces develop. At this stage the lung is known by the descriptive term 'honeycomb lung'. Fibrosing alveolitis causes a restrictive ventilation defect with severe reduction in gas transfer across the alveolar walls.

The radiological features of idiopathic fibrosing alveolitis are:

1. Hazy shadowing at the lung bases leading to a lack of clarity of the vessel outlines. Later, ill-defined nodules with connecting lines become discernible.

2. Lung volume decreases often markedly and circular translucencies are seen producing a pattern known as 'honeycomb lung' (Fig. 2.74). Eventually, the heart and pulmonary arteries enlarge due to increasingly severe pulmonary hypertension.

Determining the cause of diffuse pulmonary fibrosis

The distribution of the pulmonary shadowing may give a clue to its aetiology. In idiopathic fibrosing alveolitis

the lung shadowing is often maximal at the bases (Fig. 2.74), whereas in sarcoidosis it is usually maximal in the mid zone and in extrinsic allergic alveolitis it may be maximal in the upper zones. In scleroderma changes are often limited to the extreme lower zones, whereas in rheumatoid arthritis, they are either predominant in the lower zones or fairly uniformly distributed.

The combination of pulmonary fibrosis with other signs may lead to a specific diagnosis:

1. Past or present lymphadenopathy suggests sarcoidosis (Fig. 2.73).
2. Coexistent conglomerate masses in the mid and upper zones are virtually diagnostic of silicosis or coal miners' pneumoconiosis (Fig. 2.77).
3. Coexistent bilateral pleural thickening and calcification are diagnostic of asbetosis (Fig. 2.78).
4. Past or present pleural effusions are highly suggestive of rheumatoid arthritis.

RADIATION PNEUMONITIS (Fig. 2.75)

Radiation pneumonitis is seen following x-ray therapy for intrathoracic neoplasms and breast carcinoma with lymphatic spread. The response of the lung to radiation varies from patient to patient. Initially, there is no radio-

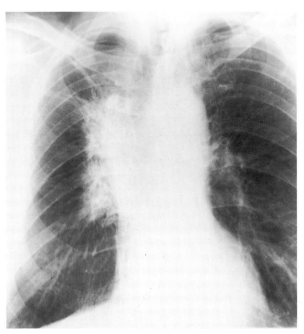

Fig. 2.75 Postradiation fibrosis. The patient had a carcinoma in the right upper lobe centrally for which he had received radiation therapy. Notice the geometric outline to the shadowing corresponding to the radiation field. There has been contraction of the irradiated lung resulting in mediastinal deviation and distortion of the pulmonary vessels.

logical change, but within a few weeks ill-defined small shadows, indistinguishable from infective consolidation, are seen in the radiation field. If the inflammatory change goes on to fibrosis there is dense coarse shadowing which may be sharply demarcated from the normal lung in a geometric fashion, conforming to the field of radiation but ignoring the lobar boundaries of the lung. There is loss of volume of the fibrosed areas. Extensive pleural thickening is also sometimes seen.

COLLAGEN DISEASES

This group of diseases includes rheumatoid arthritis, systemic lupus erythematosus, polyarteritis nodosa, systemic sclerosis, dermatomyositis and Wegener's granulomatosis. All affect the chest directly to varying degrees, but often the effects are secondary to involvement of other organs, e.g. the heart and kidneys with resultant heart failure, pulmonary oedema and pericardial effusion. Various radiological signs occur.

Rheumatoid lung

The most common finding in the chest is pleural effusion. Pulmonary fibrosis, often indistinguishable from that seen in idiopathic fibrosing alveolitis, is another important finding.

An interesting feature of pulmomary involvement in rheumatoid arthritis is the development of rounded granulomas in the periphery of the lung, similar histologically to the subcutaneous nodules seen in this disease. These spherical nodules which may be single or multiple, rarely exceed 3 cm in size. Eventually, many cavitate and resolve.

Coal miners with rheumatoid arthritis may suffer from *Caplan's syndrome*, in which numerous granulomas are seen in the lung.

Systemic lupus erythematosus

The chest x-ray is usually normal. The commonest abnormality is pleural effusion and cardiac enlargement due to pericardial effusion. Patchy consolidation in the lungs is occasionally seen.

Scleroderma and dermatomyositis

The cardinal feature is basal reticulonodular shadows due to pulmonary fibrosis similar to that seen in fibrosing alveolitis (Fig. 2.74). Usually, the fibrosis is sharply confined to the bases but, occasionally, it is more widespread. Pleural effusion is rare.

Polyarteritis nodosa

Patchy consolidations, some of which are believed to be infarcts, are the important signs of pulmonary involvement. The consolidations may be fleeting and repetitive. Pleural effusions are not uncommon.

Wegener's granulomatosis

The lungs may show one or more well-defined consolidations, usually in the mid zones which may cavitate.

PNEUMOCONIOSIS

The pneumoconioses are a group of conditions caused by the inhalation of a variety of dusts. Of the inorganic mineral dusts some are inert and although they may give rise to widespread small nodules on the chest film the dust causes no symptoms, e.g. siderosis in iron foundry workers due to the inhalation of iron or iron oxide, and stannosis in tin smelters due to the inhalation of stannous oxide. On the other hand other types of pneumoconiosis notably silicosis and coal workers' pneumoconiosis are much more serious as the inhaled dust can cause of fibrosis in the lungs. Significant pulmonary fibrosis can occur in asbestosis even with relatively minor exposure to asbestos.

Coal workers' pneumoconiosis

Simple pneumoconiosis is due to dust retention in the lungs with minor fibrosis. It is recognised radiologically by many small nodules initially in the mid and upper zones, eventually involving the whole of the lung fields (Fig. 2.76). Septal lines may also be seen. Simple pneumoconiosis does not give rise to symptoms and the diagnosis is made on the basis of the chest x-ray appearances.

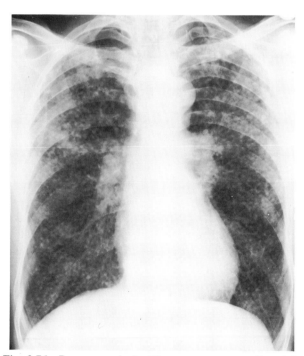

Fig. 2.76 Pneumoconiosis. There are many small nodules involving the whole of the lung fields. The patient was a coal miner.

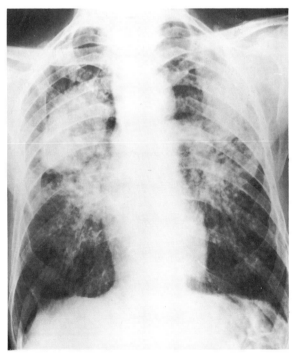

Fig. 2.77 Progressive massive fibrosis (PMF). Note the large oval shadows in the upper halves of both lungs. A nodular pattern is present elsewhere in the lung fields.

For reasons that are not entirely clear progressive massive fibrosis (PMF) may supervene. It can be recognised by homogeneous shadows, which are often ovoid in shape in the upper halves of the lungs. The shadows may be unilateral or bilateral and there is usually nodular shadowing in the rest of the lungs (Fig. 2.77). Calcification of the hilar lymph nodes may occasionally be seen. PMF can cause breathlessness and it may result in cor pulmonale.

Asbestosis

Inhalation of asbestos fibres may lead to:

1. Pleural fibrosis and calcification. Localised plaques of pleural thickening, some of which are calcified, are seen along the lateral chest wall (Fig. 2.78). The plaques in themselves are harmless, but they are a useful pointer to asbestos exposure particularly when calcified.

The differential diagnosis from postinflammatory and posttraumatic pleural thickening and calcification is made by noting the extent of the pleural disease: usually, it is bilateral in asbestosis whereas it is frequently unilateral in the other conditions. If the costophrenic angles are sharp, healed inflammatory disease or old haemothorax is very unlikely, whereas in asbestosis the costophrenic angles may be clear.

2. Pulmonary fibrosis. Pulmonary fibrosis in asbestosis is symmetrically bilateral and maximal at the bases. It produces very fine reticulonodular shadowing. Sometimes the shadowing is so fine that only a haze over the lung bases can be appreciated.

Each of the above signs may exist in isolation, but when all are seen together the diagnosis of asbestosis is certain.

DISEASES OF THE AIRWAYS

Asthma

The chest film in asthma is usually normal or shows only low flat diaphragms due to air trapping. Bronchial wall thickening may be seen. The main purpose of the chest x-ray in asthma is:

1. To determine complications, e.g. atelectasis, pneumothorax, etc.
2. To detect underlying pneumonia.
3. To exclude other causes of acute dyspnoea, e.g. pulmonary oedema or rarely tracheal obstruction.

Allergic bronchopulmonary aspergillosis results from hypersensitivity to *Aspergillus fumigatus*. Asthma is the cardinal clinical feature of this disease. The radiological signs are allergic consolidations in the lung and bronchiectasis, particularly in the mid and upper zones. The thickened walls of the dilated bronchi may be visible on a plain chest film.

Bronchiolitis

Young children with severe bronchiolitis, even when life-threatening, may show surprisingly little change on the chest film. The signs are overinflation of the lungs leading to a low position of the diaphragms. Some children show widespread small ill-defined areas of consolidation, but in many the lungs are clear.

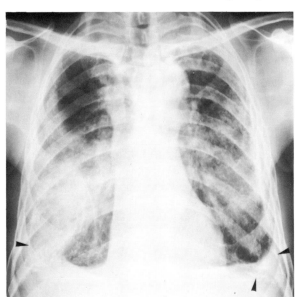

Fig. 2.78 Asbestosis. There is extensive bilateral pleural thickening and pleural calcification (arrows) best appreciated along the lateral chest wall.

Acute bronchitis

Acute bronchitis in adults and older children does not produce any radiological abnormality unless complicated by pneumonia.

Chronic obstructive pulmonary disease

Chronic obstructive pulmonary disease is an imprecise but convenient term which covers several common diseases; chronic bronchitis, emphysema and bronchiectasis.

Chronic bronchitis and emphysema

Chronic bronchitis is a clinical diagnosis based on productive cough for at least 3 consecutive months in 2 successive years. Pathologically, there is hypertrophy of the mucous glands throughout the bronchial tree with a great increase in the number of goblet cells. There is thickening of the mucous membranes and airways obstruction occurs. Bronchopneumonia is a common complication.

Emphysema is defined pathologically as 'a condition of the lung characterised by increase beyond normal size of air spaces distal to the terminal bronchiole with destructive changes in their walls'. Two types are recognised—centrilobular and panacinar depending on the extent of acinar destruction. Chronic bronchitis and emphysema often coexist though pure forms of each are seen.

Chest x-ray in chronic bronchitis. The chest film in uncomplicated chronic bronchitis is normal. Indeed, patients may die from respiratory failure due to chronic bronchitis and have a normal chest film. If the film is abnormal a complication such as emphysema, pneumonia or cor pulmonale has occurred, and the radiological features are then those of the complication in question.

Chest x-ray in emphysema. Centrilobular emphysema cannot be recognised radiologically. Its presence can be deduced when the signs of pulmonary hypertension and

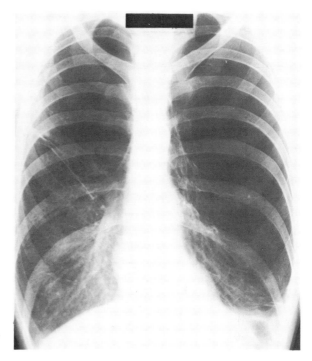

Fig. 2.79 Panacinar emphysema. The diaphragms are low and flat and the ribs are widely spaced indicating overinflation of the lungs. The peripheral vessels in most of the left lung and the upper half of the right lung are small and attenuated indicating lung destruction.

congestive heart failure are seen in a patient known to have chronic bronchitis; since post-mortem studies have shown such patients have extensive centrilobular emphysema. The precise significance of the centrilobular emphysema is still under debate.

The signs of panacinar emphysema are (Fig. 2.79):

1. Increased lung volume. The lungs increase in volume because of the combined effect of airways obstruction on abnormally compliant lungs. The diaphgrams are pushed down and become low and flat. The heart is elongated and narrowed. The ribs are widely spaced and more lung lies in front of the heart and mediastinum. (Overinflation of the lungs can be said to be present if the

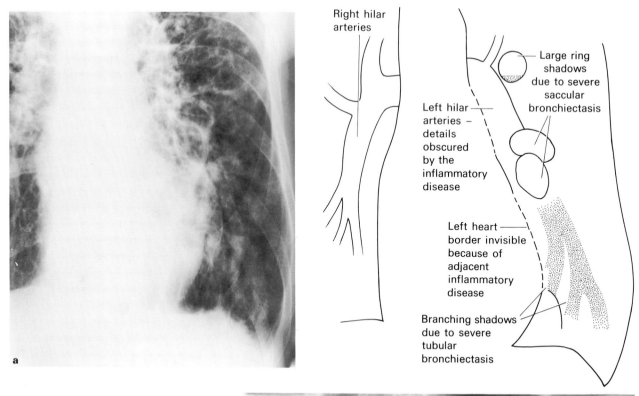

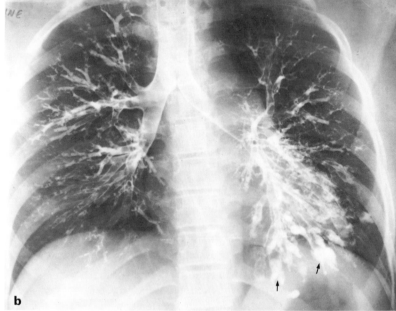

Fig. 2.80 Bronchiectasis. (a) Plain film showing a mixture of saccular and tubular bronchiectasis. The branching ectatic bronchi resemble large blood vessels but should not be confused with them; (b) bronchogram in another patient, showing bronchiectasis in the left lower lobe. Note that the affected bronchi (arrowed) are occluded as well as dilated. The remaining bronchi in the left upper lobe and right lung are normal.

diaphragms at their midpoint are below the seventh rib anteriorly or the twelfth rib posteriorly.)

2. Attenuation of the vessels. The reduction in the size and number of the blood vessels can be generalised or localised. If severe the involved area is called a bulla. The edge of a bulla may be indistinct or may be quite sharply limited by a line shadow. In some cases the normal lung adjacent to the bulla is compressed and appears opaque.

Bronchiectasis

Bronchiectasis is defined as irreversible dilatation of the bronchi. Depending on the extent and location of the disease there may be impairment of drainage of bronchial secretions leading to persistent infection.

The conditions which cause bronchiectasis include pulmonary infection in childhood, cystic fibrosis and long-standing bronchial obstruction.

Pulmonary tuberculosis is a common cause but since this usually affects the upper lobes the drainage of secretions is good and such patients rarely have symptoms due to their bronchiectasis.

The radiological features on plain film are (Fig. 2.80a):

1. Visibly dilated bronchi. If these contain air the thickened walls of the dilated bronchi may be seen as tubular or ring shadows. If filled with fluid the dilated bronchi will either be opaque or air fluid levels will be present. Since these fluid levels are very short they have to be looked for very carefully.

2. Persistent consolidation often containing dilated bronchi.

3. Loss of volume of the affected lobe or lobes is almost invariable.

A proportion of patients with symptomatic bronchiectasis have normal chest films.

Bronchography (see p. 17), which involves outlining the lumen of the bronchi with contrast media, is the definitive method of making the diagnosis (Fig. 2.80b). Bronchiectasis is frequently divided into sacular and cylindrical forms, but this serves little purpose since there is no difference in the cause or treatment of these different varieties.

The role of bronchography in known or suspected bronchiectasis is:

1. To confirm the diagnosis of bronchiectasis in those few cases where there is significant clinical doubt. This is virtually confined to cases of haemoptysis when no cause can be found.

2. To define the true extent of the disease prior to surgical resection of bronchiectatic segments. Significant bronchiectasis can be present in areas which appear totally normal on the plain film.

Cystic fibrosis

Cystic fibrosis is an inherited disorder of exocrine glands resulting in secretion of viscid mucus from mucus-secreting glands. In cystic fibrosis the movement of the

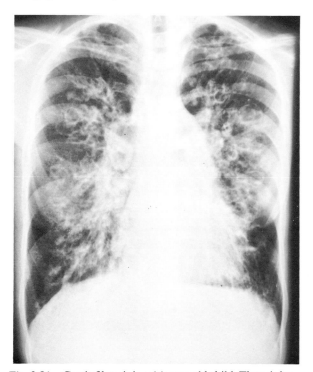

Fig. 2.81 Cystic fibrosis in a 14-year-old child. There is bronchial wall thickening, ring shadows of bronchiectasis and widespread ill-defined, shadowing. All these phenomena tend to be maximal in the mid and upper zones. The diaphragms are somewhat low due to obstructive airways disease.

viscid mucus along the airways is diminished. The small airways become blocked and secondary infections supervene. The finding of a high sodium chloride concentration in the sweat is diagnostic of the condition.

The radiological findings are (Fig. 2.81):

1. Ill-defined nodules of consolidation, maximal in the upper zones, some of which show cavitation.
2. Bronchial wall thickening and the signs of bronchiectasis; both usually maximal in the upper zones. Parallel line shadows radiating from the hila due to bronchial wall thickening are seen, whereas in normal patients the walls of the bronchi cannot be seen beyond the hilum except where they are travelling directly along the line of the x-ray beam. Dilated bronchi may be seen; these may be fluid-filled and therefore opaque, or air-filled thereby revealing the thickened bronchial walls, alternatively they may contain air and fluid and show air-fluid levels.
3. Evidence of airway obstruction. The diaphragms are low and flat and the heart is narrow and vertical until cor

pulmonale develops when cardiac enlargement may occur.

RESPIRATORY DISTRESS IN THE NEWBORN

There are many causes of clinical respiratory distress in the first few days of life. Abnormalities are visible on the chest x-ray in the majority; only two conditions are discussed here.

Hyaline membrane disease, one of the commonest abnormalities, is a disease of the premature infant, due to deficiency of surfactant in the lungs, in which the alveoli are collapsed, preventing gas exchange. The chest x-ray appearance is one of the most important criteria in making the diagnosis. The basic signs are widespread very small pulmonary opacities and visible air bronchograms (Fig. 2.82). The bronchi are visible because they are surrounded by airless alveoli. The changes are nearly

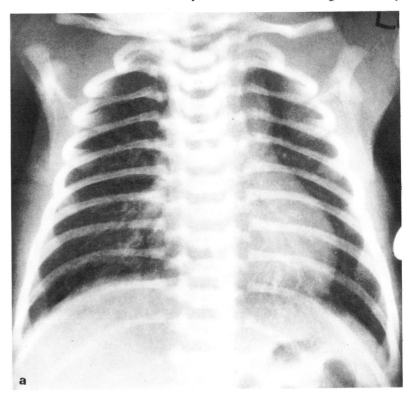

Fig. 2.82 The neonatal respiratory distress syndrome (hyaline membrane disease). (a) A normal premature neonatal chest film for comparison.

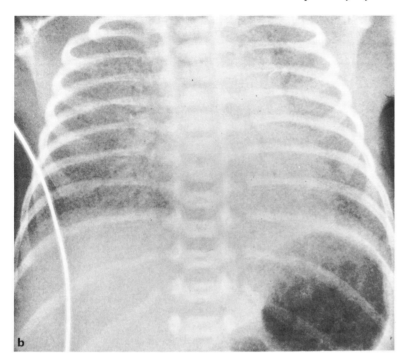

Fig. 2.82 (b) this shows the general granular opacity of the lungs typical of hyaline membrane disease. The vessels, the heart borders and the diaphragm outlines are indistinct.

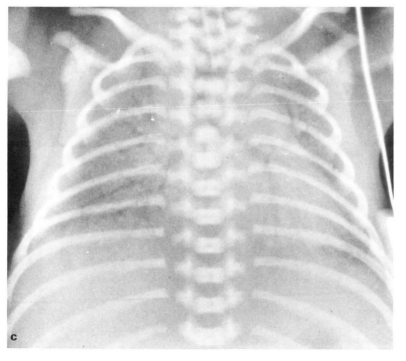

Fig. 2.82 (c) the air bronchogram sign in another baby with hyaline membrane disease. Note the uniformity of distribution of the changes in the lungs, an important diagnostic feature of hyaline membrane disease.

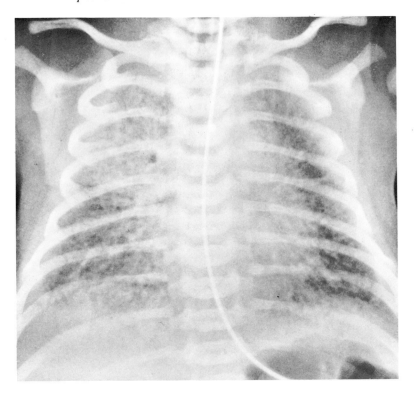

Fig. 2.83 Meconium aspiration. This baby born at term had fetal distress during delivery and was born through meconium-stained liquor. The film shows patchy consolidations rather than the uniform changes seen in hyaline membrane disease. The diaphragms are lower than normal in position which is another differentiation from hyaline membrane disease.

always uniform in distribution. In the milder forms the nodules are small and the air bronchogram may be the most obvious and easily recognised sign. In the more severe forms the pulmonary opacities become more obvious and may be confluent, the lungs appearing almost opaque, except for air bronchograms.

Meconium aspiration (Fig. 2.83). In meconium aspiration the pulmonary shadowing is more patchy and often distinctly streaky. Air bronchograms are not as obvious. In meconium aspiration the diaphragms are often lower than normal due to the airways obstruction associated with sticky meconium in the bronchi.

In addition to establishing the initial diagnosis in babies with the various causes of respiratory distress in the newborn the plain chest film is vital in detecting complications of therapy. These include lobar collapse, pneumonediastinum and pneumothorax.

PULMONARY EMBOLISM AND INFARCTION

Pulmonary embolism from thrombi originating in the veins of the legs and pelvis is very common in patients confined to bed, particularly those with heart disease and those who have had major surgery. The clinical and radiological manifestations depend on the size and number of the emboli.

Plain film abnormalities

Massive embolism. Massive embolism to the main pulmonary arteries and their lobar divisions, sufficient to produce circulatory disturbances, may produce a visible reduction in the size of the arteries beyond the occlusion on plain chest films. This is a very difficult sign to recognise and it is usually not possible to be sure that the smallness of the vessels is due to embolism rather than to

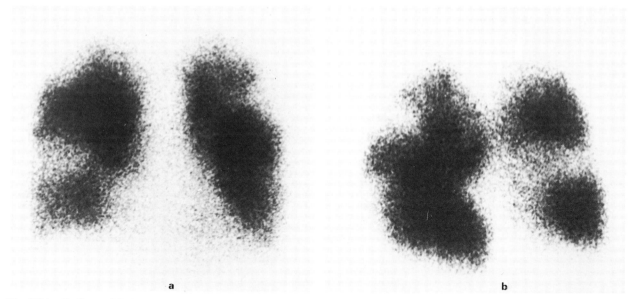

Fig. 2.84 Radionuclide lung scans in pulmonary embolism. The perfusion scan: (a) anterior view; (b) left oblique view showing several wedge-shaped filling defects in both lungs. The ventilation scan was normal.

pre-existing pulmonary disease (emphysema or old infection). In most cases the chest film in massive embolism is unremarkable.

Large emboli. In many cases there is no infarction even with large emboli but in some patients, particularly those with heart disease, infarction does occur. Radiologically, an infarct appears as an area of consolidation, indistinguishable from pneumonia. Infarcts often involve both lung bases and cause elevation of the diaphragm and pleural effusion. This distribution and combination is less common in pneumonia but there are no radiological signs specific to either pneumonia or infarction. The differentiation between these two causes of consolidation depends on clinical factors and, if necessary, an isotope lung scan (see below).

Small emboli. Small emboli do not produce any radiological abnormality unless they occur over a long period of time and cause pulmonary hypertension (see p. 95 for the signs of pulmonary hypertension).

Radionuclide lung scans

If a large number of appropriately sized particles of albumin labelled with technetium-99m are injected intravenously they become trapped in the arterioles of the lung (perfusion scan). The distribution of the radioactivity reflects the pulmonary blood flow. In the normal there is a uniform gradation from the apices to the bases. In pulmonary embolism those pulmonary arteries that are blocked prevent the radioactive isotope reaching the underperfused lung and 'filling defects' will be seen in the scan (Fig. 2.84).

Many different pathological processes diminish the blood supply to parts of the lung. For example pneumonia, tumours, bronchiectasis and emphysema all lead to underperfusion, and isotope lung scans in these conditions may be indistinguishable from one another. Only if the isotope defect is in an area which is radiologically normal is the diagnosis likely to be pulmonary embolism. If the perfusion scan is compared to a ventilation scan the diagnosis becomes even more certain. Poor perfusion in a normally ventilated area of the lung is virtually diagnostic of pulmonary embolism.

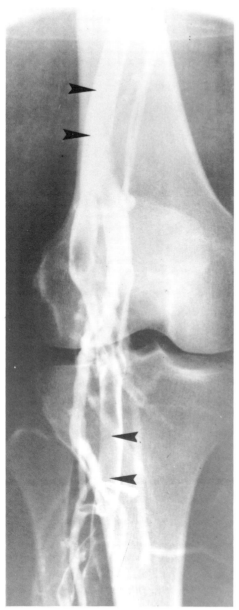

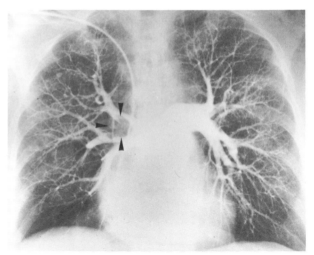

Fig. 2.86. Pulmonary arteriogram in pulmonary embolism. There is a large embolus causing a filling defect at the bifurcation of the right pulmonary artery (arrows). Note also the reduction in branches of the lower lobe arteries due to obstruction by emboli (The catheter can be seen passing from the right arm into the heart.)

Leg and pelvic phlebography

Deep venous thrombosis in the legs or thrombi in the external and common iliac veins can be demonstrated by leg and pelvic phlebography. A large volume of water soluble contrast is injected into a small vein in each foot. The contrast is forced into the deep venous system of the calf by means of tourniquets or by performing the examination with the patient half upright. Provided enough contrast is used it is possible to demonstrate the pelvic veins and the inferior vena cava as well as the veins in the legs. Thrombi may cause complete blockage of the veins or they may be seen as filling defects in the opacified veins (Fig. 2.85).

Since leg and pelvic phlebography demonstrate the source of pulmonary emboli, it is a frequent early investigation in suspected pulmonary embolism.

Fig. 2.85 Leg venogram showing deep vein thrombosis. The lower arrows point to the filling defect of the thrombus in the popliteal vein. Compare the uniform opacity of the normal vein higher up (upper arrows).

Pulmonary angiography (Fig. 2.86)

Pulmonary angiography is the most accurate method of diagnosing pulmonary emboli. Its use is rarely indi-

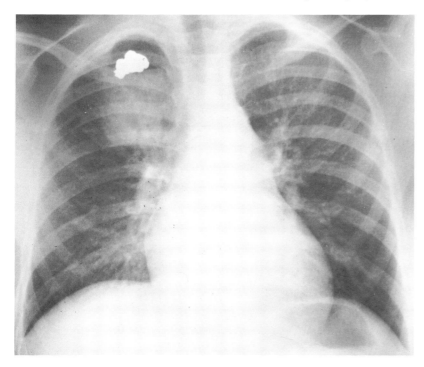

Fig. 2.87 Pulmonary contusion from a gunshot wound. The ill-defined consolidation represents haemorrhage and oedema in the right upper lobe. The deformed metallic fragments of the bullet are clearly visible.

cated, however, and it requires definite indications; e.g. to confirm massive embolism prior to thoracotomy, or when there is significant doubt regarding the diagnosis after comparing the chest x-ray with isotope perfusion and ventilation lung scans.

TRAUMA TO THE CHEST

Rib fracture can be diagnosed by noting a break or step in the cortex of a rib. Special views of the ribs may be necessary, since rib fractures are often invisible in the standard projections; particularly if the fracture lies below the diaphragm. Extrapleural soft tissue swelling due to bruising, or frank haematoma may be visible and guide the observer toward the site of the fracture.

Rib fractures are frequently multiple and may result in a flail segment.

Pleural effusion often accompanies rib fractures, the fluid frequently being blood.

Pneumothorax may occur if the lung is punctured by direct injury or by the sharp edge of a rib fracture. An air-fluid level in the pleural cavity is common in such situations, due to the associated haemorrhage.

Surgical emphysema of the chest wall may indicate the escape of air from the lungs. The presence of mediastinal emphysema may indicate the unusual phenomena of rupture of a bronchus or rupture of the oesophagus.

Pulmonary contusion, alveolar haemorrhage and oedema (Fig. 2.87)—may be seen whether or not a rib fracture can be identified. The resulting pulmonary consolidation is indistinguishable from other forms of pulmonary exudates; the relationship to the injury being important in establishing the diagnosis. Pneumatoceles are sometimes seen with pulmonary contusion.

Shock lung is an imprecise term used to describe a hypoxaemic patient with diffuse alveolar oedema and exudates following trauma. There is usually a latent interval of 12–24 hours after the traumatic episode. The

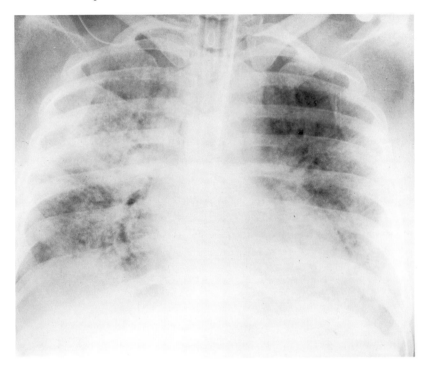

Fig. 2.88 Adult respiratory distress syndrome (ARDS) There is widespread consolidation of the lungs. This patient had suffered extensive trauma to the limbs.

trauma may be to any part of the body. A similar clinical and radiological syndrome is seen with septicaemia or other types of haemodynamic shock. A number of conditions are known to cause such a combination of events, e.g. fat embolism, but in many the mechanism is unclear and may well be multifactorial. The condition is also known by the term adult respiratory distress syndrome (ARDS). Radiologically, there is diffuse widespread air space filling often closely resembling pulmonary oedema, the precise pattern varying from patient to patient (Fig. 2.88). The radiological abnormality only develops some 12–24 hours after the onset of tachypnoea, dyspnoea and hypoxaemia.

Rupture of the diaphragm from either penetrating injury or compression trauma on the abdomen may permit herniation of the stomach or intestines into the chest. Such herniation is much commoner on the left than the right. There is usually a pleural effusion and the 'diaphragm' outline becomes indistinct during the acute stage. Gas shadows of the stomach or intestines are seen above the presumed position of the diaphragm, the diaphragm itself often being invisible. Barium meal and follow through may be indicated to establish the diagnosis.

Rupture of the aorta is a particularly serious consequence of rapid deceleration injuries. In patients that survive, the injury to the aorta is usually at the level of the ligamentum teres. Bleeding into the mediastinum may cause visible mediastinal widening, and bleeding into the pleural cavity may occur.

Aortography is usually indicated in patients with unexplained mediastinal widening following trauma, to establish the diagnosis of aortic rupture, since venous bleeding which does not require emergency surgery can cause similar signs. Widening of the mediastinum can be a very difficult sign to assess, particularly on the portable AP films that are often the only films that can be taken in these severely injured patients.

Aortic rupture can occur with surprisingly mild trauma, and though fractures of the ribs or sternum are usually present there are many cases on record without visible damage to the thoracic cage.

In some patients the diagnosis of aortic rupture is only made several months or years after the injury when the development of an aneurysm is noted.

NEOPLASTIC DISEASE

Carcinoma of the bronchus

Carcinoma of the bronchus is the commonest primary malignant tumour in the body. It is much commoner in men than women and has a clear association with cigarette smoking. With the exception of some squamous cell carcinomas and the rare alveolar cell carcinoma, it is not possible to predict with any certainty the cell type from the chest x-ray appearances.

The majority of bronchial carcinomas arise in larger bronchi at, or close to, the hilum. The remainder arise

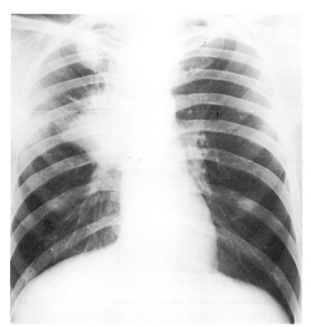

Fig. 2.89 Right hilar mass due to carcinoma of the bronchus. There is also a patch of consolidation in the right upper lobe laterally, due to the central obstruction.

peripherally. It is convenient to consider the radiological features of central and peripheral tumours separately.

Signs of a central tumour

1. The tumour itself may present as a hilar mass (Fig. 2.89) and/or narrowing of a major bronchus (Fig. 2.90). Tomography can be of great value; not only does it show the outline of the mass projecting into the lung but it is usually the simplest way to demonstrate the bronchial narrowing. The narrowing may be irregular or smooth. An irregular stricture is virtually diagnostic of carcinoma.
2. The effect of obstruction by the tumour (Fig. 2.91) is usually a combination of collapse and consolidation. The alveoli collapse because air is absorbed beyond the obstructed bronchus and cannot be replaced, whereas consolidation is the consequence of retained secretions and secondary infection. Obstructive emphysema is occasionally seen with a centrally situated carcinoma.

Signs of a peripheral tumour

A peripheral tumour usually presents as a solitary pulmonary mass. There are several causes for such a mass and these are discussed on p. 29.

The signs of a peripheral primary carcinoma are:

1. A rounded shadow with an irregular border, lobulation, notching and infiltrating edges are the common patterns (Fig. 2.92).
2. Cavitation within the mass (Fig. 2.32). Peripheral squamous cell carcinomas show a particular tendency for cavitation. The walls of the cavity are classically thick and irregular, but thin-walled smooth cavities due to carcinoma do occur.

Tomography may be required to demonstrate cavitation or the edge of lesions suspected of being bronchial carcinomas.

Signs of spread of bronchial carcinoma

1. Hilar and mediastinal lymph node enlargement (Fig. 2.93) due to lymphatic spread of tumour. The subcarinal

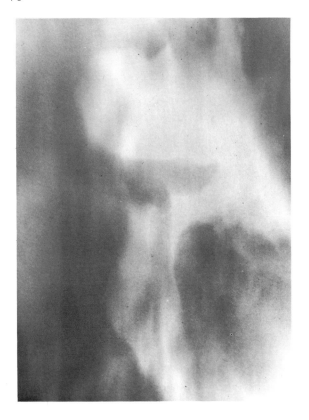

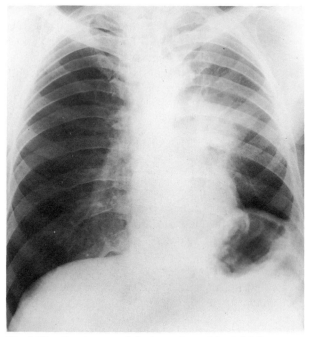

Fig. 2.91 Carcinoma of the bronchus of the left hilum causing collapse of the left upper lobe and paralysis of the left phrenic nerve. The elevated left diaphragm is too high to be due to the lobar collapse; it is due to phrenic nerve involvement by the tumour at the left hilum.

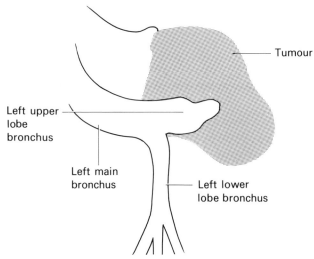

Fig. 2.90 Carcinoma of the bronchus causing narrowing of the left upper lobe bronchus (shown on tomography).

nodes are a frequent site of involvement, but enlargement of this group is difficult to diagnose radiologically.

2. Pleural effusion may be due to malignant involvement of the pleura, or may be secondary to associated infection of the lung. A large effusion or an effusion on the opposite side to the primary tumour are strong pointers to malignant spread.

3. Elevation of the diaphragm. If one diaphragm is elevated the phrenic nerve may have been involved by tumour (Fig. 2.91). Confirmation of paralysis of the diaphgram is best made by fluoroscopy where paradoxical movement will be seen.

4. Invasion of the chest wall (Fig. 2.94). Destruction of a rib immediately adjacent to a pulmonary shadow is virtually diagnostic of primary bronchial carcinoma with chest wall invasion. Recognising the rib destruction can be difficult and one has to make a conscious effort to

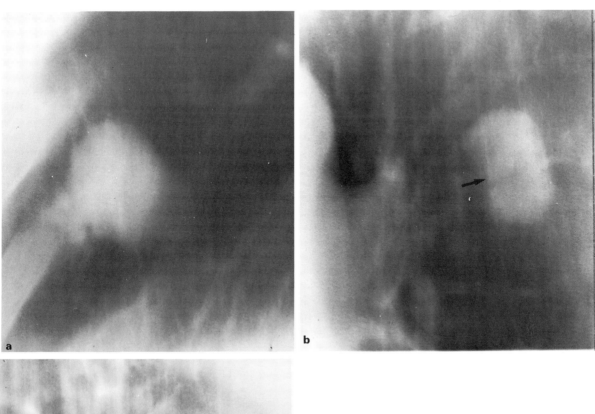

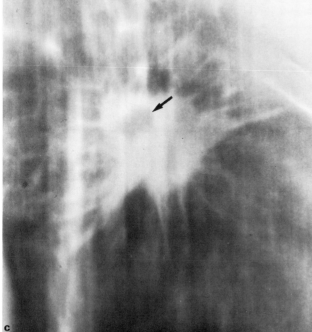

Fig. 2.92 Shape of peripheral lung carcinomas (shown on tomography). The common patterns are: (a) lobulated outline; (b) notched outline (the arrow points to the notch); (c) infiltrating edges—fine strands can be seen radiating from the edge of the cancer into the surrounding lung. A small eccentric cavity is also present in this example (arrow).

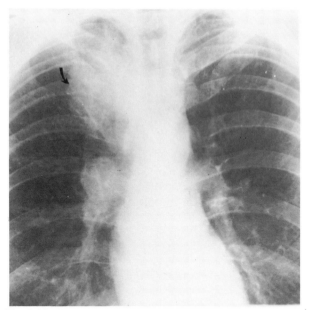

Fig. 2.93 Large carcinoma of the right upper lobe (arrowed) with lymph-node metastases at the right hilum.

look at the ribs directly. If a portion of rib is not seen it should not be assumed that this is due to lack of penetration of the x-rays through the lung shadow. Look at the other ribs; if they are visible through the shadow then the penetration is adequate and rib destruction must be present. Further films, such as oblique views and tomography, may be necessary to confirm or exclude this important sign.

5. Rib metastases. Carcinoma of the lung frequently metastasises to the ribs where it produces bone destruction. Sclerotic secondary deposits from lung carcinoma are rare.

6. Pulmonary metastases. Primary lung carcinoma not infrequently metastasises to other parts of the lungs. The rounded shadows that result are indistinguishable from secondary deposits from other primary tumours.

7. Lymphangitis carcinomatosa (Fig. 2.95) is the term applied to blockage of the pulmonary lymphatics by carcinomatous tissue. Lymphangitis carcinomatosa can be due to spread from abdominal and breast cancers as well as from carcinoma of the lung. The lymphatic

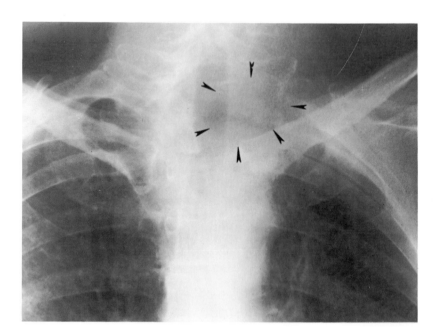

Fig. 2.94 Pancoast's tumour. The carcinoma arising at the apex of the left lung has invaded and destroyed the adjacent ribs and spine. Note that no bone is visible within the area indicated by the arrows.

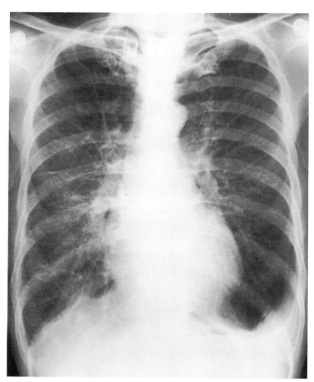

Fig. 2.95 Lymphangitis carcinomatosa. There is widespread ill-defined pulmonary shadowing with numerous septal lines. A small left pleural effusion is also present. The patient had a carcinoma of the stomach.

vessels become grossly distended and the lungs become oedematous. The signs can be identical to those seen in interstitial pulmonary oedema (septal lines, loss of vessel clarity and peribronchial thickening), but if the heart is normal in size and there is hilar adenopathy and/or lobar consolidation, the diagnosis of lymphangitis carcinomatosa becomes more certain. The clinical story is very helpful, since if the changes are due to pulmonary oedema, the patient will usually complain of sudden onset of breathlessness, whereas the patient with lymphangitis carcinomatosa will give a story of slowly increasing dyspnoea over the preceding weeks or months.

Metastatic neoplasms

Metastases from extrathoracic primary tumours may be seen in the lungs, the pleura or the bones of the thoracic cage. Hilar and mediastinal lymph node enlargement due to metastases is uncommon, other than from carcinoma of the bronchus.

Pulmonary metastases produce one or more rounded shadows within the lung (Fig. 2.96). Typically, metastases are spherical and well defined, although irregular borders are occasionally seen. Usually, they are multiple and vary in size. As with primary tumours, metastases have to be almost a centimetre in diameter or larger to be visible at all.

Lymphangitis carcinomatosa (see above) is another form of metastatic involvement.

Pleural metastases usually give rise to pleural effusion. The individual pleural metastases are only occasionally seen.

Metastases to ribs are common with those primary tumours that metastasise to bone, namely bronchus, breast, kidney, thyroid and prostate. All except prostatic and breast cancers produce mainly or exclusively

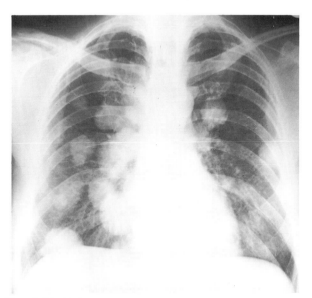

Fig. 2.96 Pulmonary metastases. There are numerous rounded shadows of varying sizes in both lungs.

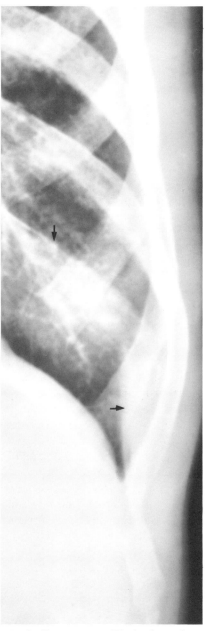

lytic metastases. Sclerotic metastases in an elderly man suggests a prostatic primary cancer. Sclerotic or mixed lytic and sclerotic deposits in a woman suggest that the primary carcinoma is in the breast.

With lytic metastases the best sign is destruction of the cortex, particularly of the upper border of the rib (Fig. 2.95). Be wary of diagnosing destruction of the lower borders of the posterior portions of the ribs, since these regions are indefinite even in the normal. When in doubt it is always wise to compare with the opposite side. Another pitfall in the diagnosis of rib metastases is that blood vessels in the lungs may cause confusing shadows. This cannot arise at the edges of the chest where there is no lung over the ribs, so this is a useful place to look for bone destruction. Soft tissue swelling is frequently seen adjacent to the rib deposits so it is a good rule to look at the outer margin of the lung for soft tissue swelling as a clue to the presence of rib metastases (Fig. 2.97).

When sclerotic metastases are suspected the best place to look is in the medulla of the ribs, particularly where the lung does not overlie the ribs.

Lymphoma

The common manifestations of thoracic malignant lymphoma are mediastinal and hilar adenopathy and pleural effusion. Pulmonary involvement by lymphoma occurs in approximately 10–15% of patients. It may take the form of large areas of infiltration of the lung parenchyma, resembling pulmonary consolidation, or occasionally it is seen as one or more mass lesions, which may cavitate. Pleural masses are a rare feature.

Since pulmonary infection is a common complication in patients with malignant lymphoma it may be impossible to decide on radiological grounds whether the pulmonary consolidation is due to lymphomatous tissue or due to infection.

Mesothelioma

This topic is discussed on p. 47.

Fig. 2.97 Lytic rib metastases. The horizontal arrows points to the soft tissue swelling adjacent to a lytic rib metastasis. Note that when the affected rib (the vertical arrow points to a normal portion of this rib) is traced downward there is extensive destruction of its axillary portion.

3

Heart Disease

When a patient with suspected heart disease is referred for x-ray films the referring doctor usually already knows the diagnosis from the physical examination and the electrocardiogram, e.g. mitral stenosis or ventricular septal defect, and wants evidence of their haemodynamic severity. Sometimes he is uncertain of the precise diagnosis but has a short list of possibilities, which can be further reduced by information from the x-ray films. Pulmonary stenosis and atrial septal defect, for example, may have similar clinical features, but radiologically they are quite different. Occasionally, the diagnosis of heart disease is first made from the chest film; atrial septal defect and coarctation of the aorta are the commonest diagnoses in this category because both are usually asymptomatic and, therefore, may be picked up on a routine chest film.

As with all chest x-rays, whatever their indication, unsuspected diseases may be discovered.

Standard films and additional views

The standard films for the evaluation of cardiac disease are the posteroanterior view (PA) and the lateral chest film (Fig. 3.1).

The following views may be added:

1. A penetrated PA film. This allows one to look at the shadows *within* the heart, e.g. the double contour of the left atrium and valve calcification. In many conventionally exposed chest films the heart shadow is uniformly white, so any 'shadow' within it would be invisible.

2. Lateral and PA films with barium in the oesophagus. The posterior border of the heart is often difficult to define on the plain lateral view. Since the oesophagus is so closely applied to the back of the heart, bulges of the cardiac contour will indent the oesophagus and these can be demonstrated once the oesophagus is outlined by barium.

3. Oblique views. In some centres these views are routine but in others they are rarely done. They add very little, if anything, to the information obtained from frontal and lateral projections.

Special techniques

Fluoroscopy. Cardiac fluoroscopy (screening) enables the movement of the cardiac outline to be observed. In particular, areas of abnormal pulsation, e.g. paradoxical pulsation in ventricular aneurysms, can be identified. Another major use is the detection and localisation of intracardiac calcification. Calcification is more easily seen at fluoroscopy than on static films. Certain calcifications show characteristic movements which enable them to be accurately localised. The best example of this is aortic valve calcification, which moves quite differently from calcification in the mitral valve.

Cardiac catheterisation and angiography is almost invariably carried out under fluoroscopic control. Cardiac angiography is a specialised topic which will not be discussed further.

Routine approach to chest films in cardiac disease

When interpreting chest films in patients with cardiac problems the following features should always be assessed:

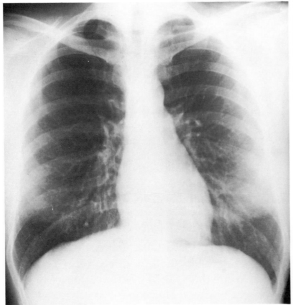

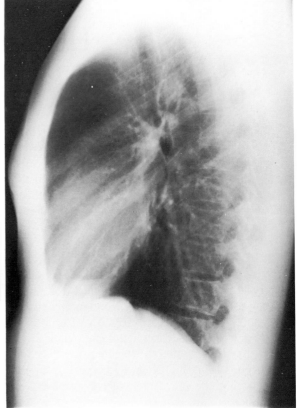

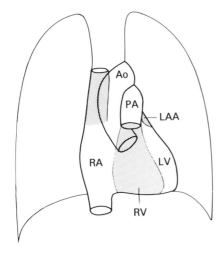

Fig. 3.1 Outline of heart in PA and lateral views. Ao—aorta;
LAA—left atrial appendage; LA—left atrium; LV—left ventricle; PA—pulmonary artery; RH—right atrium; RPA—
right pulmonary artery; RV—right ventricle.

The heart

1. Is the overall size of the heart enlarged?
2. Does the shape of the heart suggest enlargement of the individual cardiac chambers?
3. Is there any pericardial or intracardiac calcification?

The aorta

1. What is the size of the ascending aorta?
2. Is the aortic arch on the left or the right?
3. Are there any signs of coarctation of the aorta?

The main pulmonary artery

What is the size of the main pulmonary artery?

The lungs

Are the blood vessels at the hilum and in the lungs normal in size, or are they larger or smaller than normal?

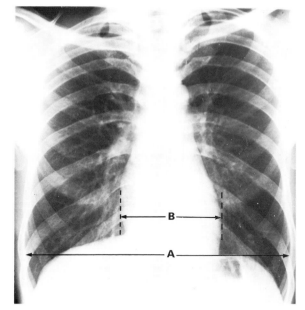

Fig. 3.2 Measurement of heart size. The transverse diameter of the heart is the distance between the two vertical tangents to the heart outline. When calculating the cardiothoracic ratio (CTR) the maximum diameter of the heart (B) is divided by the maximum internal diameter of the chest (A).

BASIC SIGNS IN HEART DISEASE

THE HEART

The size of the heart

There are a variety of ways of measuring heart size. None is entirely satisfactory. A widely used method is to express heart size as the cardiothoracic ratio (CTR); in normal people the transverse diameter of heart is usually less than half the internal diameter of the chest (Fig. 3.2). This ratio has some value in children, but in adults it may be better simply to measure the transverse cardiac diameter. In most normal men the heart measures less than 15·5 cm and in most normal women less than 15 cm. Since many normal people have heart diameters around 11 cm it is clear that in such individuals substantial dilatation needs to take place before these measurements become 'abnormal'. The best method of evaluating heart size is comparison with previous films; when such films are available.

An increase in heart size may be due to:

1. Dilatation of one or more cardiac chambers.
2. A pericardial effusion

The shape of the heart

It has always been the wish of radiologists and cardiologists to be able to diagnose which chambers are enlarged. This wish has been frustrated by the fact that only one, or at most two, of the borders of any chamber are visible on plain films and that enlargement of one of the ventricles will affect the shape of the other. It is possible, however, to make a reasonably accurate assessment of the size of the left atrium, and this has become one of the most valuable aspects of the analysis of the cardiac shadow.

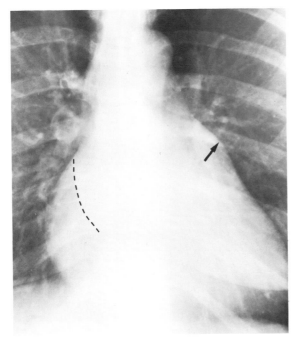

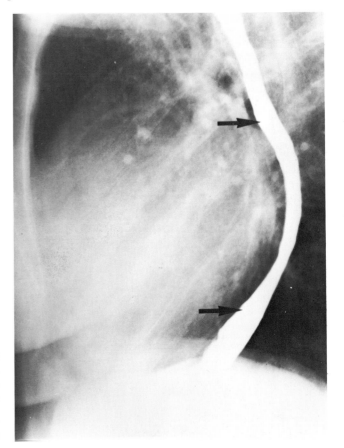

Fig. 3.3   Left atrial enlargement in a patient with mitral valve disease showing the 'double contour sign' (the left atrial border has been drawn in) and dilatation of the left atrial appendage (LAA) (straight arrow). The enlarged LAA should not be confused with dilatation of the main pulmonary artery. The main pulmonary artery is the segment immediately below the aortic knuckle. The LAA is separated from the aortic knuckle by the main pulmonary artery (compare with Fig. 3.13). The lateral view shows how the enlarged left atrium displaces the barium-filled oesophagus between the two arrows.

Left atrial enlargement

The *signs* of left atrial enlargement are (Fig. 3.3):

1. 'Double contour' on frontal projection. The wall of the enlarged left atrium is seen as a separate contour adjacent to the right heart border, usually within the cardiac shadow.

2. Enlargement of the left atrial appendage occurs as part of the general enlargement of the left atrium. A bulge is seen below the pulmonary artery.

3. Backward displacement of the oesophagus. In the normal patient the barium-filled oesophagus on the lateral view follows a smooth curve down to the diaphragm, with no local indentation as it passes behind the left atrium. When the left atrium is enlarged there is an indentation which starts abruptly just below the left main bronchus. In mitral stenosis, where the left atrium but not the left ventricle is enlarged, the indentation stops where the left atrium joins the left ventricle (Fig. 3.3). If the left ventricle is also enlarged, as in mitral incompetence, the lower limit of the indentation is often invisible.

4. Elevation of the left main bronchus. The left atrium is situated immediately beneath the bifurcation of the trachea. With marked left atrial enlargement the angle between the main bronchi is widened, the left main bronchus being pushed upward.

The *causes* of left atrial enlargement include:

1. Mitral stenosis.
2. Mitral incompetence.
3. Intra-atrial obstruction, e.g. left atrial myxoma.

Right atrial enlargement (Fig. 3.4)

Right atrial enlargement causes an increase of the curvature of the right heart border. It is often accompanied by enlargement of the superior vena cava, the border of which forms the right side of the superior mediastinum. Right atrial enlargement occurs in right ventricular failure, and in tricuspid stenosis and incompetence.

Ventricular enlargement

Ventricular hypertrophy without dilatation is difficult to recognise; the increasing wall thickness encroaches on the cavity of the chamber so that little change occurs in the external contour of the heart. Those conditions demanding an increase in pressure lead to hypertrophy without dilatation until decompensation occurs, so 'pressure overload' situations such as aortic or pulmonary stenosis and systemic hypertension do not lead to a readily identifiable change in the shape of the left or right ventricle on plain chest films, unless the relevant ventricle fails.

'Volume overload' situations such as ventricular failure, valvular incompetence, left to right shunts and severe chronic anaemia result in both hypertrophy and dilatation of the relevant ventricle. Ventricular enlargement results in an overall increase in the transverse cardiac diameter and may lead to an alteration in car-

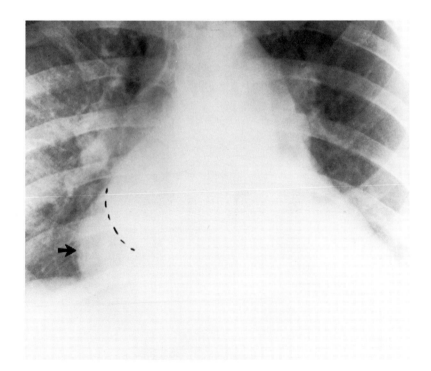

Fig. 3.4 Right atrial enlargement. The right heart border is displaced to the right and is much more curved than normal. In this patient the right atrial enlargement (arrow) is due to tricuspid incompetence. The patient also has mitral valve disease causing enlargement of the left atrium (dotted line) and pulmonary arterial hypertension.

diac contour that permits one to identify whether the left or right ventricle is enlarged.

Left ventricular enlargement

If the left ventricle is the only chamber that is significantly enlarged the shape of the left heart border is as shown in Fig. 3.5, with the apex displaced downwards and outwards. On the lateral view the lower third of the posterior contour of the heart (and the adjacent oesophagus) is bulged backwards.

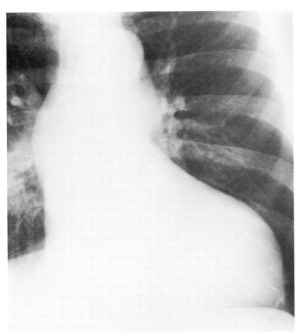

Fig. 3.5 Left ventricular enlargement in a patient with aortic incompetence. The cardiac apex is displaced downwards and to the left. Note also that the ascending aorta causes a bulge of the right mediastinal border; a feature that is almost always seen in significant aortic valve disease.

The commoner causes of enlargement of the left ventricle are:

1. Aortic valve disease, particularly aortic incompetence.
2. Mitral incompetence.

3. Systemic hypertension, once cardiac decompensation has occurred.
4. Ischaemic heart disease, once substantial muscle damage has occurred.
5. Cardiomyopathy of many types.
6. Patent ductus arteriosus and ventricular septal defect with large shunts.

Left ventricular aneurysm may cause a localised bulge in the left heart border usually involving the apex of the heart (Fig. 3.23).

Right ventricular enlargement (Fig. 3.6)

This is the most difficult chamber enlargement to diagnose. No portion of the right ventricle is visible on the frontal view so on this view the right ventricle has to displace the left ventricle for the enlargement to be recognised. The apex may be displaced upwards and outwards, rather than downwards and outwards as happens with left ventricular enlargement. On the lateral view only the lower third of the right ventricle is normally against the sternum, but when the right ventricle enlarges more of its outline lies in contact with the sternum. The common causes of right ventricular enlargement are:

1. Atrial septal defect.
2. Pulmonary hypertension of any cause.
3. Pulmonary stenosis.
4. Fallot's tetralogy.

Pericardial effusion (Fig. 3.7)

The appearance of the heart shadow with pericardial effusion may be indistinguishable from other forms of cardiac enlargement (Fig. 3.8), but there are certain useful signs:

1. A marked change in the transverse cardiac diameter within a week or two, particularly if no pulmonary oedema occurs, is virtually diagnostic.
2. With very large effusions the heart may assume a characteristic shape, bulged inferiorly with a well-defined outline.

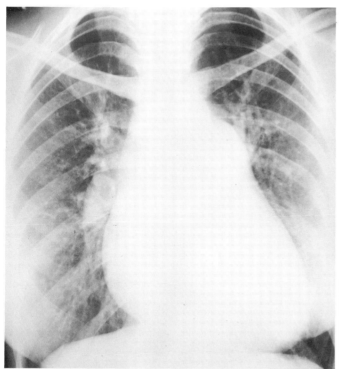 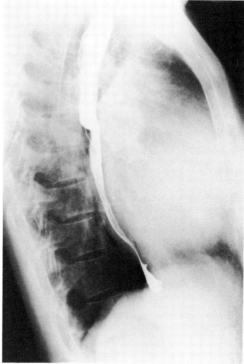

Fig. 3.6 Right ventricular enlargement in an adult with primary pulmonary hypertension. The heart is enlarged with the apex of the heart somewhat lifted off the diaphragm. On the lateral view the enlargement is all anterior, there being no displacement of the lower esophagus, and more of the right ventricle than normal is in contact with the sternum. Note also the features of pulmonary arterial hypertension-enlargement of the main pulmonary artery and hilar arteries with normal vessels within the lungs.

It is important to realise, however, that a patient may have a pericardial effusion causing life-threatening tamponade, but have very little cardiac enlargement and an otherwise normal contour. Ultrasound is now the best non-invasive method of detecting a pericardial effusion.

Depressed sternum (pectus excavatum)

In patients with a severely depressed sternum the heart is rotated and displaced into the left chest. This so alters the cardiac outline that heart disease may be suspected unless the chest wall deformity is recognised (Fig. 3.9).

Cardiac calcification

There is normally no visible calcification in the heart.

When calcification is identified the first step is to decide its site; after this the diagnosis is usually easy.

Valvular calcification occurs in the aortic and/or mitral valves in rheumatic heart disease and in the aortic valve in adults with congenital aortic stenosis. The calcification is seen as clumps, sometimes vaguely conforming to a ring, in the centre of the heart shadow. Calcification confined to the mitral valve ring is seen in the elderly; providing the heart is otherwise normal, and the calcification is confined to the mitral valve ring, it is of no haemodynamic significance.

Valve calcification is best evaluated in the lateral view. On the frontal view it is often projected over the spine and difficult to see. The easiest method of deciding

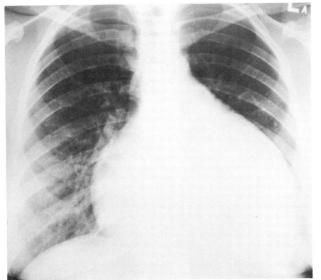

Fig. 3.7 Pericardial effusion. The heart is greatly enlarged. (3 weeks before, the heart had been normal in shape and size.) The outline is well defined and the shape globular. The lungs are normal. The cause in this case was a viral pericarditis. This appearance of the heart, though highly suggestive, is not specific to pericardial effusion. (Compare with Fig. 3.8.)

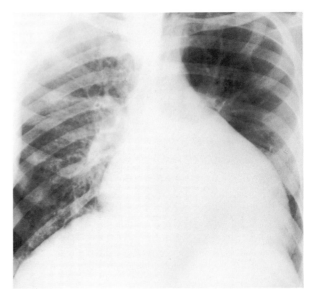

Fig. 3.8 Congestive cardiomyopathy causing generalised cardiac dilatation. This appearance can easily be confused radiologically with a pericardial effusion.

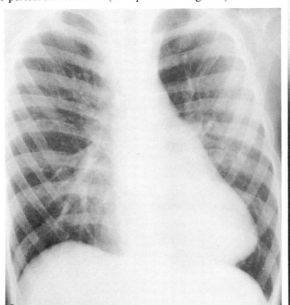

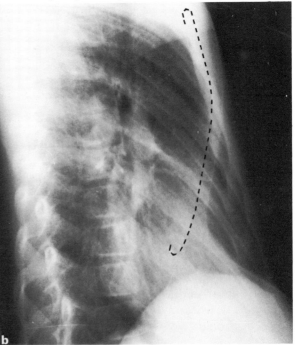

Fig. 3.9 Pectus excavatum. (a) PA view: note how the heart is displaced and altered in shape by the depressed sternum; (b) lateral film: the edge of the sternum has been traced in on this film. There was no cardiac disease in this patient.

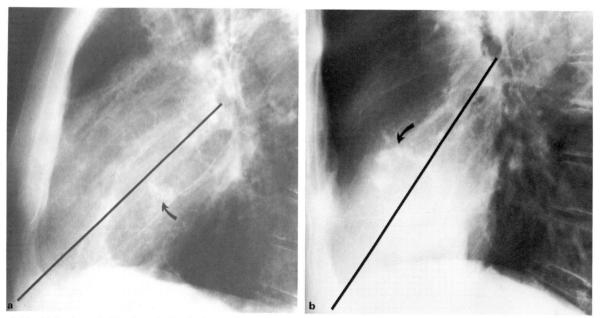

Fig. 3.10 Valvular calcification. A line is drawn from the sternodiaphragmatic junction to the left main bronchus. (a) Mitral valve calcification: the calcification lies below this line; (b) aortic valve calcification: the calcification lies above this line.

which valve is calcified is shown in Fig. 3.10. A line is drawn from the junction of the diaphragm and sternum obliquely upwards and backwards to the left main bronchus. This line will be at approximately 45° to the horizontal. Calcification in the mitral valve lies behind and below the line, whereas aortic valve calcification lies in front and above it. Should the line pass through the calcification the smaller portion is ignored, but should it bisect the calcification serious consideration should be given to the possibility that both valves are calcified.

Coronary artery calcification always indicates atheroma, but it is a poor indicator of the severity of the disease. The calcification is easily recognised as arterial in nature because it is composed of parallel lines and usually seen near the origin of the coronary arteries.

Pericardial calcification is seen in approximately 50% of patients with constrictive pericarditis. Calcific constrictive pericarditis is usually postinfective in aetiology, tuberculosis and Coxsackie infections being the common known causes. In many cases no infecting agent can be identified. The calcification occurs patchily in the pericardium, even though the pericardium is thickened and rigid all over the heart. It may be difficult or even impossible to see the calcification on the frontal view. On the lateral film it is usually maximal along the anterior and inferior pericardial borders (Fig. 3.11). Pericardial calcification is a particularly important sign because, if present, it makes the diagnosis of constrictive pericarditis certain.

THE AORTA

The appearance of the aorta varies with age. In the infant it lies close to the midline and is difficult to identify distinct from the other mediastinal shadows. With increasing age the aorta elongates, but since it is fixed at the aortic valve and at the aortic hiatus in the diaphragm elongation necessarily involves unfolding. This unfolding results in the ascending aorta deviating

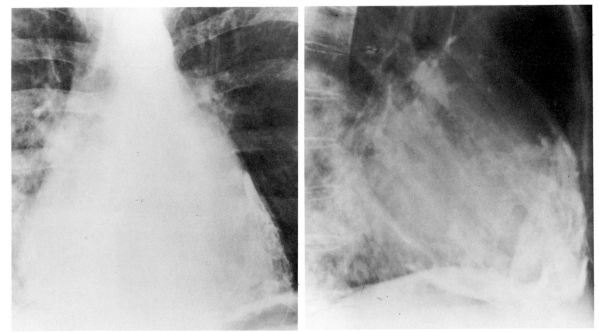

Fig. 3.11.   Pericardial calcification in a patient with severe constrictive pericarditis. The distribution of the calcification is typical. It follows the contour of the heart and is maximal anteriorly and inferiorly. As always, it is more difficult to see the calcification on the PA film. (This patient also had pneumonia in the right lower lobe.)

to the right and the descending aorta deviating to the left. The only problem with aortic unfolding is to be sure that one is not dealing with an aneurysm.

The ascending aorta dilates in aortic incompetence, aortic stenosis (see Figs 3.5 & 3.22), systemic hypertension and aneurysm formation, but substantial dilatation is necessary before a bulge of the right mediastinal border can be recognised.

True dilatation of the aortic arch, as opposed to unfolding, occurs with systemic hypertension and aneurysms (Fig. 3.12). The appearance of the aortic arch in coarctation of the aorta is described on p. 103.

THE PULMONARY VESSELS IN CARDIAC DISEASE

The pulmonary vasculature

The plain chest film provides a simple method of assessing the pulmonary vasculature. By observing the size of the main pulmonary artery, the hilar arteries and the vessels in the lungs it is sometimes possible to draw valuable diagnostic conclusions from the plain chest film.

It is not possible to measure the true diameter of the main pulmonary artery on plain film, since less than half of its circumference is visible and the amount seen in profile is variable. Nevertheless, there are degrees of bulging that permits one to say that the pulmonary artery is enlarged (Fig. 3.13). Conversely, the artery may be recognisably small.

The assessment of the hilar vessels can be more objective since the diameter of the right lower lobe artery can be measured; the diameter at its midpoint (Fig. 3.14) is normally between 9 and 16 mm.

The size of the vessels within the lungs is crucial in deciding pulmonary blood flow. Surprisingly, there are no generally accepted measurements of normality, so

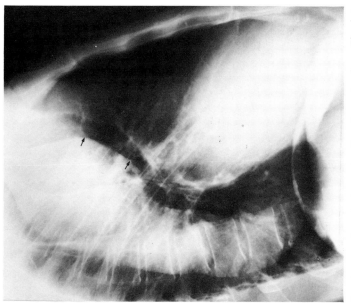

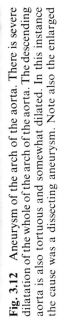

Fig. 3.12 Aneurysm of the arch of the aorta. There is severe dilatation of the whole of the arch of the aorta. The descending aorta is also tortuous and somewhat dilated. In this instance the cause was a dissecting aneurysm. Note also the enlarged heart secondary to long-standing hypertension. The lateral view confirms the dilatation and demonstrates forward displacement of the trachea (arrows).

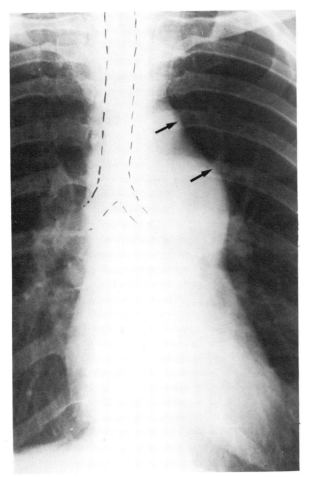

Fig. 3.13 Enlarged main pulmonary artery in a patient with pulmonary valve stenosis. The bulge of the main pulmonary artery (lower arrow) is clearly greater than normal and at first glance one might be deceived into diagnosing enlargement of the aorta. However, the aortic knuckle is the first 'bump' on the left mediastinal border (upper arrow). It projects $2\frac{1}{2}$–3 cm lateral to the trachea. The pulmonary artery forms the segment immediately below the aortic knuckle.

the diagnosis is based on experience with normal films.

Certain valuable conclusions can be drawn from observing these signs. They usually are not specific diagnoses, but assessments of the haemodynamic situation. It may be possible to suggest the following:

1. Pulmonary valve stenosis (Fig. 3.14). The main pul-

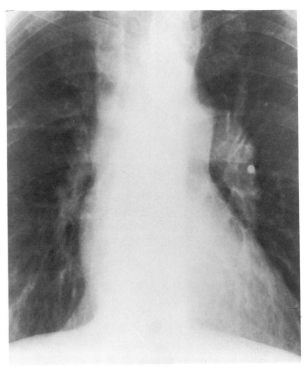

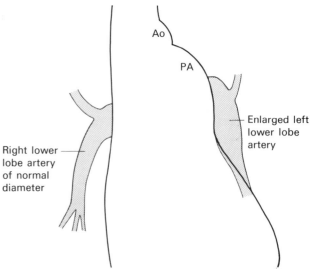

Fig. 3.14 Pulmonary valve stenosis. The heart is normal in size; the main pulmonary artery is enlarged. The left lower lobe artery is enlarged, whereas the right lower lobe artery is of normal diameter. The appearances are otherwise normal. PA—pulmonary artery; Ao—aorta.

monary artery is large due to poststenotic dilatation. The left pulmonary artery may also be large. If so, this combination is diagnostic of stenosis of the pulmonary valve. The remainder of the pulmonary vessels are normal or sometimes small.

2. Increased pulmonary blood flow. (Fig. 3.15) Atrial septal defect, ventricular septal defect and patent ductus arteriosus are the common anomalies in which there is

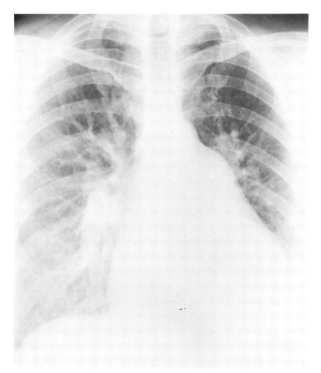

Fig. 3.15 Increased pulmonary blood flow in an atrial septal defect. Note the large heart and enlargement of the pulmonary vessels from the main pulmonary artery to the periphery of the lungs.

shunting of blood from the systemic to the pulmonary circuits, thereby increasing pulmonary blood flow. The severity of the shunt varies greatly. In patients with a haemodynamically significant left to right shunt (2 : 1 or more), all the vessels from the main pulmonary artery to the periphery of the lungs are large. This radiographic

appearance is known as *pulmonary plethora*. There is a reasonably good correlation between the size of the vessels on the chest film and the degree of shunting.

3. Pulmonary arterial hypertension. (Figs. 3.16 & 3.6). The pressure in the pulmonary arteries is dependent on cardiac output and pulmonary vascular resistance. The conditions that cause significant pulmonary arterial hypertension all increase the resistance to blood flow through the lungs. There are many such conditions including various lung diseases (cor pulmonale), pulmonary emboli, pulmonary arterial narrowing in response to mitral valve disease or left to right shunts, and idiopathic.

Pulmonary arterial hypertension needs to be severe before it can be reliably diagnosed radiologically. It is difficult to quantify in most cases.

The radiological features common to most forms of pulmonary hypertension are; enlargement of the main pulmonary artery and hilar arteries, the vessels within the lungs being normal or small. When the pulmonary hypertension is part of Eisenmenger's syndrome (greatly raised pulmonary arterial resistance in association with a pulmonary-systemic communication, e.g. ASD, VSD or PDA, leading to a right-to-left shunt through these defects) the vessels within the lungs may also be large, but there is still disproportionate enlargement of the central vessels (Fig. 3.16).

The reason for the pulmonary arterial hypertension may be visible on the chest film; e.g. in cor pulmonale the lung disease is often radiologically obvious, and in mitral valve disease the other features described on p. 99 will also be seen, particularly left atrial enlargement.

4. Pulmonary venous hypertension. (Fig. 3.17). Mitral valve disease and left ventricular failure are the common causes of an elevated pulmonary venous pressure. In the normal upright person the lower zone vessels are larger than those in the upper zones. In raised pulmonary venous pressure the upper zone vessels enlarge and in severe cases become larger than those in the lower zones. This is a very useful sign, since there is no other non-invasive method of assessing the pulmonary venous pressure.

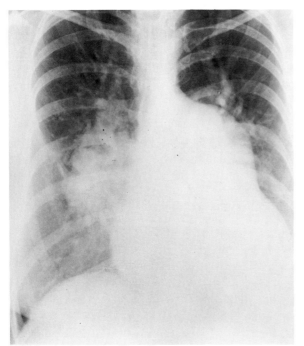 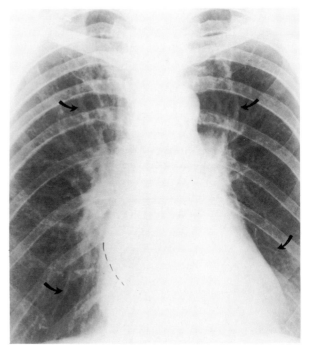

Fig. 3.16 Pulmonary arterial hypertension due, in this case, to an atrial septal defect with Eisenmenger's syndrome. The main pulmonary artery and hilar arteries are massive with an abrupt change in calibre of the vessels at the level of the segmental arteries. Note that the heart is also large.

Fig. 3.17 Pulmonary venous hypertension in a patient with mitral valve disease. The upper zone vessels (arrowed) are larger than the equivalent vessels in the lower zones (arrowed). This is the reverse of the normal situation. (The left atrial border has been drawn in.)

5. Decreased pulmonary blood flow. To be recognisable radiologically the reduction in pulmonary blood flow must be substantial. The commonest cause is the tetralogy of Fallot (see Fig. 3.26), where there is obstruction to the right ventricular outflow and a ventricular septal defect which allows right to left shunting of the blood.

Pulmonary valve stenosis only causes oligaemia in extremely severe cases, and these are virtually confined to babies and very young children.

PULMONARY OEDEMA

Pulmonary oedema occurs in many different situations, e.g. cardiac disease, renal failure and fluid overload. The common cardiac conditions causing pulmonary oedema are mitral stenosis and left ventricular failure from many different causes. Pulmonary oedema occurs when the pulmonary venous pressure rises above 24 to 25 mmHg (the osmotic pressure of plasma). Initially, pulmonary oedema is confined to the interstitial tissues of the lung, but if it becomes more severe the oedema will also collect in the alveoli. Both interstitial and alveolar pulmonary oedema are recognisable on plain chest films.

Interstitial oedema (Fig. 3.18)

There are many septa in the lungs which are invisible in the normal chest film because they consist of little more than a sheet of connective tissue and very small blood and lymph vessels. When thickened by oedema the peripherally located septa may be seen as line shadows. These lines, known as Kerley B lines, named after the

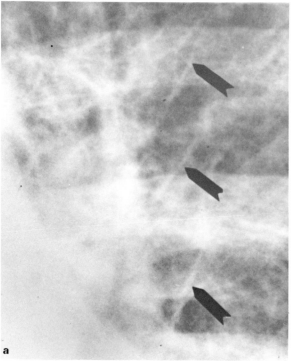

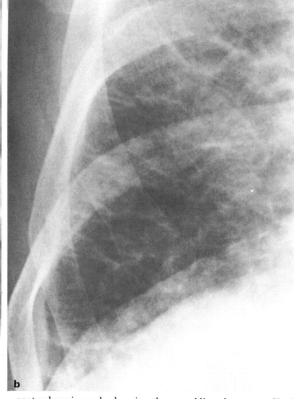

Fig. 3.18 Septal lines in interstitial pulmonary oedema. (a) Left upper zone showing the septal lines known as Kerley A lines (arrowed) in a patient with acute left ventricular failure following a myocardial infarction. Note that these lines are narrower and sharper than the adjacent blood vessels; (b) right costophrenic angle showing the septal lines known as Kerley B lines in a patient with mitral stenosis. Note that these oedematous septa are horizontal non-branching lines which reach the pleura.

man who first described them, are horizontal lines never more than 2 cm long seen laterally in the lower zones. They reach the lung edge and are, therefore, readily distinguished from blood vessels, which never extend into the outer centimetre of the lung. Other septa radiate towards the hila in the mid and upper zones (Kerley A lines). These are much thinner than the adjacent blood vessels, and are only 3–4 cm in length.

The outline of the blood vessels may become indistinct due to oedema collecting around them. This loss of clarity is a difficult sign to evaluate and it may only be recognised by looking at follow-up films after the oedema has cleared.

Alveolar oedema (Fig. 3.19)

This is a more severe form of oedema where the fluid collects in the alveoli. Alveolar oedema is almost always bilateral involving all the lobes. Shadowing is maximal close to the hilar and fades out peripherally leaving a relatively clear zone, which may contain septal lines around the edge of the lobes. This pattern of oedema is sometimes known as the 'butterfly pattern' or 'bat's wing pattern'.

HEART FAILURE

One or more of the following general signs of heart

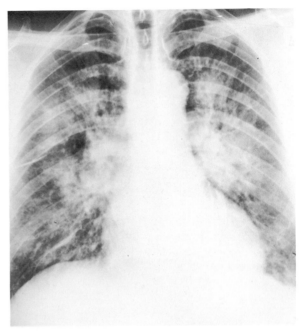

Fig. 3.19 Alveolar oedema in a patient with acute left ventricular failure following a myocardial infarction. The oedema fluid is concentrated in the more central portion of the lungs leaving a relatively clear zone peripherally. Note that all the lobes are fairly equally involved.

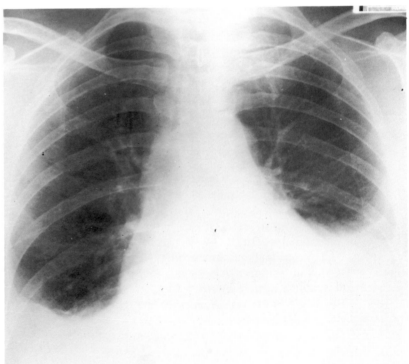

Fig. 3.20 Congestive cardiac failure. There are large bilateral pleural effusions. The heart is enlarged although it is difficult to measure it precisely because the pleural fluid obscures its borders.

failure may be seen (Fig. 3.20). (These signs may coexist with the specific disorders described in the remainder of this chapter.)

1. Cardiac enlargement with or without specific chamber enlargement.
2. Evidence of raised pulmonary venous pressure; namely enlargement of the vessels in the upper zones of the lungs.
3. Evidence of pulmonary oedema.
4. Pleural effusions.

Pleural effusions due to heart failure are usually bilateral, often larger on the right than the left, and if unilateral almost always right-sided. In acute left ventricular failure small effusions are seen in the costophrenic angles running up the lateral chest wall. These may in fact be due to oedema in the lungs rather than true pleural effusions.

SPECIFIC CARDIAC DISORDERS

Valvular heart disease

Valve stenosis and incompetence often coexist. It is, however, convenient to describe each of these phenomena separately.

Mitral stenosis (Fig. 3.21)

The three most important features are:

1. Enlargement of the left atrium. The degree is very variable and correlates poorly with the severity of the mitral stenosis.
2. Evidence of raised pulmonary venous pressure and pulmonary oedema. Interstitial oedema is commoner than the alveolar form. These signs correlate well with the severity of the mitral stenosis.
3. Calcification of the mitral valve.

Unless pulmonary hypertension develops the transverse cardiac diameter is often normal. Pulmonary hypertension leads to several additional abnormalities: dilatation of the central pulmonary arteries, right ventricular enlargement and, if functional tricuspid incompe-

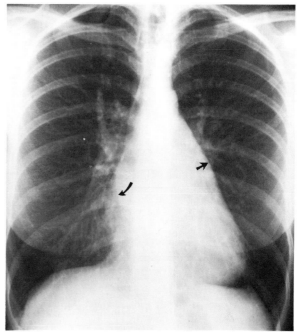

Fig. 3.21 Mitral stenosis. The enlarged left atrium shows as a double contour at the right heart border (curved arrow). The left atrial appendage is also enlarged (straight arrow). Note how the upper zone vessels are larger than those in the lower zones. The overall size of the heart is not increased.

tence occurs, enlargement of the right atrium (Fig. 3.4).

Widespread fine nodules may be seen in the lungs in patients who develop haemosiderosis and, occasionally, multiple small foci of bone are seen in the mid and lower lung fields in long-standing cases.

Mitral incompetence

As with mitral stenosis left atrial enlargement and raised pulmonary venous pressure are important signs. In most cases of mitral incompetence the size of the left atrium correlates well with the severity of the incompetence. An important difference from mitral stenosis is the presence of left ventricular enlargement.

Aortic stenosis (Fig. 3.22)

The major features are aortic valve calcification and

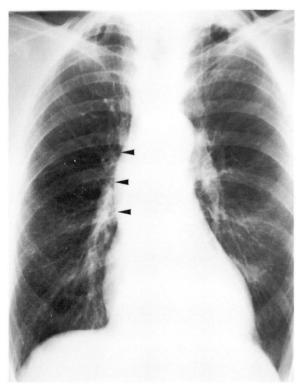

Fig. 3.22 Aortic stenosis showing poststenotic dilatation of the aorta (arrows). Despite the presence of a severe gradient there is little if any cardiac enlargement. The lateral view (see Fig. 3.10b) showed heavy calcification of the aortic valve—a feature not visible in the frontal view of this instance.

poststenotic dilatation of the ascending aorta. It is rare for a patient to have a severe gradient across the aortic valve unless calcification is seen, but it may be necessary to fluoroscope the heart to be certain of the presence of calcification. Recognisable left ventricular enlargement and raised pulmonary venous pressure are late signs.

Aortic incompetence (Fig. 3.5)

Aortic incompetence leads to enlargement of the left ventricle early in the course of the disease. The ascending aorta is dilated. As the severity of the aortic incompetence increases the left atrium enlarges and the changes of raised pulmonary venous pressure develop.

Pulmonary stenosis

This condition is considered on p. 101.

Tricuspid stenosis and incompetence

Both of these give rise to enlargement of the right atrium and superior vena cava. They are almost never seen as isolated abnormalities and the features of coexistent mitral valve disease or pulmonary hypertension often dominate the radiological picture.

Ischaemic heart disease

Most patients with angina or an acute myocardial infarct have a normal chest film. The abnormalities which may be seen include:

1. Signs of raised pulmonary venous pressure and pulmonary oedema. The chest x-ray is a sensitive method of detecting these phenomena; indeed, it is often more reliable than the physical examination in this respect.
2. Cardiac enlargement and aneurysm formation.

It is only when there is substantial muscle damage that cardiac enlargement occurs. Usually, there are no specific features to the cardiac contour but an aneurysm may be recognised on plain film if it involves the left heart border or apex (Fig. 3.23). Myocardial infarcts occasionally calcify. Usually, this is seen in association with aneurysm formation. Fluoroscopy is more sensitive than plain films in detecting aneurysm formation and myocardial calcification.
3. Atheromatous calcification may be seen in the coronary arteries but, as elsewhere in the body, arterial calcification, though it indicates the presence of atheroma, is a poor indicator of its severity.

Hypertensive heart disease and other myocardial problems

As discussed previously, ventricular hypertrophy without dilatation does not cause recognisable abnormality on plain chest films. Therefore, in systemic hypertension and various forms of cardiomyopathy the chest x-ray only becomes abnormal once ventricular dilatation has

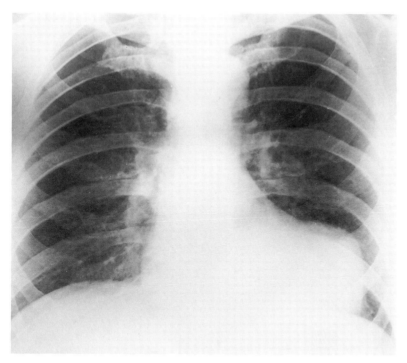

Fig. 3.23 Left ventricular aneurysm. The transverse diameter of the heart is moderately enlarged. There is a bulge of the lower half of the left heart border extending down to the apex. This bulge is due to the aneurysm itself.

taken place. If the left ventricular end-diastolic pressure and hence left atrial pressure rises the signs of moderate left atrial enlargement and elevation of the pulmonary venous pressure may be seen.

The shape of the heart is the same regardless of the causes of the myocardial disorder. The aorta is, however, large in systemic hypertension.

Congenital heart disease

There are a large number of congenital malformations of the heart and great vessels. Frequently, they are multiple. Only a few selected disorders are discussed here.

Atrial septal defect, ventricular septal defect and patent ductus arteriosus

All these give rise to cardiac enlargement, enlargement of the main pulmonary artery, large hilar arteries and pulmonary plethora (Fig. 3.24). These signs are usually

only visible when the left-to-right shunt through the defect is 2:1 or more. The value of plain film radiology is that it provides a simple method of estimating the degree of shunting in a patient known to have a left-to-right shunt. It is, however, of limited value in making a diagnostic distinction between the three major causes of left-to-right shunt. In cases of clinical diagnostic doubt (e.g. pulmonary stenosis can be confused clinically with atrial septal defect, and mitral incompetence can be confused with ventricular septal defect) the presence of plethora indicates that the patient has a left-to-right shunt.

Pulmonary valve stenosis

This can give rise to a specific radiological appearance (Fig. 3.14). There is enlargement of the main pulmonary artery and there may be enlargement of the left pulmonary artery, the remainder of the lung vasculature being normal. The dilatation of these vessels is due to the phenomenon of poststenotic dilatation. Usually, the

Fig. 3.24 Ventricular septal defect in a child. The heart is enlarged and there is obvious enlargement of the pulmonary vessels. The left-to-right shunt in this case was 3:1.

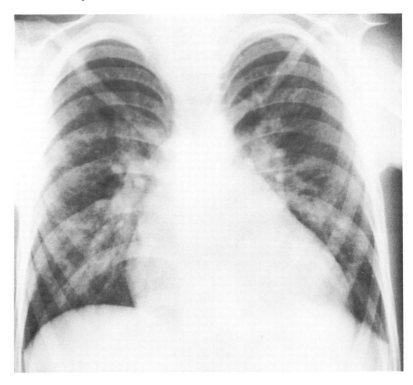

Fig. 3.25 (*below*) Coarctation of the aorta. (a) Abnormal aortic knuckle: the lower arrow points to the poststenotic dilatation of the aorta immediately below the coarctation. The upper arrow indicates the dilated left subclavian artery above the coarctation; (b) rib notching: portion of the ribs in another patient showing notching on the under surfaces.

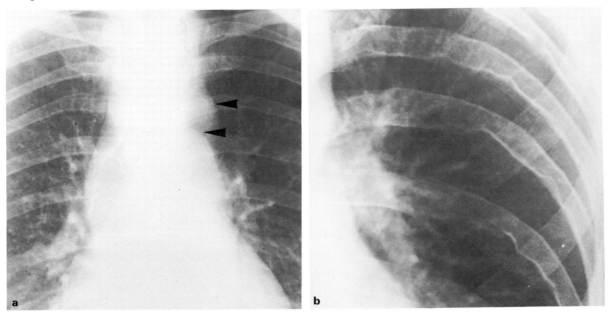

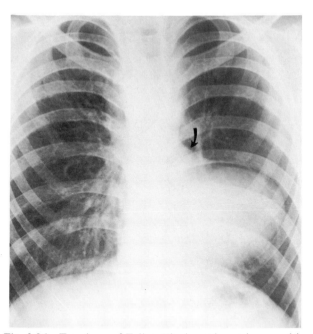

Fig. 3.26 Tetralogy of Fallot. The heart is not increased in overall size but shows a very abnormal outline. The apex is lifted up and there is a bay in the region of the main pulmonary artery (arrow). This combination is known as 'coeur en sabot' (boot-shaped heart). The aorta is right-sided as it is in 25% of cases. The hilar vessels are very small and the vessels in the lungs are also small.

heart is not enlarged. In severe cases presenting in early childhood the heart may be large and the lungs oligaemic.

Coarctation of the aorta (Fig. 3.25)

An abnormal aortic knuckle is the commonest finding. The site of narrowing of the aorta can be seen as an indentation and there is frequently a bulge above the coarctation due to dilatation of the left subclavian artery and a dilatation below it, due to poststenotic dilatation of the aorta.

Over half the cases in adults show rib notching. This sign is due to the development of anastomatic pathways via the intercostal arteries. One or more small corticated indentations are seen on the inferior margins of the posterior halves of the ribs from the third or fourth ribs downward. It is not seen until late childhood.

The heart is often enlarged, as is the ascending aorta, due to the long-standing hypertension.

The tetralogy of Fallot

This consists of a ventricular septal defect, right ventricular outflow obstruction, usually subvalvar or valvar stenosis with consequent right ventricular hypertrophy. The aorta overrides the ventricular septal defect.

Table 3.1 Commoner congenital heart disease: diagnosis based on blood vessel pattern.

	Plethora	*Normal vascularity*	*Oligaemia*
Not cyanosed	Atrial septal defect Ventricular septal defect Patent ductus arteriosus	Left to right shunts <2:1 Pulmonary stenosis Coarctation of the aorta	Severe pulmonary stenosis
Cyanosed	Transposition of the great arteries Total anomalous pulmonary venous drainage Eisenmenger ASD, VSD and PDA Truncus arteriosus	Fallot's tetralogy Pulmonary atresia	Fallot's tetralogy Pulmonary atresia Tricuspid atresia

Approximately half the cases have a normal chest film. The abnormal radiological signs, when present, are (Fig. 3.26); an upturned cardiac apex and a bay in the region of the main pulmonary artery. (This shape is sometimes referred to as a 'boot-shaped heart'.) The aorta is right-sided in 25% of patients. An important feature is oligaemia of the lungs.

4

The Plain Abdomen

The standard plain films of the abdomen are the supine AP view and the erect AP view. An alternative to the erect AP view in patients unable to sit or stand is to take a lateral decubitus view (i.e. an AP film taken with the patient lying on his side). This view, like the erect view, utilises a horizontal x-ray beam. The main purpose of horizontal beam films is to detect air-fluid levels and free intraperitoneal air.

How to look at a plain abdominal film (Fig. 4.1)

In a normal patient relatively large amounts of gas are usually present in the stomach and colon, with only a small amount in the small intestine. The stomach can be readily identified by its location above the transverse colon, by the band-like shadows of the gastric rugae in the supine view, and by the gas-fluid level beneath the diaphragm in the erect view.

The duodenum often contains air and shows a fluid level. There may be some gas in the normal small bowel, but it is rarely sufficient to outline the whole of a loop. Short fluid levels in the small and large bowel are normal. Fluid levels become abnormal when they are seen in dilated loops of bowel or when they are very numerous. If the bowel is dilated it is important to try and decide which portion is involved.

Look for any gas outside the lumen of the bowel. Its location and pattern often gives valuable diagnostic information.

Identify the liver. The liver is seen as a homogeneous opacity in the right upper quadrant usually extending into the left upper quadrant. Occasionally, there is a tongue-like extension of the right lobe into the right iliac fossa. This is a normal variant known as a Reidl's lobe and should not be confused with generalised liver enlargement. The lower border of the liver is often difficult to see but its position can be predicted by the position of the gas in the hepatic flexure and transverse colon.

Identify the borders of the spleen, kidneys, bladder and psoas muscles. A layer of extraperitoneal fat indicates the position of the peritoneum (Fig. 4.1b). The psoas outline is often obscured with a retroperitoneal abnormality.

Look for any soft tissue masses in the abdomen and pelvis.

If there is any calcification try to locate exactly where it lies.

Dilatation of the bowel (Figs 4.2, 4.3 & 4.4)

The distinction between dilatation of the large and small bowel can be difficult. It depends on: the appearance of the dilated bowel, the position and number of the bowel loops, and the presence of solid faeces. The presence of solid faeces is a useful and reliable indication of the position of the colon.

The colon can be recognised by its haustra, which usually form incomplete bands across the colonic gas shadows. They are always present in the ascending and transverse colon but may be absent distal to the splenic flexure. When the jejunum is dilated the valvulae conniventes can be identified. They are always closer together than the colonic haustra, and cross the width of the bowel often giving rise to the appearance known as 'a stack of coins' (Fig. 4.2). Problems may be encountered in distinguishing the lower ileum from the sigmoid colon since both may be smooth in outline. The radius of curvature of the loops is sometimes helpful in this

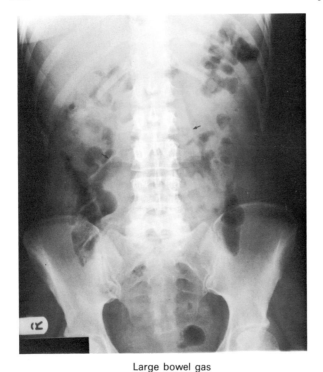

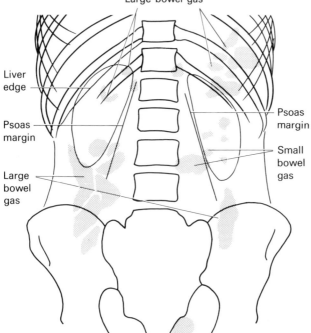

Large bowel gas

Liver edge

Psoas margin

Large bowel gas

Psoas margin

Small bowel gas

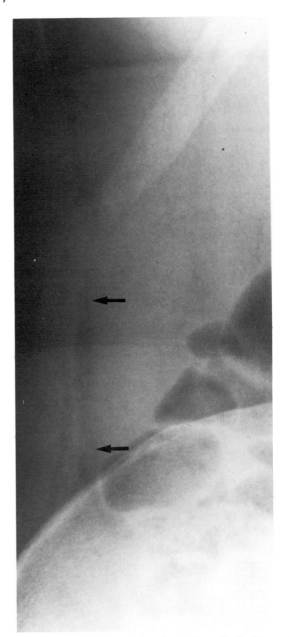

Fig. 4.1 Normal plain abdominal film. (*left*) Normal abdomen: the arrows point to the lateral borders of the psoas muscles. The renal outlines are obscured by the overlying colon; (*above*) Normal extraperitoneal fat stripe: part of the right flank showing the layer of extraperiton fat (arrows) which indicates the position of the peritoneum.

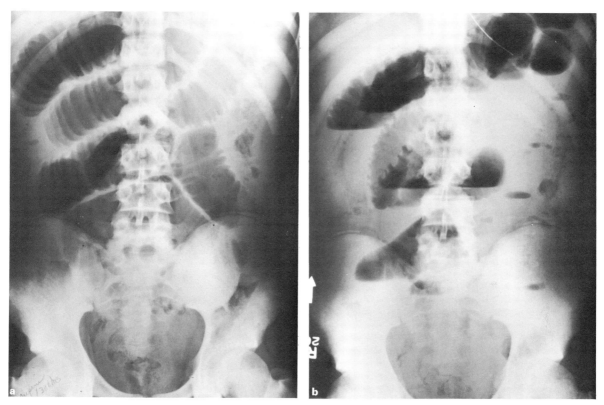

Fig. 4.2 Small bowel obstruction due to adhesions. (a) supine; (b) erect. The jejunal loops are markedly dilated and show air-fluid levels in the erect film. The jejunum is recognised by the presence of valvulae conniventes. Note the large bowel contains less gas than normal.

respect; the tighter the curve the more likely it is to be a dilated loop of small bowel.

The small bowel usually lies in the centre of the abdomen within the 'frame' of the large bowel, but the sigmoid and transverse colons are frequently very redundant and may also lie in the centre of the abdomen particularly when dilated.

The number of dilated loops is a valuable distinguishing feature between small and large bowel dilatation, because even with a very redundant colon one does not see the numerous layered loops that are so often present with small bowel dilatation.

Dilatation of the bowel occurs in mechanical obstruction, paralytic ileus and inflammatory bowel disease. The radiological differential diagnosis of these phenomena depends mainly on the distribution of the dilated loops. The following patterns can be recognised:

1. Mechanical obstruction of the small bowel (Fig. 4.2) causes small bowel dilatation with a normal or reduced calibre to the large bowel.

2. Obstruction of the large bowel (Fig. 4.3) causes dilatation of the colon down to the point of obstruction, and may be accompanied by small bowel dilatation if the ileocaecal valve becomes incompetent.

3. In generalised paralytic ileus (Fig. 4.4) both the large and the small bowel will be dilated. The dilatation often extends down into the sigmoid colon and gas may be present in the rectum. It may be difficult to differentiate such cases from low large bowel obstruction.

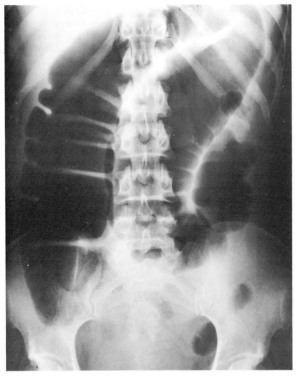

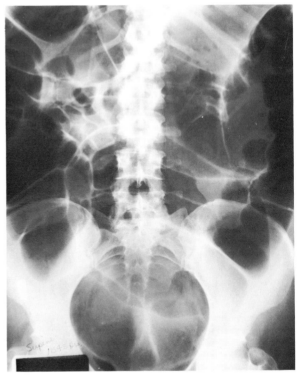

Fig. 4.3 Large bowel obstruction due to carcinoma at the splenic flexure. There is marked dilatation of the large bowel from the caecum to the splenic flexure.

Fig. 4.4 Paralytic ileus. There is considerable dilatation of the whole of the large bowel extending well down into the pelvis. Small bowel dilatation is also seen.

4. Local peritonitis often results in dilatation of the loops adjacent to the inflammatory process, giving rise to the so-called 'sentinel loops' which may be seen, for example, in appendicitis and pancreatitis.

5. Patients with gastroenteritis may show a number of patterns: some have a normal film and some show excess fluid levels without dilatation, whereas some mimic paralytic ileus and others small bowel obstruction.

6. Small bowel infarction may mimic both obstruction of the small bowel and obstruction of the large bowel.

7. Closed loop obstructions. The diagnosis depends on whether the loop in question contains air. If it does, as for example in a caecal volvulus or sigmoid volvulus, the dilated loop is seen filled with gas in a characteristic shape (Fig. 4.5). If the closed loop is filled with fluid it

may not be visible, the common situation in most obstructed hernias.

8. Toxic dilatation of the colon. Should this occur in patients with ulcerative colitis or more rarely Crohn's disease, the large bowel becomes distended (Fig. 4.6). In most of the patients maximal dilatation occurs in the transverse colon; indeed, the descending colon may be narrower than normal. The haustra are lost or grossly abnormal and the swollen islands of mucosa between the ulcers can be recognised as polypoid shadows. If the transverse colon is more than 6 cm in diameter in a patient with colitis, toxic dilatation should be strongly suspected. One cannot, however, make the diagnosis of 'toxic' dilatation from radiological findings alone since not all patients with these findings are toxic on clinical examination.

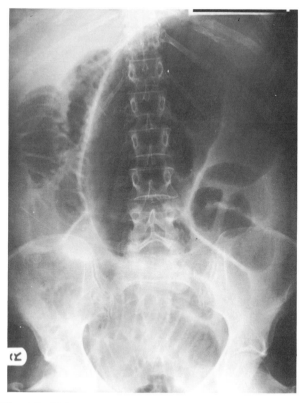

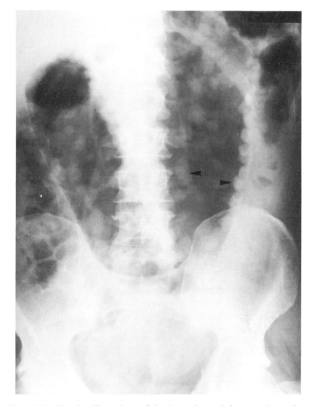

Fig. 4.5 Volvulus of the caecum. The twisted obstructed caecum and ascending colon now lie on the left side of the abdomen and appear as a large gas shadow. There is also extensive small bowel dilatation due to obstruction by the volvulus.

Fig. 4.6 Toxic dilatation of the large bowel due to ulcerative colitis. The dilatation is maximal in the transverse colon. Note the loss of haustrae and islands of hypertrophied mucosa. Two of these pseudopolyps are arrowed.

Gas outside the lumen of the bowel

Gas outside the lumen of the bowel is abnormal. Its location can usually be assessed:

1. Gas in the peritoneal cavity (Fig. 4.7) is almost always due to perforation of the gastrointestinal tract, or follows surgical intervention in the abdomen. The most common cause of spontaneous pneumoperitoneum is a perforated peptic ulcer and two-thirds of such cases are recognisable radiologically. The largest quantities of free gas are seen after colonic perforation and the smallest amounts with leakage from the small bowel. A pneumoperitoneum is very rare in acute appendicitis even if the appendix is perforated.

Free intraperitoneal air is a normal finding after a laparotomy. In adults all the air is usually absorbed within 7 days. In children, the air absorbs very much faster, usually within 24 hours. An increase in the amount of air on successive films indicates continuing leakage of air.

Air under the right diaphragm is usually easy to recognise on an erect abdominal or chest film as a curvilinear collection of gas between the line of the diaphragm and the opacity of the liver. Free gas under the left diaphragm is more difficult to identify because of the overlapping gas shadows of the stomach and splenic flexure of colon. Gas in these organs may mimic free intraperitoneal air when none is present. Gas under the

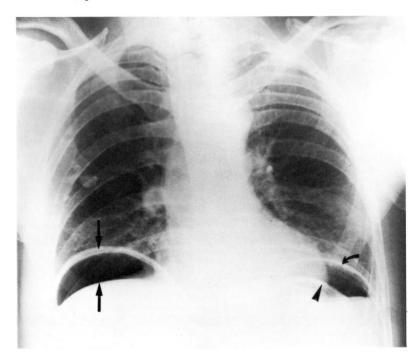

Fig. 4.7 Free gas in the peritoneal cavity. On this chest x-ray air can be seen under the domes of both diaphragms. The curved arrow points to the left diaphragm and the arrow head to the wall of the stomach. The two vertical arrows on the right point to the diaphragm and upper border of the liver.

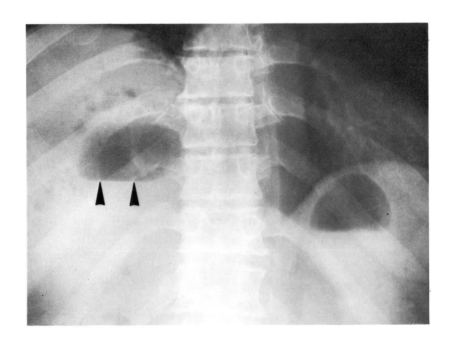

Fig. 4.8 Gas in a right subphrenic abscess. There are several collections of gas within the abscess. The largest of these contains a fluid level (arrow). The air-fluid level under the left diaphragm is normal. It is in the stomach.

diaphragm is much easier to diagnose on an erect chest film than on an upright abdominal film. If there is doubt about the presence of a pneumoperitoneum a lateral decubitus film will show the air collected beneath the flank. It is important to realise that when the patient is lying flat the gas collects centrally beneath the abdominal wall and is very difficult to identify on the conventional supine film.

2. Gas in an abscess (Fig. 4.8) produces a very variable pattern. It may form either small bubbles or larger collections of air, both of which could be confused with gas within the bowel. Fluid levels in abscesses may be seen on a horizontal ray film. Since abscesses are mass lesions they displace the adjacent structures, e.g. the diaphragm is elevated with a subphrenic abscess, and the bowel is displaced by pericolic and pancreatic abscesses. Pleural or pulmonary shadows are very common in association with subphrenic abscess. Although fluoroscopy of the diaphragm is extremely popular, it is of limited value in the diagnosis of subphrenic abscess, partly because the diaphragm is so frequently obscured by pleural fluid and partly because inflammatory processes at the lung bases also reduce diaphragmatic movement.

3. Gas in the wall of the bowel (*pneumatosis intestinalis*). Numerous spherical or oval bubbles of gas are seen in the wall of the large of small bowel in adults in the benign condition known as *pneumatosis cystoides intestinalis*. Linear streaks of intramural gas have a more sinister significance as they usually indicate infarction of the bowel wall. Gas in the wall of the bowel in the neonatal period, whatever its shape, is diagnostic of necrotising enterocolitis (Fig. 4.9) a disease that is fairly common in premature babies with respiratory problems.

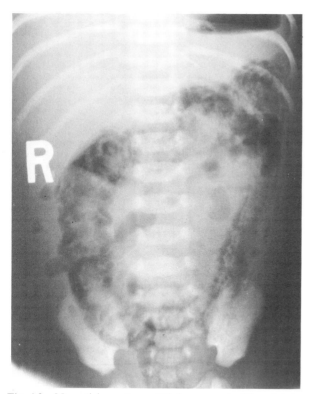

Fig. 4.9 Necrotising enterocolitis in a neonate. There is intramural gas throughout the colon.

Ascites

Small amounts of ascites cannot be detected on plain films. Larger quantities separate the loops of bowel from one another and displace the ascending and descending colon from the fat stripes which indicate the position of the peritoneum along the lateral abdominal walls. The loops of small bowel float to the centre of the abdomen (Fig. 4.10).

In practice, plain films are of very limited value in the diagnosis of ascites, since the signs are so difficult to interpret confidently except when large amounts of ascites are present.

Abdominal calcifications

An attempt should always be made to determine the nature of any abdominal calcification. The first essential is to localise the calcification; for this a lateral or oblique view may be necessary. Once the organ of origin is known the pattern or shape of the calcification will usually limit the diagnosis to just one or two alternatives.

The most common calcifications are of little or no significance to the patient. These include phleboliths,

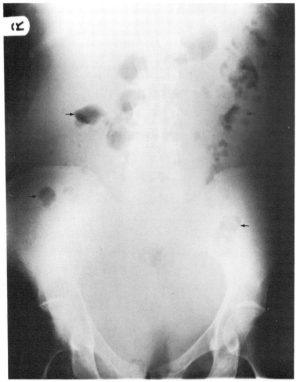

Fig. 4.10 Ascites. Note how the gas in the ascending and descending colon (arrows) is displaced by the fluid away from the side walls of the abdomen.

calcified lymph nodes, costal cartilages and arterial calcification.

1. Pelvic vein phleboliths (Fig. 4.12) are very common; the only problem they cause is that they may be difficult to distinguish from urinary calculi and faecaliths. As with phleboliths calcified *mesenteric lymph nodes*, due to old tuberculosis, are important only in that they may be difficult to differentiate from other more important calcifications. Their pattern is often specific: they are irregular in outline and very dense and since they lie in the mesentery they are often mobile. It is usually possible to see that they are composed of a conglomeration of smaller rounded calcifications.

2. Vascular calcification occurs in association with atheroma, but there is no correlation with the haemodynamic severity of the vascular disease.

Calcification is frequently present in the walls of *abdominal aortic aneurysms*. It is usually easier to assess the size of such aneurysms on the lateral projection (Fig. 4.11).

3. Uterine fibroids (Fig. 4.12) may contain numerous irregularly shaped well-defined calcifications conforming to the spherical outline of fibroids. Again the calcification is by itself of no significance to the patient.

4. Malignant ovarian masses occasionally contain visible calcium. The only benign ovarian lesion that is visibly calcified is the *dermoid cyst*, which may contain various calcified components, of which teeth are the commonest.

5. Adrenal calcification (Fig. 4.13) occurs after adrenal haemorrhage, after tuberculosis and occasionally in adrenal tumours. However, the majority of patients

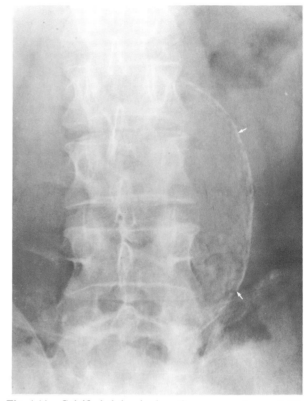

Fig. 4.11 Calcified abdominal aortic aneurysm (arrows). The aneurysm measured 8 cm in diameter on the lateral view.

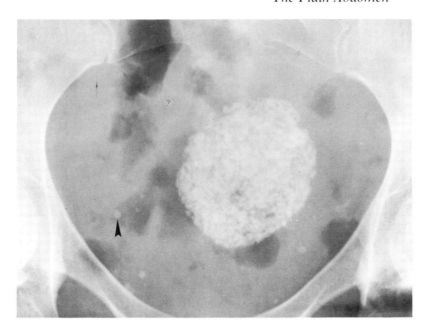

Fig. 4.12 Calcification in a large uterine fibroid. There are also several phleboliths, one of which is arrowed.

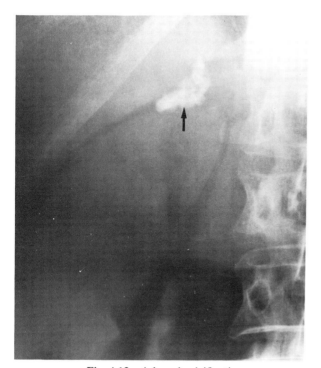

Fig. 4.13 Adrenal calcification.

with adrenal calcification are asymptomatic healthy people in whom the cause of the calcification is unclear. Only a minority of patients with Addison's disease have adrenal calcification.

6. Liver calcification occurs in hepatomas and rarely in other tumours. Hydatid cysts, old abscesses and tuberculosis may also calcify. Gall-stones, renal stones and costal cartilage are common causes of calcification projected over the liver shadow.

7. Splenic calcification is rarely of clinical significance. It is seen in cysts, infarcts, old haematomas and following tuberculosis.

8. Pancreatic calcification occurs in chronic pancreatitis. The calcifications are mainly small calculi within the pancreas. The position on the frontal and lateral views usually enables the diagnosis to be made without difficulty (Fig. 4.14).

9. Faecaliths. Calcified faecaliths may be seen in diverticula of the colon or in the appendix (Fig. 4.15). Appendiceal faecaliths are an important radiological observation since the presence of an appendolith is a strong indication that the patient has acute appendicitis, often with gangrene and perforation. However, only a

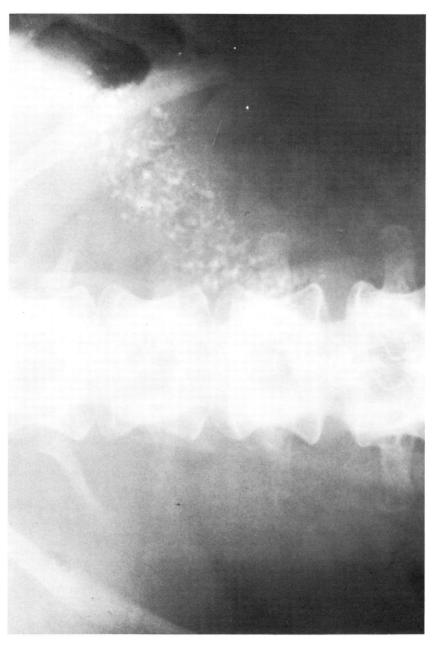

Fig. 4.14 Pancreatic calcification.

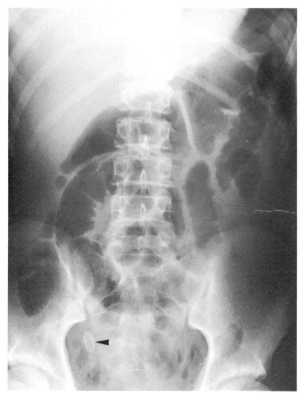

Fig. 4.15 Appendolith. The oval calcified shadow (arrow) is a faecalith in the appendix. The patient had perforated appendicitis. Note the dilated loops of small bowel in the centre of the abdomen due to peritonitis—the so-called 'sentinel loops'.

small proportion of patients with appendicitis have a radiologically visible appendolith.

10. Soft tissue calcification in the buttocks may be seen following injection of certain medicines. These shadows can at times be confused with intra-abdominal calcifications.

11. Calcification of the urinary tract is discussed on p. 191.

The liver and spleen

Substantial enlargement of the liver has to occur before it can be recognised on a plain abdominal film. As the liver enlarges it extends well below the costal margin displacing the hapatic flexure, transverse colon and right kidney downwards and displacing the stomach to the left. The diaphragm may also be elevated.

As the spleen enlarges the tip becomes visible in the left upper quadrant below the lower ribs. Eventually, it may fill the left side of the abdomen and even extend across the midline into the right lower quadrant. The splenic flexure of the colon and the left kidney are displaced downwards and the stomach is displaced to the right.

Following trauma, rupture of the spleen occurs more frequently than of the liver, but the principles underlying the plain film diagnosis are the same in both cases. As a haematoma forms the plain films may show a mass in the upper abdomen displacing adjacent structures. In the case of a ruptured spleen the stomach will be displaced medially and the splenic flexure of the colon displaced downwards. There may be paralytic ileus. Fractures of the lower ribs may also be present. These signs, though helpful if present, are often not seen even with significant lacerations of the liver and spleen.

5

The Gastrointestinal Tract

The chapter on the abdomen (p. 105) showed that the plain film can be very informative in patients with an acute abdomen, but for most other intestinal disorders some form of contrast examination is necessary. Barium sulphate is the best contrast medium for the gastrointestinal tract. It produces excellent opacification, good coating of the mucosa and is completely inert. Its only major disadvantage is that when water is reabsorbed in the colon the barium may solidify and impact proximal to a colonic or rectal stricture. The other available contrast is a water-soluble medium, Gastrografin. It has several disadvantages: it is hypertonic and soon becomes diluted; it is irritant should it inadvertently enter the lungs; and it is less radio-opaque than barium.

Gastrointestinal contrast examinations are carried out under fluoroscopic control so that the passage of contrast can be observed on a television monitor. By watching the television screen the radiologist is able to position the patient so that any abnormality is shown clearly. Films are taken to show fine detail and to serve as a permanent record. One of the values of fluoroscopy is to ensure that an abnormality has a constant appearance. Peristaltic waves are transitory and so can be easily distinguished from a true narrowing which is constant.

The double-contrast examination of the stomach and colon has recently been gaining in popularity. In the single-contrast method the bowel is filled only with barium. In the double-contrast technique the mucosa is coated with barium and the stomach or colon distended by introducing gas, often in combination with an injection of a short-acting smooth-muscle relaxant to paralyse the bowel. The double-contrast method is a little more time consuming but shows the mucosa to advan-

tage and demonstrates small abnormalities which would be obscured by a large volume of barium.

It is important to understand some basic terms applicable to the radiology of the gastrointestinal tract which are often used in rather a loose way.

The wall of the bowel is never seen as such. What is seen is the outline of the lumen and from this one has to draw conclusions about the state of the wall. Usually, the most reliable information is obtained when the bowel is fully distended.

Mucosal folds are seen when the bowel is in a contracted state so that the mucosa becomes folded (Fig. 5.7b). When the bowel is distended these mucosal folds disappear. The normal mucosal fold pattern may be altered by smoothing out or by abnormal irregularity.

Filling defect is a term used to describe something occupying space within the bowel thereby preventing the normal filling of the lumen with barium. This creates an area of total or relative radiolucency within the barium column. There are three types of filling defects, each having distinct radiological signs:

1. An intraluminal filling defect is entirely within the lumen of the bowel (e.g. food), and has barium all around it (Fig. 5.1a).
2. An intramural filling defect arises from the wall of the bowel (e.g. a carcinoma or leiomyoma). It causes an indentation from one side only, making a sharp angle with the wall of the bowel and is not completely surrounded by barium (Fig. 5.1b).
3. An extramural filling defect arises outside the bowel

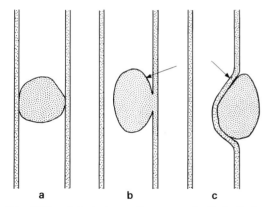

Fig. 5.1 Filling defects shown diagrammatically in the bowel. (a) Intraluminal; (b) intramural: note the sharp angle (arrow) made with the wall; (c) extramural: there is a shallow angle (arrow) with the wall of the bowel.

but compresses it, e.g. enlarged pancreas or lymph nodes. It also gives a narrowing from one side only but makes a shallow angle with the wall of the bowel. The mucosa is preserved but stretched over the filling defect (Fig. 5.1c).

A stricture is a circumferential or annular narrowing. A stricture must be differentiated from the transient narrowing which occurs with normal peristalsis. A stricture may have tapering ends (Fig. 5.2a) or it may end abruptly and have overhanging edges giving an appearance known as 'shouldering' (Fig. 5.2b). Shouldering is a feature of malignancy.

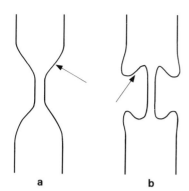

Fig. 5.2 Stricture. (a) Tapering ends (arrow); (b) overhanging edges or shouldering (arrow).

Ulceration. An ulcer is a breach of a mucosal surface which becomes visible when the crater contains barium. When viewed in profile it appears as an outward projection from the barium-filled lumen (Fig. 5.3a). When viewed *en face* the ulcer crater appears as a rounded collection of barium (Fig. 5.3b).

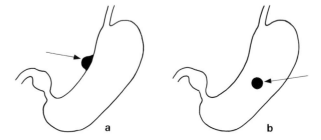

Fig. 5.3 Ulceration. (a) In profile the ulcer is seen as an outward projection (arrow); (b) *en face* the ulcer appears rounded (arrow).

THE SALIVARY GLANDS

Calculi occur most commonly in the submandibular duct or gland and as they normally contain calcium they can be seen on plain films.

Sialography

A sialogram is performed by injecting contrast into the ducts of the salivary glands. Only the submandibular and parotid glands have ducts which can be cannulated (Fig. 5.4). Stones and strictures in the ducts can be identified. There may be dilatation of small ducts which is known as sialectasis and this may occur with obstruction to the main duct (Fig. 5.5) but may also be seen without obvious obstruction. Tumours of the salivary gland may cause stretching and displacement of the ducts within the gland.

THE OESOPHAGUS

Plain films do not normally show the oesophagus unless it is very dilated (e.g. achalasia), but they are of use in

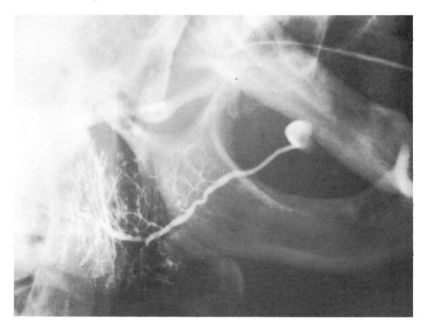

Fig. 5.4 Normal parotid sialogram.
Note the long duct of even calibre and
the fine branching of the ducts within
the gland.

demonstrating an opaque foreign body such as a bone
lodged in the oesophagus (Fig. 5.6).

The barium swallow is the contrast examination
employed to visualise the oesophagus. The patient
drinks some barium and its passage down the oesophagus is observed on a television monitor. Films are
taken in an oblique position to project the oesophagus
clear of the spine with the oesophagus both full of
barium, to show the outline and empty to show the
mucosal pattern.

Normal barium swallow

The oesophagus when full of barium should have a
smooth outline. When empty and contracted barium
normally lies in between the folds of mucosa which
appear as three or four long, straight parallel lines (Fig.
5.7).

The aortic arch gives a clearly visible impression on
the left side of the oesophagus, which is more pronounced in the elderly as the aorta becomes tortuous

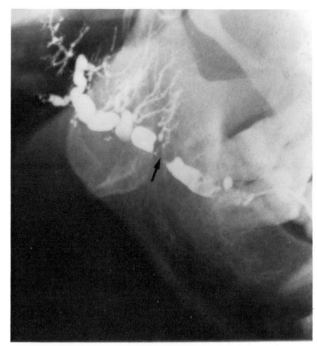

Fig. 5.5 Sialectasis. There is dilatation of the ducts due to a
stone (arrow) in the main parotid duct.

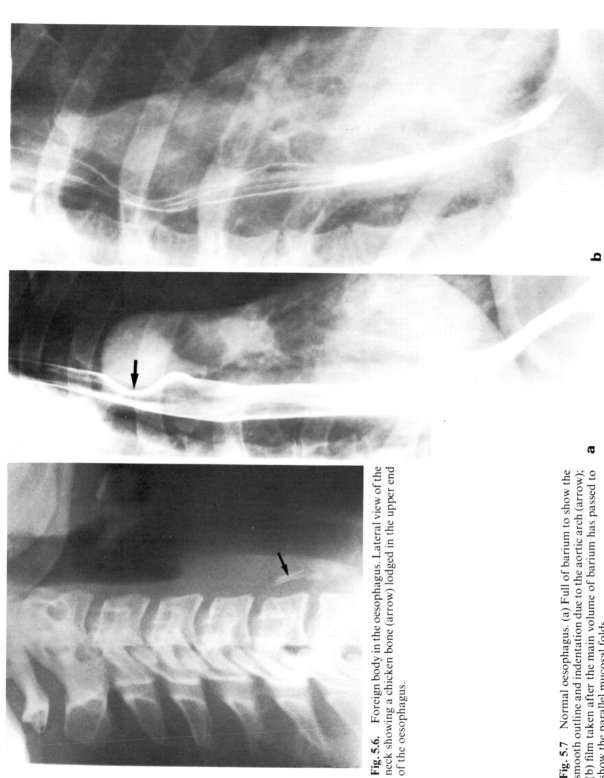

Fig. 5.6. Foreign body in the oesophagus. Lateral view of the neck showing a chicken bone (arrow) lodged in the upper end of the oesophagus.

Fig. 5.7 Normal oesophagus. (a) Full of barium to show the smooth outline and indentation due to the aortic arch (arrow); (b) film taken after the main volume of barium has passed to show the parallel mucosal folds.

b

a

and elongated. Below the aortic impression there is often a smaller impression due to the left main bronchus. The lower part of the oesophagus sweeps gently forward closely applied to the back of the left atrium and left ventricle.

Peristaltic waves can be observed during fluoroscopy. They move smoothly along the oesophagus to propel the barium rapidly into the stomach even if the patient swallows when lying flat. It is important not to confuse a contraction wave with a true narrowing; a narrowing is constant whereas a contraction wave is transitory. Sometimes the contraction waves do not occur in an orderly fashion but are pronounced and prolonged to give the oesophagus an undulated appearance known as a 'corkscrew oesophagus' (Fig. 5.8). These so-called tertiary contractions usually occur in the elderly and in most instances they do not give rise to symptoms.

Abnormal barium swallow

Strictures

Strictures are an important cause of dysphagia. There are four main causes; carcinoma, peptic, achalasia and corrosive. In order to distinguish between these possibilities it is useful to answer the following questions:

1. Where is the stricture?
2. What is its shape?
3. How long is it?
4. Is there a soft tissue mass?

Carcinomas rarely arise from only one wall but usually involve the full circumference to form strictures. The stricture, which may occur anywhere in the oesophagus, shows an irregular lumen with shouldered edges and is normally several centimetres in length (Fig. 5.9). A soft tissue mass may be visible.

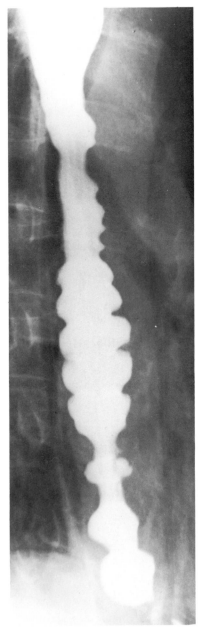

Fig. 5.8 Tertiary contractions (corkscrew oesophagus) giving the oesophagus an undulated appearance.

Peptic strictures are found at the lower end of the oeso-phagus and are almost invariably associated with a hiatus hernia and gastro-oesophageal reflux and, there-fore, the stricture may be some distance above the dia-phragm. Peptic strictures are characteristically short and have smooth outlines with tapering ends (Fig. 5.10).

An ulcer may be seen in close proximity to the stricture.

Achalasia is a neuromuscular abnormality resulting in failure of relaxation at the cardiac sphincter which pre-sents radiologically as a smooth, tapered narrowing,

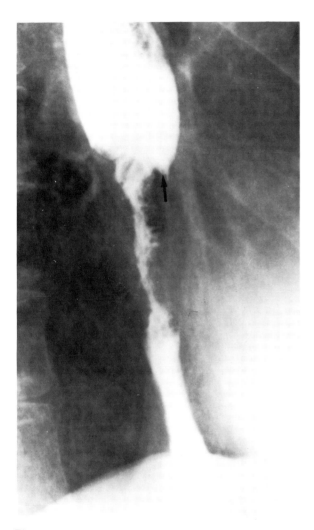

Fig. 5.9 Carcinoma. There is an irregular stricture with shouldering (arrow) at the upper end.

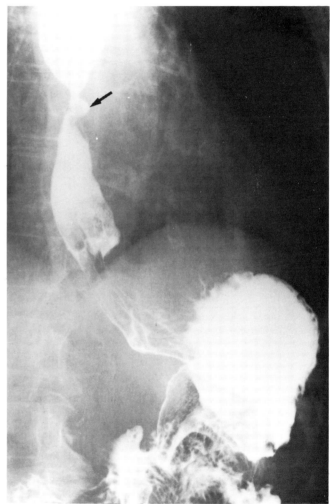

Fig. 5.10 Peptic stricture due to grastro-oesophageal reflux in a patient with a hiatus hernia. There is a short smooth stricture at the oesophagogastric junction with an ulcer crater within the stricture (arrow).

which is always at the lower end of the oesophagus (Fig. 5.11). There is associated dilatation of the oesophagus which often shows absent peristalsis. The dilated oesophagus usually contains food residues and may be visible on the plain chest radiograph. The lungs may

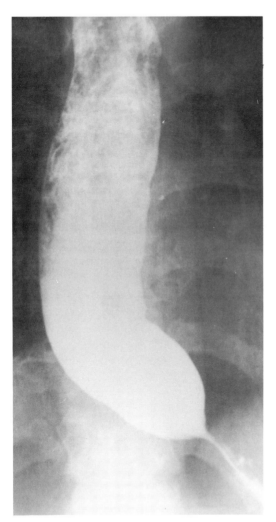

Fig. 5.11 Achalasia. The very dilated oesophagus containing food residues shows a smooth narrowing at its lower end.

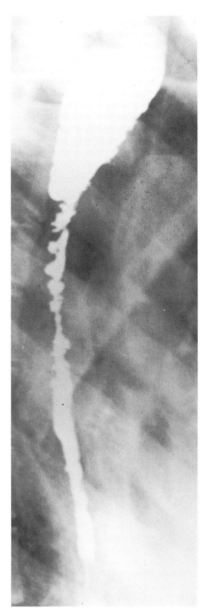

Fig. 5.12 Corrosive stricture.

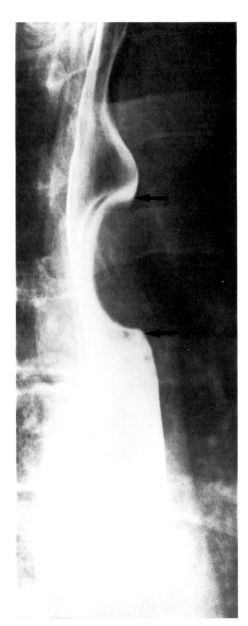

show consolidation and bronchiectasis due to aspiration of the oesophageal contents. The stomach gas bubble is usually absent because the oesophageal contents act as a water seal, but this sign is not diagnostic of achalasia as it is seen in other causes of oesophageal obstruction and can occasionally be observed in normal people.

Corrosive strictures are the result of swallowing corrosives such as acids or alkalis. They are long strictures which begin at the level of the aortic arch. As with the other benign strictures they are usually smooth with tapered ends (Fig. 5.12).

Filling defects

Filling defects may be caused by a tumour arising in the wall of the oesophagus, by a lesion arising from outside the oesophagus or by objects in the lumen of the oesophagus.

An intramural filling defect is likely to be a leiomyoma (Fig. 5.13). A leiomyoma causes a smooth, rounded indentation into the lumen of the oesophagus. A soft tissue mass may be seen in the mediastinum indicating extraluminal extension.

A carcinoma may cause an irregular filling defect, but as mentioned above carcinomas usually present as strictures.

Extramural lesions compressing the oesophagus include carcinoma of the bronchus, enlarged mediastinal lymph nodes and an aneurysm of the aorta (Fig. 5.14). In all these conditions the chest film will usually show the underlying pathology.

Fig. 5.13 Leiomyoma. There is an intramural filling defect in the oesophagus below the aortic arch (arrows). The sharp angle this makes with the wall of the oesophagus indicates that the filling defect is due to a mass arising in the wall of the oesophagus.

An anomalous right subclavian artery, which instead of coming from the inominate artery, arises as the last major branch from the aortic arch, gives rise to a characteristic short, smooth narrowing as it crosses behind the upper oesophagus (Fig. 5.15).

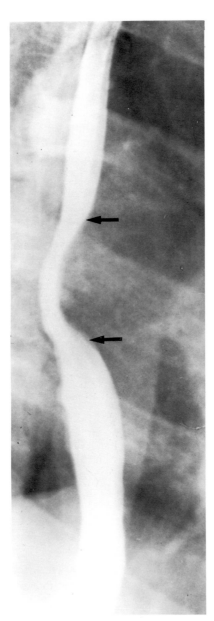

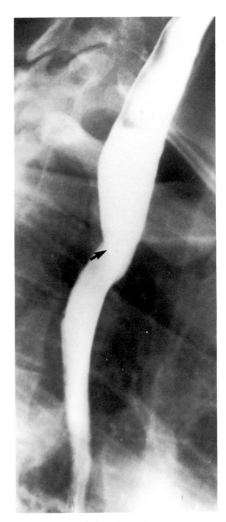

Fig. 5.14 Extrinsic compression of oesophagus by metastatic carcinoma of the bronchus (arrows). Note the shallow angle made with the wall of the oesophagus, indicating extraluminal compression.

Fig. 5.15 Anomalous right subclavian artery. There is a localised indentation caused by the anomalous artery as it passes behind the oesophagus (arrow).

Intraluminal filling defects. A lump of food may impact in the oesophagus and may cause a complete obstruction. This is usually associated with a stricture.

Dilatation of the oesophagus

There are two main types—obstructive and non-obstructive.

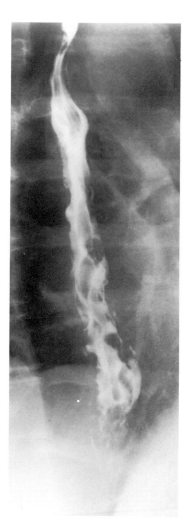

Fig. 5.16 Oesophageal varices. Tortuous worm-like filling defects are seen in the lower half of the oesophagus.

1. Dilatation due to obstruction is associated with a visible stricture and the diagnosis becomes that of the stricture (see p. 118). The patient with a carcinoma usually presents with dysphagia before the oesophagus becomes very dilated. On the other hand a markedly dilated oesophagus indicates a very long-standing condition, usually achalasia or occasionally a benign stricture.

2. Dilatation without obstruction occurs in scleroderma. The disease involves the oesophageal muscle resulting in dilatation of the oesophagus, which resembles an inert tube with no peristaltic movement so that barium does not flow from the oesophagus into the stomach unless the patient stands upright.

Varices

Oesophageal varices appear as lucent tortuous worm-like filling defects which distort the mucosal pattern so that the folds are no longer parallel (Fig. 5.16). When the oesophagus is distended with barium the varices are often obliterated and they are, therefore, best demonstrated after the main bolus of barium has passed. Small varices are difficult to spot but may become obvious if the patient performs the Valsalva manoeuvre.

Oesophageal web

A web is a thin, shelf-like projection arising from the anterior wall of the cervical portion of the oesophagus. To demonstrate it, that part of the oesophagus must be full of barium (Fig. 5.17). A web may be an isolated finding but the combination of a web, dysphagia and an iron deficiency anaemia is known as the Plummer Vinson syndrome.

Diverticula

Diverticula are saccular outpouchings which are often seen as chance findings in the intrathoracic portion of the oesophagus. One type of diverticulum, the pharyngeal pouch or Zenker's diverticulum (Fig. 5.18), is important as it may give rise to symptoms due to retention of food and pressure upon the oesophagus. A pharyngeal pouch arises through a congenital weakness in the

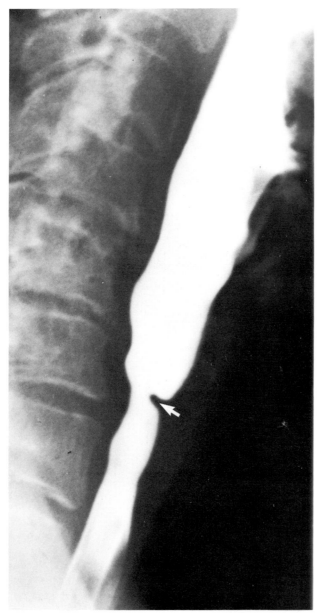

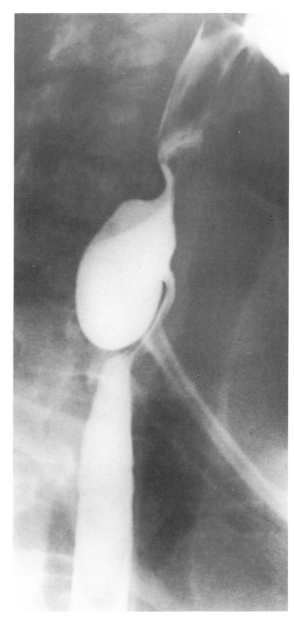

Fig. 5.17 Oesophageal web. There is a shelf-like indentation (arrow) from the anterior wall of the upper oesophagus.

Fig. 5.18 Pharyngeal pouch (Zenker's diverticulum). The pouch is lying behind the oesophagus which is displaced forward.

inferior constrictor muscle of the pharynx and comes to lie behind the oesophagus near the midline. It may reach a very large size and can cause displacement and compression of the oesophagus.

Oesophageal atresia

In atresia the oesophagus ends as a blind pouch in the upper mediastinum. Several different types exist (Fig. 5.19) but the commonest is for the upper part of the oesophagus to be a blind sac with a fistula between the lower segment of the oesophagus and the tracheobronchial tree. A plain abdominal film will show air in the bowel if a fistula is present between the tracheobronchial tree and the oesophagus distal to the atretic segment.

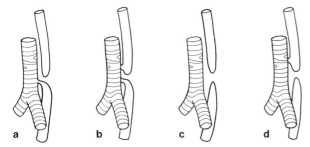

Fig. 5.19 Diagram of the various types of oesophageal atresia. The first two types also have an oesophagotracheal fistula distal to the atretic segment and will show air in the stomach.

To confirm the diagnosis of oesophageal atresia it is usually sufficient to pass a soft tube into the oesophagus when it is held up, or coils in the blindly ending pouch. Oily contrast, such as the bronchographic medium Dionosil (0·5–1·0 ml), injected through the tube has been used to outline the oesophagus (Fig. 5.20), but this is a hazardous procedure because the viscous contrast may cause respiratory obstruction if it spills over into the trachea.

THE STOMACH AND DUODENUM

The barium meal is the standard contrast examination to examine the stomach and duodenum. For this the

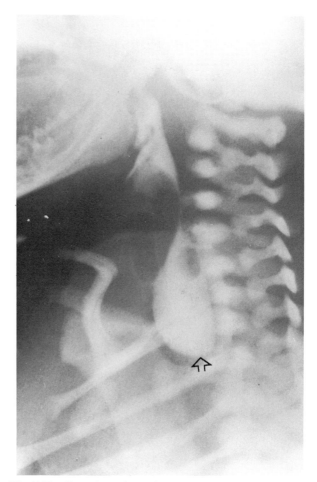

Fig. 5.20 Oesophageal atresia. The dilated oesophagus ends blindly at the thoracic inlet (arrow). There was a fistula from the oesophagus below the atretic segment to the trachea. This is the commonest type of oesophageal atresia.

patient drinks about 200 ml of barium. Each radiologist has his own routine but the aim is to take films in various positions with the patient both erect and lying flat (Figs. 5.21 a & b) so that each part of the stomach and duodenum is shown distended by barium and also distended with air but coated with barium to show the mucosal pattern. A double-contrast technique is often used to provide better mucosal detail: the stomach is coated with barium and distended by giving gas tablets and an

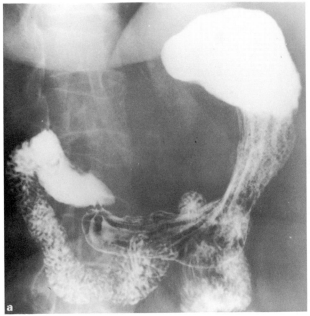

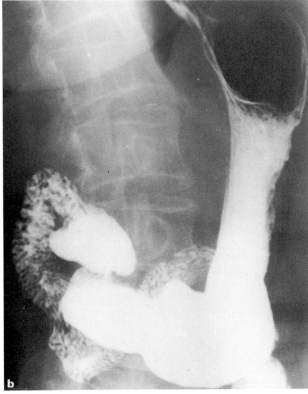

Fig. 5.21 Normal stomach and duodenum. (a) Supine view (with the patient lying flat): the barium outlines the mucosal folds of the antrum and body of the stomach but mostly collects in the fundus. The duodenal cap has a smooth outline. Note how the fourth part of the duodenum and duodeno-jejunal flexure are superimposed on the body of the stomach; (b) erect view (same patient): the barium collects in the body and antrum of the stomach.

intravenous injection of a short-acting smooth muscle relaxant (Fig. 5.22).

Food residues in the stomach produce pictures which are very difficult to interpret. For this reason it is most important that the patient fasts for at least 6 hours prior to the examination.

Normal barium meal

Each part of the stomach and duodenum should be checked to ensure that no abnormal narrowing is present. A transient contraction wave must not be confused with a constant pathological narrowing. The outline of the lesser curve of the stomach is smooth with no filling defects or projections visible but the greater curve is nearly always irregular due to prominent mucosal folds. In the stomach the mucosa is thrown up into a number of smooth folds and barium collects in the troughs between the folds. There should be no effacement of the folds or rounded collections of barium.

The duodenal cap or bulb should be approximately triangular in shape. It arises just beyond the short pyloric canal and may be difficult to recognise if deformed due to chronic ulceration.

The duodenum forms a loop around the head of the pancreas to reach the duodenojejunal flexure. Diverti-

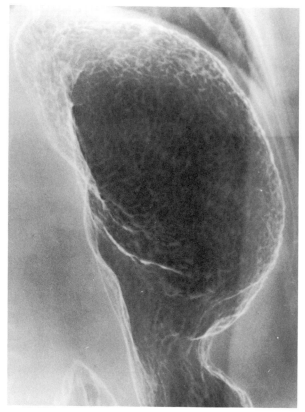

Fig. 5.22 Normal stomach. With a double contrast technique the mucosa is coated with barium and the stomach distended with air to show fine mucosal detail in this case in the fundus of the stomach.

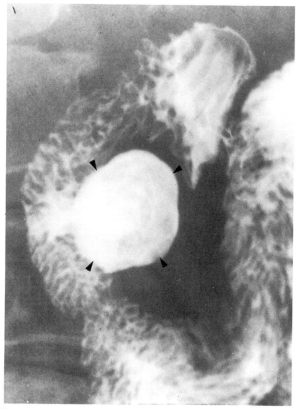

Fig. 5.23 Duodenal diverticulum arising from the third part of the duodenum (arrows).

cula arising beyond the first part of the duodenum are a common finding (Fig. 5.23) and are usually without significance.

Abnormal barium meal

Gastric ulcers

Depending on the projection ulcers may be seen either *en face* as a collection of barium occupying the ulcer crater, or in profile as a projection from the lumen of the stomach (see p. 117).

Gastric ulcers may be benign or malignant. Table 5.1 below gives some of the distinguishing features between benign and malignant gastric ulcers. Although it may be possible to diagnose an ulcerating carcinoma with confidence it is never possible to be sure that a gastric ulcer is benign until it has healed.

Filling defects in the stomach

These may arise from the wall of the stomach or be due to masses that press in from outside the stomach. A few are entirely intraluminal. It should be remembered that carcinoma is by far the commonest cause of a filling defect in an adult.

Carcinoma usually produces an irregular filling defect with alteration of the normal mucosal pattern. The tumour is often larger than is readily appreciated from

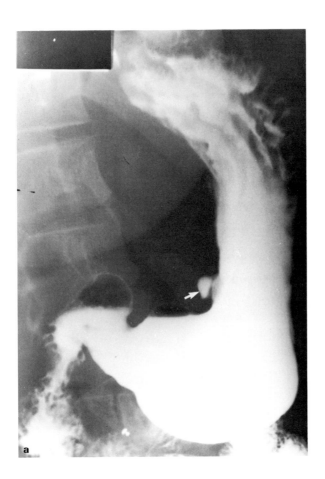

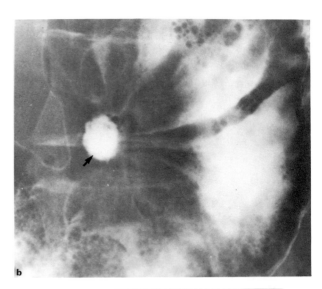

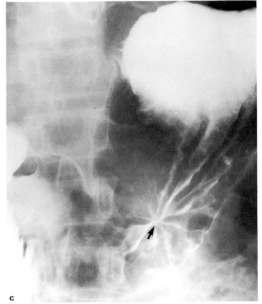

Fig. 5.24 Benign ulcer. (a) In profile the ulcer (arrow) projects from the lesser curve of the stomach; (b) en face the ulcer (arrow) is seen as a rounded collection of barium; (c) radiating mucosal folds reach the ulcer crater (arrow).

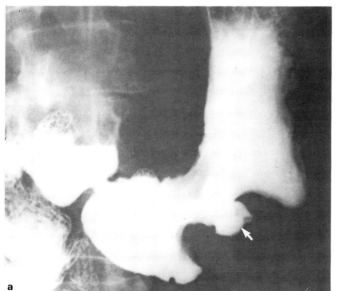 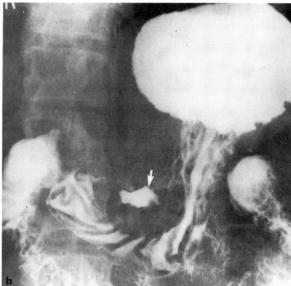

Fig. 5.25 Malignant ulcer. (a) An ulcer (arrow) appears with a large filling defect; (b) in this instance the filling defect is not obvious and the ulcer (arrow) does not project from the lumen of the stomach.

Table 5.1 Distinguishing features between benign and malignant gastric ulcers

Benign (Fig. 5.24)	*Malignant* (Fig. 5.25)
1. The ulcer projects beyond the lumen of the stomach	*1.* The ulcer occurs within an irregular filling defect, which may be difficult to demonstrate, therefore the ulcer appears not to project beyond the original lumen
2. The edge of the ulcer is regular and round when seen *en face*	*2.* The edge of the ulcer is irregular
3. Usually on lesser curve, rarely on greater curve	*3.* May occur anywhere but ulcers in the antrum and on greater curve are particularly suspicious
4. Radiating mucosal folds reach the edge or near to the edge of the ulcer crater	*4.* The mucosal folds obliterated some distance from the edge of the ulcer
5. Usually heals with reduction in size after 4 weeks' medical treatment	*5.* Usually little change with medical treatment but, occasionally, a malignant ulcer may temporarily get a little smaller

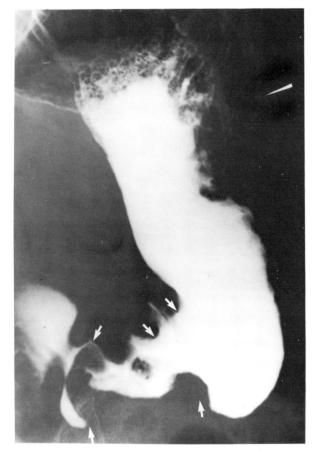

Fig. 5.26 Carcinoma. There is a large filling defect in the antrum and body of the stomach with overhanging edges (arrows).

the barium meal. Overhanging edges or shouldering may be seen at the junction of the tumour and the stomach wall (Fig. 5.26). A carcinoma at the fundus may obstruct the oesophagus while one in the antrum may cause gastric outlet obstruction (Fig. 5.32). Carcinoma diffusely involving the stomach is known as linitis plastica and is discussed below.

Recently, because of the much better prognosis, emphasis has been placed on the diagnosis of early gastric cancer which is confined to the mucosa. It may be flat or appear as a shallow ulcer indistinguishable from a benign ulcer. This stage of cancer is difficult to detect

and is usually only diagnosed with a double contrast barium meal.

Leiomyoma. A smooth, round filling defect arising from the wall of the stomach may be caused by a benign tumour such as a leiomyoma (Fig. 5.27). A leiomyoma is a submucosa tumour which as well as projecting into the lumen of the stomach may show a large extraluminal extension. One of the characteristic features of a leiomyoma is that it may have an ulcer on its surface.

Polyps may be single or multiple. They may be sessile or have a stalk. Even with high quality radiographs it is often impossible to distinguish benign from malignant polyps. For this reason gastroscopy with biopsy or

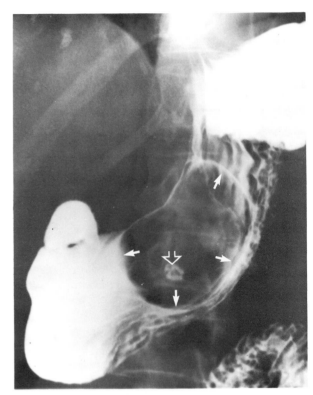

Fig. 5.27 Leiomyoma. There is a large filling defect in the stomach with smooth borders (outer arrows). An ulcer crater (central arrow) is present within the filling defect—a characteristic feature of a leiomyoma.

operative removal is invariably carried out on all suspected polyps.

Intraluminal defects are completely surrounded by barium and are often mobile in the stomach. Examples are food or blood following a haematemesis. Sometimes ingested fibrous material such as hair may intertwine forming a ball or bezoar (Fig. 5.28).

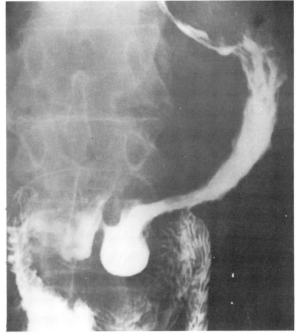

Fig. 5.29 Linitis plastica. The stomach is narrowed by an extensive carcinoma converting it to a rigid tube with obliteration of mucosal folds.

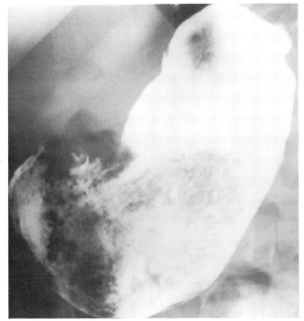

Fig. 5.28 Bezoar. Masses of hair in the stomach have caused irregular filling of the stomach with barium.

Narrowing of the stomach

If constant, narrowing is an important feature as it may herald malignancy.

When the whole stomach is involved the narrowing is due to a carcinoma. This is the so-called leather bottle stomach or linitis plastica (Fig. 5.29). The stomach behaves as a thick rigid tube lacking peristalsis with obliteration of the mucosal folds. Rapid gastric emptying takes place because the cardia and pylorus are held open by the rigid stomach wall.

It is when the narrowing is localised that diagnostic difficulties arise as it may be due to either an infiltrating carcinoma, an active ulcer causing spasm or an ulcer which has healed with scarring and fibrosis (Fig. 5.30).

Displacement of the stomach

The stomach may be displaced by masses adjacent to it. The pattern of displacement of the stomach may help to decide which structure is enlarged.

Enlargement of the liver displaces the stomach to the left stretching the lesser curve. The spleen stretches and displaces the greater curve of the stomach. The stomach may be pushed forward by retroperitoneal masses arising in the left kidney, aorta or pancreas (Fig. 5.31).

Thick gastric mucosal folds

Enlarged mucosal folds are associated with a high acid secretion. They are seen in patients with duodenal ulcers

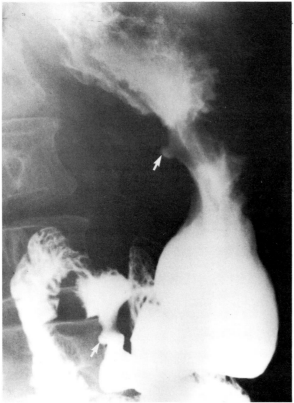

Fig. 5.30 Narrowing of the stomach due to an ulcer. The antrum is narrowed due to spasm resulting from the antral ulcer (arrow). Note the second ulcer on the lesser curve of the stomach (arrow).

and also in the Zollinger–Ellison syndrome which comprises a high acid secretion, gastrin secreting tumour in the pancreas and multiple peptic ulcers.

Occasionally, diffuse infiltration of the stomach with malignant lymphoma can produce generalised thickening of the mucosal folds.

Gastric outlet obstruction

Emptying of the stomach can be a difficult feature to assess. In most patients barium rapidly leaves the stomach to enter the duodenum, but in others this only occurs after the patient has been lying on the right side for several minutes. In gastric outlet obstruction less than 50% of the barium leaves the stomach after 4 hours and some may still be present after 24 hours. The stomach will be large and will contain food residues.

It is not sufficient merely to diagnose gastric outlet obstruction; an attempt must be made to find the underlying cause, which may be situated in the duodenum, pylorus or antrum. The term 'gastric outlet obstruction' is preferred to the older term 'pyloric stenosis' as the pylorus is often not the site of the obstruction.

In adults the causes of gastric outlet obstruction are:

Chronic duodenal ulceration. The diagnosis depends on demonstrating a very deformed stenosed duodenal cap. It may or may not be possible to identify an actual ulcer crater.

Carcinoma of the antrum may cause narrowing. The diagnosis is made by recognising an irregular filling defect in the antrum of the stomach (Fig. 5.32).

Adult hypertrophic pyloric stenosis is a rare condition giving rise to an elongated pyloric canal with a smooth, narrow channel through it.

In infants pyloric stenosis is by far the commonest cause of gastric outlet obstruction. Often the diagnosis is made clinically and a barium meal is only carried out in doubtful cases. The pyloric canal is narrowed and elongated but the stomach is usually not dilated.

Hiatus hernia

A hiatus hernia is a herniation of the stomach into the mediastinum through the oesophageal hiatus in the diaphragm. It is a common finding. Two main types of hiatus hernia exist: sliding and rolling or para-oesophageal (Fig. 5.33).

The commoner type is the sliding hiatus hernia where the gastro-oesophageal junction and a portion of the stomach are situated above the diaphragm (Fig. 5.34). The cardiac sphincter is usually incompetent, so reflux from the stomach to the oesophagus occurs readily and this may cause oesophagitis, ulceration or peptic stricture. A small sliding hernia may be demonstrated in

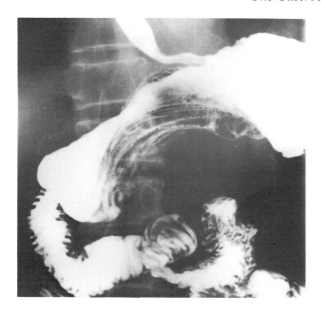

Fig. 5.31 Displacement of the stomach by a large pancreatic mass (a pseudocyst).

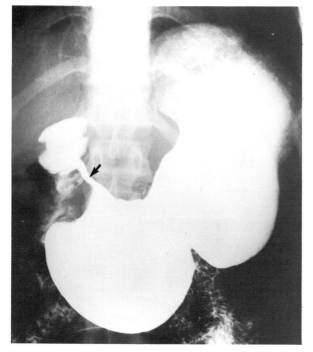

Fig. 5.32 Gastric outlet obstruction. A carcinoma is causing narrowing of the antrum (arrow). The speckled appearance in the fundus of the enlarged stomach is due to food residues.

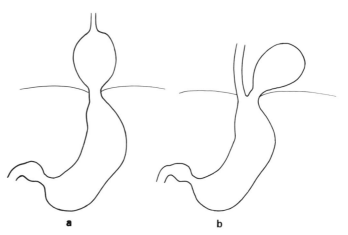

Fig. 5.33 Hiatus hernia. (a) Sliding: a portion of the stomach and the gastro-oesophageal junction are situated above the diaphragm; (b) rolling or para-oesophageal: the gastro-oesophageal junction is below the diaphragm.

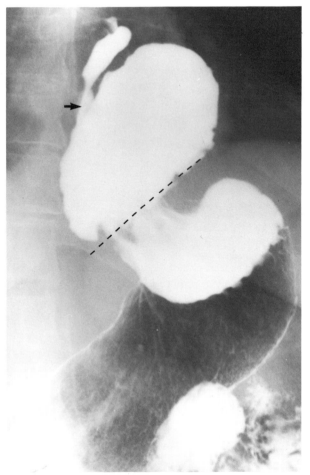

Fig. 5.34 Sliding hiatus hernia. The fundus of the stomach and the gastro-oesophageal junction (arrow) have herniated through the oesophageal hiatus and lie above the diaphragm (dotted lines).

most people during a barium meal examination, provided enough manoeuvres are undertaken to increase intra-abdominal pressure. It is, therefore, difficult to assess the significance of a small hernia with little or no reflux.

In a para-oesophageal hernia the fundus of the stomach herniates through the diaphragm but the oesophagogastric junction often remains competent below the diaphragm.

A large hernia, particularly one of the para-oesopha-geal type, may not be reduced when the patient is in the erect position. In these instances the hiatus hernia will be seen on chest films.

Duodenal ulcers

The great majority of duodenal ulcers occur in the duodenal cap (duodenal bulb) but a few are found just beyond the cap and are known as postbulbar ulcers. The ulcer crater (Fig. 5.35a) may have a surrounding lucent zone due to oedema and mucosal folds are often seen radiating towards the ulcer (Fig. 5.35b). With chronic ulceration the cap becomes deformed due to scarring. In a grossly scarred cap it is often impossible to be certain of the presence of an ulcer.

It is not worthwhile carrying out repeat barium meal examinations to assess healing of duodenal ulcers for the following reasons:

1 It is often not possible to comment on the presence or absence of an ulcer in a deformed cap.
2. Follow-up for malignant change is unnecessary as duodenal ulcers are almost invariably benign and do not undergo malignant degeneration.

Deformity of the duodenal loop

Enlargement of the head of the pancreas due to tumour or pancreatitis may press upon and widen the duodenal loop. This is difficult to appreciate unless it is gross, as there is so much variation in the shape of the duodenal loop in normal individuals and unfortunately no reliable measurements are available. A carcinoma of the head of the pancreas may also invade and destroy the duodenal mucosa which then appears irregular and spiky (Fig. 5.36).

The changes may be detected more readily with a hypotonic duodenogram which involves paralysing the duodenum with a short-acting smooth muscle relaxant and distending it with air. This may be conveniently done by passing a tube into the duodenum through which both barium and air can be injected.

Gastroscopy in relation to radiology

With the advent of flexible fibro-optic instruments gas-

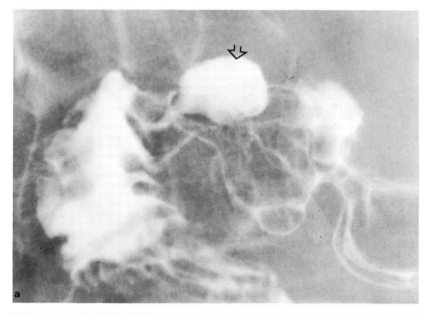

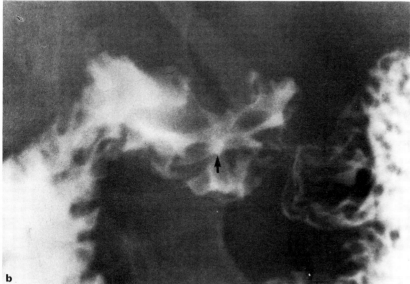

Fig. 5.35 Duodenal ulcer (two patients). (a) seen as a large collection of barium in the duodenal cap (arrow); (b) mucosal folds are radiating to a central ulcer crater (arrow).

Fig. 5.36 Carcinoma of the pancreas. The duodenal loop is widened but the lumen of the second part of the duodenum is narrowed and the normal mucosal pattern is lost due to invasion by the carcinoma.

troscopy has become a straightforward procedure which enables the mucosa of the stomach and duodenum to be directly inspected and biopsied. Gastroscopy and radiology are complementary investigations and each has its own advantages.

Although gastroscopy will not be discussed in detail the main indications for its use are:

1. Making a histological diagnosis of an abnormality shown on a barium meal.
2. Investigation of dyspepsia in a patient with a normal barium meal.
3. Assessment of ulcer healing.

4. In acute bleeding from the upper gastrointestinal tract (see below).

Radiology in acute upper gastrointestinal bleeding

The patient presenting with haematemesis and melaena is a common medical emergency. The main causes of bleeding are:
1. Peptic ulcer.
2. Gastric erosions.
3. Varices.
4. Carcinoma.

Gastroscopy, barium meal examination and arteriography are the investigations which can be used to identify the bleeding site. Which method is used depends largely on the rate of bleeding, the facilities available and on personal preferences.

Gastroscopy is superior to a barium meal in making a precise diagnosis in the bleeding patient. An emergency barium meal is a difficult examination to perform as it is often difficult to move the very ill patient into the various positions required. Blood in the stomach also makes it difficult to appreciate small lesions. Gastroscopy has the major advantage that it will identify gastric erosions which are too shallow to be reliably visualised on a barium examination.

Another great advantage of gastroscopy is that the bleeding point may be seen directly. Bleeding is not always from the lesion demonstrated radiologically. For example patients with oesophageal varices not infrequently bleed from ulcers rather than from their varices.

The bleeding site may also be detected by arteriography. Contrast medium is injected into the coeliac axis, superior mesenteric artery or branches of these vessels. The bleeding site may be shown by the presence of contrast within the bowel, providing the patient is bleeding at a rate of at least $\frac{1}{2}$–1 ml per minute at the time of the angiogram. This method is more complex than endoscopy or barium meal examination and is only successful if performed during active bleeding. If the site of haemorrhage is detected, bleeding may be arrested by infusion of a vasopressor agent directly into the artery, or an attempt can be made to block the bleeding artery by embolisation techniques.

THE SMALL INTESTINE

The standard contrast examination for the small intestine is the barium small bowel follow-through. The patient drinks about 200–300 ml of barium and its passage through the small intestine is observed by taking films at regular intervals until the barium reaches the colon. This can be a time-consuming procedure and usually takes 2–3 hours, but the transit time is very variable.

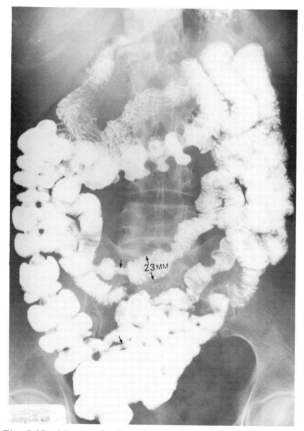

Fig. 5.37 Normal barium follow-through. The small intestine, ascending and transverse colon are filled with barium. The jejunum in the left side of the abdomen has a much more marked mucosal fold pattern than the ileum which is lying in the pelvis. When a peristaltic wave contracts the bowel the mucosal folds lie longitudinally (arrows). Note the way of measuring the diameter of the bowel. In the pelvis the loops overlap and details of the bowel become hidden.

The normal barium follow-through

The normal small intestine (Fig. 5.37) occupies the central and lower abdomen, usually framed by the colon. The terminal portion of the ileum (Fig. 5.38) enters the medial aspect of the caecum through the ileocaecal valve. As the terminal ileum may be the first site of disease this region is often fluoroscoped and observed on a television monitor so that peristalsis can be seen and films can be taken with the terminal ileum unobscured by other loops of small intestine.

The barium forms a continuous column defining the diameter of the small bowel which is normally not more than 25 mm. Transverse folds of mucous membrane project into the lumen of the bowel. Barium lies between the folds which appear as lucent filling defects normally about 2–3 mm in width. The appearance of the mucosal folds depends upon the diameter of the bowel. When distended the folds are seen as lines traversing the bar-

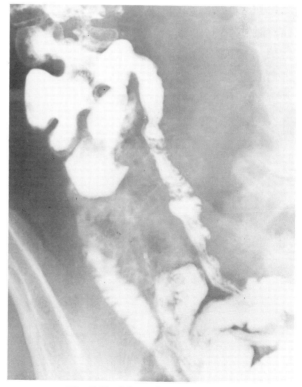

Fig. 5.38 Normal terminal ileum.

ium column known as valvulae conniventes. When the small bowel is contracted the folds lie longitudinally and when relaxed the folds assume an appearance described as feathery. The mucosal folds are largest and most numerous in the jejunum and tend to disappear in the lower part of the ileum.

The abnormal barium follow-through

The following signs should be looked for:

1. Dilatation usually indicates either malabsorption, paralytic ileus or small bowel obstruction (Fig. 5.39). If necessary measure the diameter of the bowel. A value over 30 mm is definitely abnormal but make sure you are

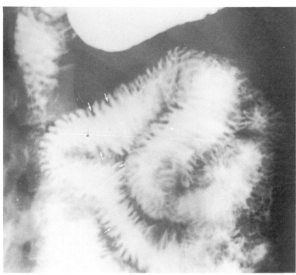

Fig. 5.40 Mucosal abnormality with infiltration of the bowel, in this case due to oedema. The mucosal folds become thickened. Some of the thickened folds are arrowed.

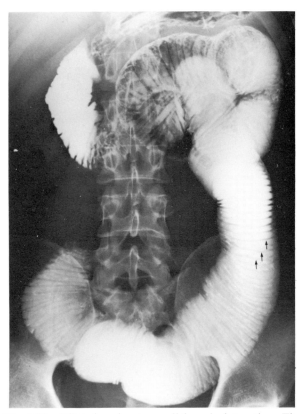

Fig. 5.39 Dilatation due to small bowel obstruction. The diameter of the bowel is greatly increased. The feathery mucosal pattern is lost and the folds appear as thin lines traversing the bowel known as valvulae conniventes (arrows).

not measuring two overlapping loops. As the bowel dilates the normal mucosal pattern becomes largely effaced and the valvulae conniventes become clearly visible.

2. Mucosal abnormality. The folds become thickened in many conditions, e.g. malabsorption states, with oedema or haemorrhage into the bowel wall, and when inflamed or infiltrated (Fig. 5.40). Since mucosal fold thickening occurs in many diseases it is not possible to make a particular diagnosis unless other more specific features are present.

3. Narrowing. The only normal narrowings are those caused by peristaltic waves. They are smooth, concentric and transient with normal mucosal folds traversing them and normal bowel proximally. The common causes of strictures are: Crohn's disease, tuberculosis and malignant lymphoma (Fig. 5.41). Strictures do not contain normal mucosal folds and usually result in dilatation of the bowel proximally.

4. Ulceration. The outline of the small bowel should be smooth apart from the indentation caused by normal mucosal folds. Ulcers appear as spikes projecting outwards which may be shallow or deep (Fig. 5.42).

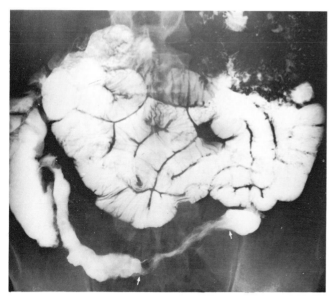

Fig. 5.41 Narrowing. There is a long irregular stricture (arrows) in the terminal ileum due to Crohn's disease. There is an abnormal mucosal pattern in the remainder of the terminal ileum. Note the contracted caecum—another feature of the disease.

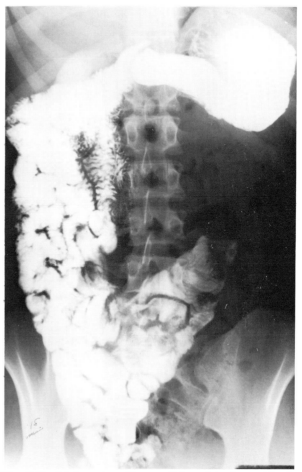

Fig. 5.43 Malrotation. The small bowel is situated in the right side of the abdomen. Later films showed the colon on the left side.

Fig. 5.42 Ulceration. Abnormal loops of bowel in Crohn's disease showing the ulcers as outward projections (arrows).

Ulceration is seen in Crohn's disease, tuberculosis and lymphoma. When there is a combination of fine ulceration and mucosal oedema a cobblestone appearance may be seen.

5. *Alteration in position.* (a) Congenital malrotation. During intrauterine life the bowel undergoes a series of rotations. Failure of the normal rotation may result in the small bowel being situated in the right side of the abdomen and the colon on the left side (Fig. 5.43). However, this state of affairs only occasionally gives rise

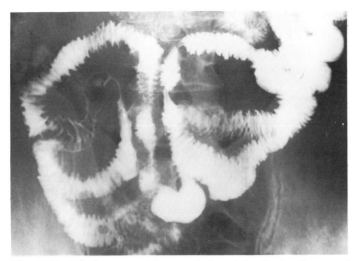

Fig. 5.44 Displacement. The small bowel is displaced around enlarged abdominal lymph nodes due to metastases from a teratoma of the testis.

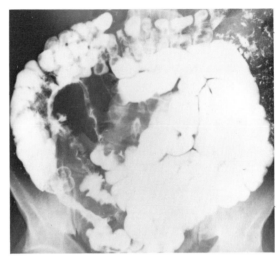

Fig. 5.45 Crohn's disease. Many of the features are illustrated in this film. There are long strictures affecting different loops of ileum. The outline of the strictures are irregular due to ulceration. Note how the affected loops lie separately, displaced from other loops because of the presence of inflammatory masses.

to problems, mainly volvulus associated with abnormal mesenteric attachments.

(b) Displacement by a mass.   Because of its mesentery the small intestine is freely mobile and will be displaced by an abdominal or pelvic mass (Fig. 5.44). The bowel will appear stretched around the mass but the mucosal pattern is usually preserved.

Crohn's disease

Crohn's disease is a disease of unknown aetiology characterised by localised areas of non-specific chronic granulomatous inflammation, which nearly always affects the terminal ileum. In addition it may cause disease in several different parts of the small and large intestine often leaving normal intervening bowel; the affected parts being known as skip lesions. The major signs on the barium follow-through are strictures and mucosal abnormality (Fig. 5.45).

The strictures are extremely variable in length. Sometimes a loop of bowel is so narrow, either due to spasm in an extensively ulcerated loop of bowel or to oedema and fibrosis in the bowel wall, that its appearance has been called 'the string sign'. The bowel proximal to a stricture is often dilated. When there is obvious disease in the terminal ileum the caecum may be contracted.

Ulcers are seen which are sometimes quite deep. Fine ulceration combined with mucosal oedema gives rise to the so-called 'cobblestone' appearance.

Owing to thickening of the bowel wall the mucosal folds may become thickened, distorted or even disappear. When this thickening of the bowel wall is severe then the loops of bowel become separated; the presence of an inflammatory mass will cause even greater displacement of the loops.

Fistulae may occur to other small bowel loops, colon, bladder or vagina. When between adjacent loops of small intestine a fistula is difficult to detect on the barium follow-through.

Crohn's disease may cause malabsorption so the radiological features of this condition may be present as well. Occasionally, the flocculation of barium that occurs with malabsorption may be so severe that it becomes impossible to demonstrate the signs of Crohn's disease.

Tuberculosis

Tuberculosis is indistinguishable from Crohn's disease on barium examination. It commonly affects the ileocaecal region and also causes contraction of the caecum.

Lymphoma

The infiltration in the wall of the bowel with lymphoma gives an appearance that is often extremely difficult to distinguish from Crohn's disease. Additional features to look for that may help differentiate the two conditions are small mucosal filling defects due to tumour nodules (Fig. 5.46), and displacement of loops caused by enlarged lymph nodes. Enlargement of the liver and spleen may also be present.

Malabsorption

A number of disorders result in defective absorption of food stuffs, minerals or vitamins. The definitive test for malabsorption is the jejunal biopsy. Radiology is no substitute for a jejunal biopsy but along with biochemical tests is an important complementary investigation. The use of the barium follow-through in malabsorption is twofold.

1. It may show a structural abnormality causing the malabsorption.
2. It may help to make the diagnosis in doubtful cases where the biochemical tests are equivocal or normal.

The signs of malabsorption in the small bowel follow-through are (Fig. 5.47):

(a) Small bowel dilatation, the jejunum being affected more than the ileum.

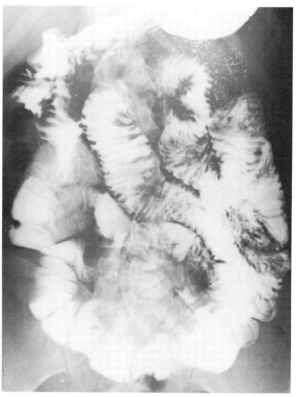

Fig. 5.47 Malabsorption. The bowel is dilated and the mucosal folds thickened. In the lower loops the barium appears less dense due to it becoming diluted. No specific cause for the malabsorption can be detected, which in this case was due to gluten enteropathy.

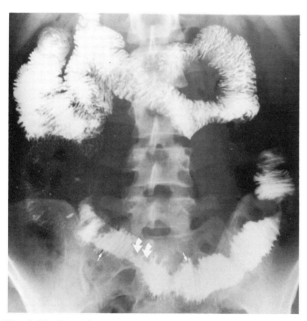

Fig. 5.46 Lymphoma. Lymphomatous infiltration has occurred in the lower loops of bowel causing thickening of the mucosal folds (small arrows) and discrete filling defects due to tumour nodules (curved arrows).

(b) Thickening of mucosal folds.

(c) Flocculation and dilution of the barium occur in advanced disease. Instead of the barium forming a continuous column there may be clumping or flocculation so that the barium column is broken up into a number of segments. The barium may become diluted by the excessive fluid in the small bowel and so appears less dense.

These signs occur with any of the causes of malabsorption. In the following conditions no clue to the cause can be obtained from a study of the barium follow-through.

1. Diffuse mucosal lesions
    (a) Gluten enteropathy (coeliac disease and idiopathic steatorrhoea).
    (b) Tropical sprue.
2. Deficiency of absorptive factors , e.g. bile or pancreatic enzymes.
3. Postgastrectomy due to rapid emptying of gastric remnant and insufficient mixing with bile and pancreatic juice.

Those conditions where the cause of malabsorption may be seen include:
Crohn's disease (see p. 142).
Lymphoma (see p. 143).
Anatomical abnormalities
1. Decreased length of small bowel available for absorption, e.g. surgical resection or a fistula short circuiting a length of small bowel.
2. Stagnation of bowel contents allowing bacterial overgrowth which utilise nutriments from the bowel lumen.
    (a) Multiple small bowel diverticula (Fig. 5.48).
    (b) A dilated loop cut off from the main stream of the bowel in which there is delayed filling and emptying (blind loop).
    (c) A dilated loop proximal to a stricture (stagnant loop).

Disaccharidase deficiency

Patients suffering from milk intolerance due to a deficiency of the enzyme lactase in the small bowel mucosa

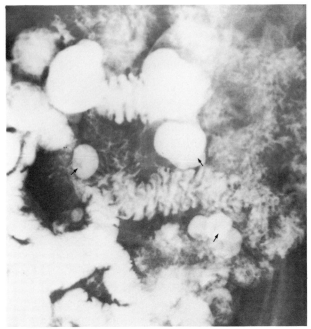

Fig. 5.48 Diverticulosis. A number of diverticula of varying size are arising from the small bowel. Some of these are arrowed.

have an abnormal appearance on a barium follow-through if lactose is added to the barium. The small bowel becomes dilated, the barium diluted and the barium rapidly reaches the colon.

Acute small bowel obstruction

A barium examination is not carried out in most cases of obstruction as the diagnosis is usually made on clinical examination with the help of plain abdominal films (p. 107). When barium is given by mouth it shows that the small bowel proximal to an obstruction is dilated, and often markedly so. The barium becomes diluted by the excess fluid in the bowel (Fig. 5.39).

Although a barium follow-through must be avoided in a colonic obstruction owing to the barium becoming solid and impacting proximal to the obstruction this danger is not present in a small bowel obstruction because the fluid in the bowel prevents the barium solidi-

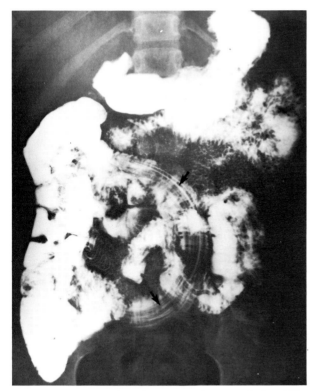

Fig. 5.49 Worm infestation. Several long tubular filling defects (arrows) due to roundworms (Ascaris) in the small bowel.

fying. However, because of the dilution of the barium in the fluid filled bowel it is often difficult to predict the nature or site of the obstruction.

Worm infestation

Roundworms (Ascaris) are the commonly encountered worms that are large enough to be seen as filling defects in the lumen of the bowel (they may grow up to 35 cm long) (Fig. 5.49). The worms themselves may ingest the barium to have their own barium meal and barium may be seen in their digestive tracts.

THE LARGE INTESTINE

The standard radiological examination of the large intestine is the barium enema. Barium is run into the colon under gravity through a tube inserted into the rectum. Films are taken in various projections so that all the loops of colon are unravelled. In the 'single contrast method' the whole colon is distended with barium. When a 'double contrast technique' is used only part of the colon is filled with barium and air is then blown in to push the barium around the colon with the result that the colon is distended with air and the mucosa coated with barium.

Prior bowel preparation by means of enemas or washouts is most important to rid the colon of faecal

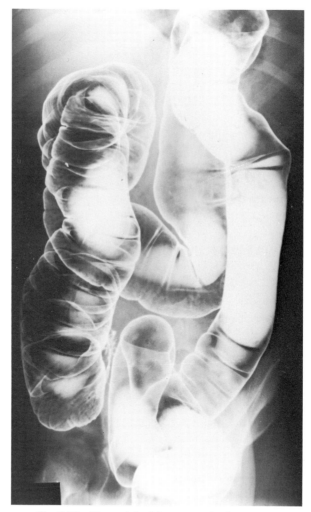

Fig. 5.50 Normal double contrast barium enema.

material, which might otherwise mask small lesions and cause confusion by simulating polyps.

Sigmoidoscopy is complementary to every barium enema as lesions in the rectum, especially mucosal abnormalities, may be missed by barium examinations. In selected cases colonoscopy will be required.

The normal barium enema

The radiological anatomy of the normal colon is shown in Fig 5.50. Certain features are worth emphasising.

The length of the colon is very variable and sometimes there are redundant loops particularly in the sigmoid and transverse colon. The calibre decreases from the caecum to the sigmoid colon.

The caecum is usually situated in the right iliac fossa but it may be seen under the right lobe of the liver or even in the centre of the abdomen if it possesses a long mesentery. The lips of the ileocaecal valve may project in to the caecum and cause a filling defect and must not be mistaken for a tumour. Filling of the terminal ileum and appendix may occur but if they do not fill no significance can be attached to this.

Haustra can usually be recognised in the whole of the colon although they may be absent in the descending and sigmoid regions. The outline of the distended colon, apart from the haustra, is smooth but when the colon is contracted as in an after evacuation film the mucosa appears as smooth regular folds (Fig. 5.51).

The abnormal barium enema

1. Narrowing of the lumen

Narrowing of the colon may be due to spasm, stricture formation or compression by an extrinsic mass.

Spasm is often seen in normal patients and providing it is an isolated finding it can be ignored. Spasm is also seen in conjunction with diverticular disease and various inflammatory disorders.

Spasm gives rise to a smooth concentric narrowing which usually varies in severity during the period under observation. It can often be abolished by the intra-

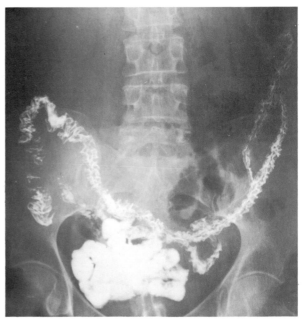

Fig. 5.51 Normal colon. After evacuation the colon contracts and the mucosa is thrown into folds. Barium has refluxed into loops of ileum in the pelvis.

venous administrations of a smooth muscle relaxant, e.g. Buscopan or Glucagon.

Strictures. The main causes of stricture formation are:
(a) Carcinoma.
(b) Diverticular disease.
(c) Crohn's disease.
(d) Ischaemic colitis.
(e) Rarer causes include tuberculosis, lymphogranuloma venereum, amoebiasis and radiation fibrosis.

When attempting to diagnose the nature of a stricture in the colon the following points should be borne in mind:

Neoplastic strictures have shouldered edges, an irregular lumen and are rarely more than 6 cm in length (Fig. 5.52), whereas benign strictures classically have tapered ends, a relatively smooth outline and may be of any length.

Ulceration may be seen in strictures due to Crohn's disease and sacculation of the colon is a feature of ischaemic strictures.

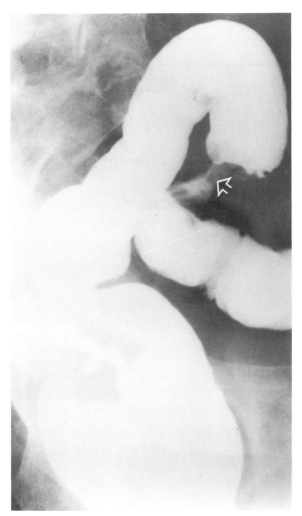

Fig. 5.52 Stricture. A short circumferential narrowing is seen in the sigmoid colon (arrow) due to a carcinoma.

Narrowing due to diverticulitis is usually accompanied by other signs of diverticular disease. It is sometimes impossible to distinguish a stricture due to a carcinoma in an area of diverticular disease from a stricture due to diverticular disease.

The *site* of the stricture might help in limiting the differential diagnosis. Strictures due to diverticular disease are almost always confined to the sigmoid colon. Ischaemic strictures are usually centred somewhere between the splenic flexure and the sigmoid colon. Crohn's disease and tuberculosis have a predilection for the caecum.

Extrinsic compression by a mass arising outside the wall of the bowel causes a smooth narrowing of colon from one side only and often displaces the colon, e.g. ovarian and uterine masses. Extrinsic compression causing a smooth indentation on the caecum may be seen with a mucocele of the appendix, appendix abscess or an inflammatory mass due to Crohn's disease (Fig. 5.53).

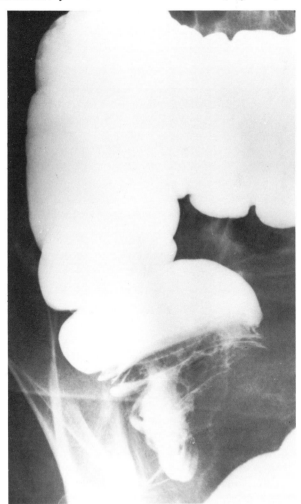

Fig. 5.53 Extrinsic compression. An appendix abscess is compressing and narrowing the caecum.

2. Dilatation

Dilatation of the colon is difficult to assess. The barium enema, particularly the double contrast examination, involves distending the colon, so that its diameter is partly dependent on the amount of barium and air introduced.

The causes of dilatation of the colon are:

(a) *Obstruction.* Here the important consideration is not the dilatation itself but the nature of the obstructing lesion. In complete obstruction the barium enema may only show one end of the stricture, so some of the valuable signs described above are lost.

(b) *Paralytic ileus.* This diagnosis is usually made on clinical grounds with the help of plain films of the abdomen (p. 107). In those few cases where it proves difficult to distinguish paralytic ileus from mechanical obstruction to the distal colon a barium enema can be undertaken. This will show a dilated but otherwise normal colon.

(c) *Volvulus* (p. 156).

(d) *Ulcerative colitis* with toxic dilatation (Fig. 4.6, p. 109).

(e) *Hirschsprung's disease and megacolon* (p. 159).

3. Filling defects

Filling defects in the colon, as elsewhere in the gastrointestinal tract, may be intraluminal, arise from the wall or be due to pressure from an extrinsic mass.

In a clean colon a localised filling defect is likely to be a polyp or a neoplasm (p. 9). Faeces will cause a filling defect and can be very difficult to distinguish from a polyp or tumour (Fig. 5.54). Faeces have no attachment to the wall of the bowel, are completely surrounded by barium or air and they move freely varying with the position of the patient. All barium enema examinations should be done with a clean colon in order to avoid diagnosing polyps that are in fact faeces.

Intramural haemorrhage, oedema or air in the wall of the colon (pneumatosis coli) all cause multiple smooth filling defects arising from the wall of the bowel.

A unique type of filling defect is seen in intussusception (p. 156).

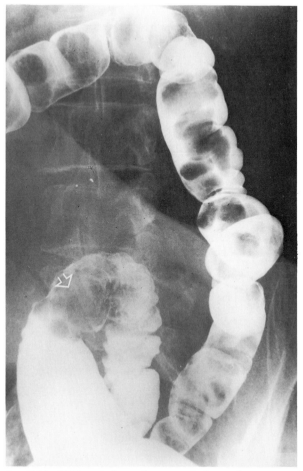

Fig. 5.54 Filling defects. Lumps of faeces have caused smooth filling defects surrounded by barium. However, in the sigmoid colon there is a large filling defect with ill-defined edges (arrow). This is a carcinoma. A clean colon is essential for a satisfactory barium enema.

4. Ulceration

Ulcers of the colonic mucosa can be recognised as small projections from the lumen into the wall of the bowel. This results in the normally smooth outline of the colon having a fuzzy or shaggy appearance (Fig. 5.55). The two major causes of ulceration are ulcerative colitis and Crohn's disease. Rarer causes include tuberculosis, amoebic and bacillary dysentery.

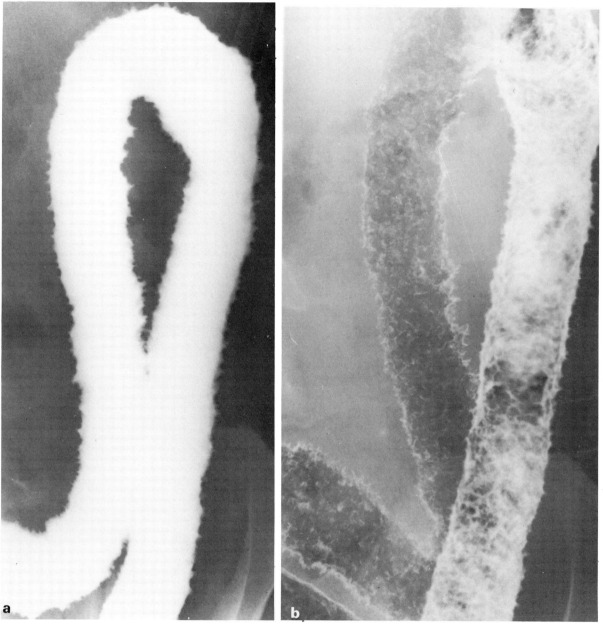

Fig. 5.55 Ulceration. (a) Single contrast; (b) double contrast: in this case of ulcerative colitis the ulceration causes the normally smooth outline of the colon to be irregular.

5. *Diverticula and muscle hypertrophy*

These are seen with diverticular disease (Fig. 5.56) and are discussed on p. 152.

6. *Displacement of the colon*

Displacement of the colon from its normal position may be caused by a variety of abdominal or pelvic masses, e.g. enlargement of the liver and spleen or ovarian cyst.

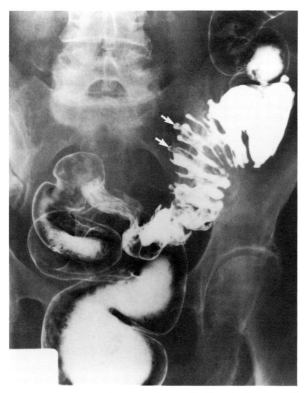

Fig. 5.56. Muscle hypertrophy and diverticula. Muscle hypertrophy gives the sigmoid colon a serrated appearance. Two small diverticula are arrowed.

These masses may also compress the colon. Scrutiny of the plain abdominal films should be made as these may show further details of the mass.

Displacement of the colon is also seen in malrotation.

Ulcerative colitis and Crohn's disease of the colon

Although classical changes are described for both ulcerative colitis and Crohn's disease it is sometimes difficult to distinguish between them. Radiology is important not only to diagnose these conditions but also to assess the extent and severity of the disease and to detect complications.

Ulcerative colitis

Ulcerative colitis is a disease of unknown aetiology characterised by inflammation and ulceration of the colon. The disease always involves the rectum. When more extensive it extends in continuity around the colon sometimes affecting the whole colon. The cardinal radiological sign is widespread ulceration (Fig. 5.55). The ulcers are usually shallow but in severe cases may be

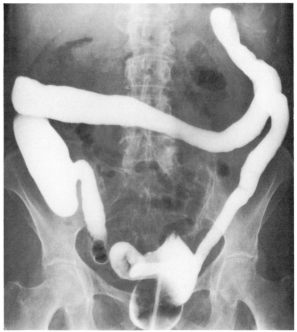

Fig. 5.57 Ulcerative colitis. With long-standing disease the haustrae are lost and the colon becomes narrowed and shortened coming to resemble a rigid tube. Reflux into the ileum through an incompetent ileocaecal valve has occurred.

quite deep. In all but the milder cases there is loss of the normal colonic haustrae in the affected portions of the colon. Oedema of the perirectal tissues causes widening of the space between the sacrum and the rectum. Narrowing and shortening of the colon giving the appearance of a rigid tube (Fig. 5.57) and pseudopolyps are seen in advanced disease. Pseudopolyps are small filling defects projecting into the lumen of the bowel formed by swollen mucosa in between the areas of ulceration. The swelling of these islands of inflamed mucosa makes it difficult to assess the true depth of the ulceration.

Strictures are rare and when present are likely to be due to carcinoma; the incidence of colonic carcinoma in long-standing ulcerative colitis is significantly increased.

When the whole colon is involved the terminal ileum may become dilated. Since the ileocaecal valve in this situation is incompetent the abnormal terminal ileum is usually demonstrated at barium enema.

Toxic dilatation (toxic megacolon) is a serious complication. The diagnosis is made on clinical grounds and on examination of the plain abdominal film. A barium enema should never be performed in the presence of toxic dilatation owing to the risk of perforating the colon.

Crohn's disease of the colon (granulomatous colitis, regional enteritis)

Crohn's disease is a chronic granulomatous conditions of unknown aetiology which may affect any part of the gastrointestinal tract, but most frequently involves the lower ileum and the colon. The colon may be the only part of the alimentary tract to be involved, but frequently the disease is confined to either the small or large bowel.

At the early stage in the disease the findings at barium enema are: loss of haustration, narrowing of the lumen of the bowel and a 'cobblestone' appearance to the mucosa (Fig. 5.58). The cobblestone appearance is due to a combination of mucosal oedema with criss-crossing fine ulceration. Later, the ulcers become deeper and may track in the submucosa (Fig. 5.59). The ulcers may be very deep, penetrating through the mucosal layer, when they are described as rose-thorn ulcers or deep fissures.

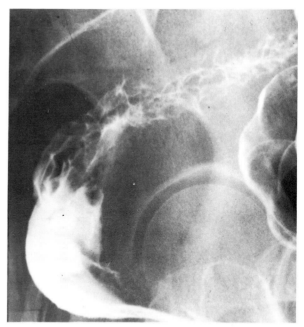

Fig. 5.58 Crohn's disease. The mucosal pattern has a cobblestone appearance due to criss-crossing fine ulceration.

The deep ulceration in Crohn's disease may lead to the formation of intra- and extramural abscesses. Fistulae are an important complication.

Strictures are a common finding in Crohn's disease (Fig. 5.60). The strictures are smooth and have tapered ends. When the caecum is involved it is usually markedly contracted. Ulcers may or may not be present in the strictured area. The disease is not always circumferential; one of the features that distinguishes it from ulcerative colitis is that it may involve only one portion of the circumference of the bowel.

Another important diagnostic feature is the presence of the so-called 'skip lesion' (Fig. 5.60), namely areas of disease with intervening normal bowel. Skip lesions are virtually diagnostic of Crohn's disease. However, the entire colon may be involved or the disease may be limited to just one segment. There is a predilection for the caecum and terminal ileum. The rectum is often spared—another important differentiating feature from ulcerative colitis.

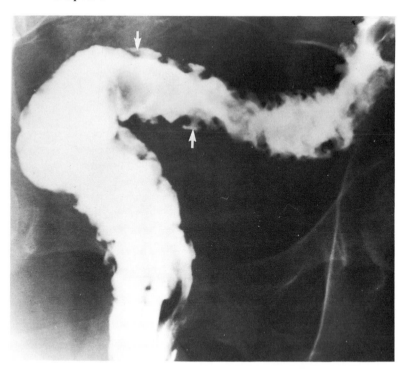

Fig. 5.59 Crohn's disease. Very deep
ulcers are present. Two examples of an
ulcer tracking in the submucosa are
arrowed.

Differences between ulcerative colitis and Crohn's disease

The only specific features are: the presence of skip
lesions and a normal rectum providing this is confirmed
sigmoidoscopically. If either of these are seen then the
diagnosis of Crohn's disease can be confidently made.
Other differences are listed in Table 5.2 below.

Diverticular disease

Diverticula are sac-like out-pouchings of mucosa
through the muscular layer of the bowel wall. They are
associated with hypertrophy of the muscle layer and are
probably due to herniation of mucosa through areas of
weakness where blood vessels penetrate the muscle.

Table 5.2

Crohn's disease	Ulcerative colitis
1. Rectum involved in half the cases	*1.* Rectum involved in all the cases
2. Colon may be affected segmentally	*2.* Colon always affected continuously
3. Ulcers deep	*3.* Ulcers shallow
4. Some cases show asymmetrical loss of haustrae	*4.* Symmetrical loss of haustrae is the rule
5. Fistulae are a feature	*5.* Fistulae very rarely occur
6. Anal or perianal lesions frequent	*6.* Anal or perianal lesions uncommon
7. Small bowel involvement common—particularly of the terminal ileum with narrowing in the region of the ileocaecal valve	*7.* Small bowel normal—dilatation of the terminal ileum may be seen

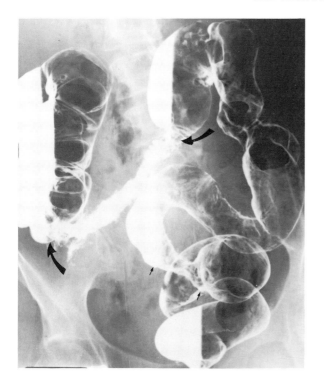

Fig. 5.60 Crohn's disease—strictures. A long stricture is present in the transverse colon (between curved arrows) and a shorter one in the sigmoid colon (between small arrows). In this case the outline of the strictures are irregular, due to ulceration. These two abnormal segments with normal intervening bowel are an example of 'skip lesions'—an important diagnostic feature of Crohn's disease.

Diverticula are very common, particularly in the elderly. They are seen in all parts of the colon but are commonest in the sigmoid colon. At one time the term 'diverticulitis' was applied when infection was thought to be causing symptoms and diverticulosis when the diverticula were considered asymptomatic. As no radiological distinction can be made between these two entities the term 'diverticular disease' is nowadays used to cover both situations.

The diverticula when filled with barium are seen as spherical out-pouchings with a narrow neck (Fig. 5.61). The colon may also show a 'saw tooth' serrated appearance due to hypertrophy of the muscle coats (Fig. 5.56). Sometimes the signs of muscle hypertrophy are seen in isolation. Some diverticula may not fill; this is particularly true when inflammation occludes the necks of the diverticula.

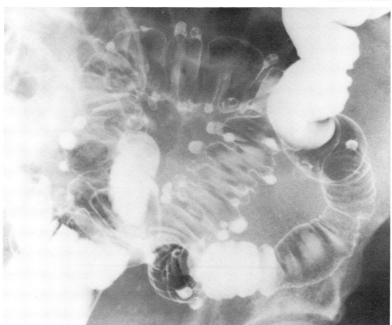

Fig. 5.61 Diverticular disease. Numerous diverticula are seen as out-pouchings from the sigmoid colon.

A diverticulum may perforate resulting in a pericolic abscess or fistula into the bladder, small bowel or vagina. This is recognised by noting barium outside the colon, either in the pericolic region (Fig. 5.62) or within the structure to which the fistula has occurred. Occasionally, diverticula perforate directly into the peritoneal cavity giving rise to peritonitis and free intraperitoneal air is recognised on a plain abdominal film.

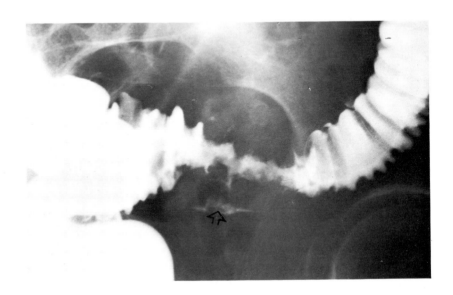

Fig. 5.62 Diverticular disease. Barium is seen outside the lumen of the bowel in a pericolic abscess (arrow). Muscle hypertrophy gives the bowel a serrated appearance.

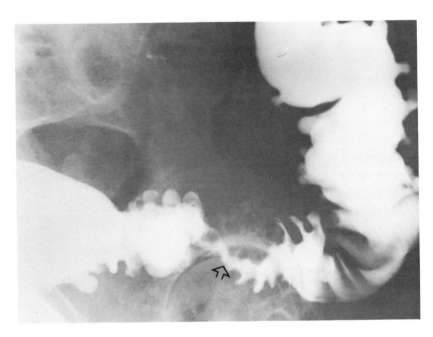

Fig. 5.63 Diverticular disease. A stricture is present (arrow). Although there is recognisable diverticular disease at both ends of the stricture it is impossible to exclude definitely a carcinoma.

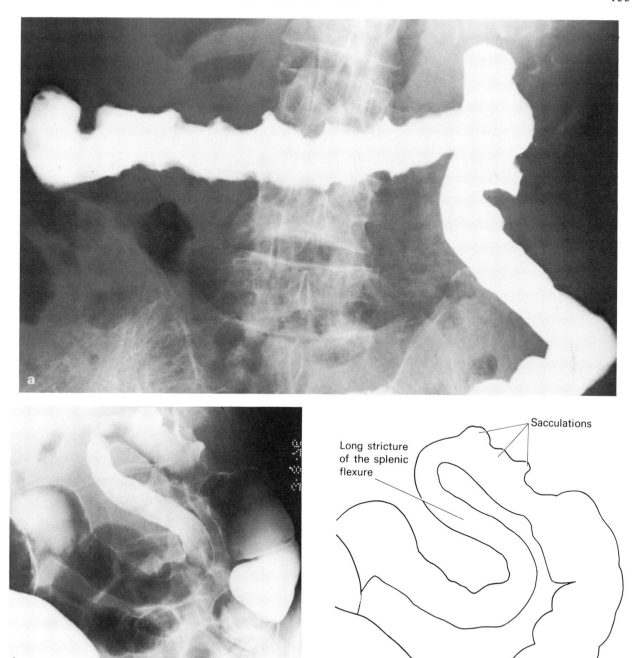

Fig. 5.64 Ischaemic colitis. (a) Mucosal haemorrhage and oedema have caused indentations resembling thumb prints in the transverse colon; (b) a long smooth stricture involving the splenic flexure with sacculation arising from one side of the colon in another patient.

A stricture with or without local abscess formation (Fig. 5.63) may occur. Usually, this is clearly within an area of recognisable diverticular disease. It is, however, often impossible to differentiate such a stricture from a carcinoma occurring coincidentally in a patient with diverticular disease.

Ischaemic colitis

Acute infarction of the large bowel is very rare. Ischaemia is usually a more chronic process giving rise, initially, to mucosal oedema and haemorrhage which may resolve. In the later stages a stricture may form. The findings on barium enema depend on the stage at which the examination is performed.

Mucosal haemorrhage and oedema may be recognised by observing multiple smooth indentations into the lumen of the bowel, resembling thumb prints (Fig. 5.64a). If stricture formation occurs the stricture will be smooth and have tapered ends. The site is usually centred between the splenic flexure and the sigmoid colon since these are the regions of the colon with the most vulnerable blood supply (Fig. 5.64b). Sacculations may be seen arising from one side of the strictured area.

Pneumatosis coli

In this unusual condition gas-filled spaces are present in the wall of the bowel. These cyst-like spaces do not communicate with the lumen. They can be identified on a plain film of the abdomen but the diagnosis is much easier with a barium enema where the cysts cause smooth translucent filling defects projecting from the wall of the bowel (Fig. 5.65). The appearance could be confused with intramural haemorrhage and oedema, or with colitis if the presence of air within the cysts is not appreciated.

Volvulus

In a volvulus a loop of bowel twists on its mesentery. This happens most frequently in the sigmoid colon, particularly when it is redundant and less often in the caecum. The twisted loop becomes greatly distended

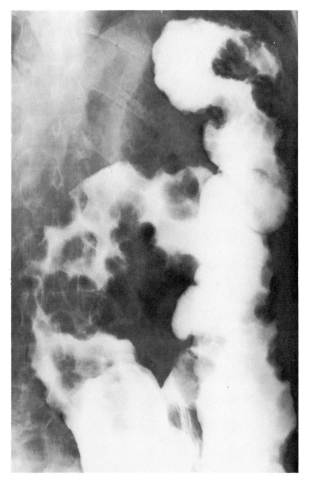

Fig. 5.65 Pneumatosis coli. Part of the colon showing numerous translucencies in the wall of the colon due to many gas-filled cysts.

and the bowel proximal to the volvulus is obstructed by the twist and may, therefore, also be dilated.

The diagnosis is usually made on the plain abdominal films (see p. 108) but a barium enema may be helpful in doubtful cases. This will show a smooth tapered narrowing (Fig. 5.66) due to twisting of the colon with marked dilatation of the bowel proximal to the twist.

Intussusception

An intussusception is the invagination of one segment of

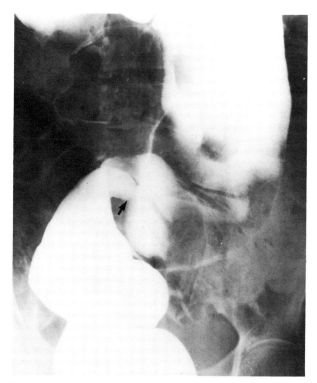

Fig. 5.66 Volvulus. A smooth narrowing is seen in the sigmoid colon where the colon has twisted (arrow). Note the dilated colon proximal to this.

the bowel into another. Infants are much more liable to intussusception than adults.

By far the commonest type is the ileum invaginating into the colon which is known as an ileo-colic intussusception. Other types are colo-colic when the colon invaginates into another part of the colon and ileo-ileal when the ileum invaginates into a more distal segment of ileum.

At barium enema the flow of barium is obstructed by the leading edge of the intussusception which causes a convex filling defect. This filling defect may show stretched mucosal folds on its surface giving the so-called 'coiled spring' appearance (Fig. 5.67). In infants and young children an intussusception can sometimes be reduced with a barium enema, so avoiding a laparotomy. If such a reduction utilising hydrostatic pressure is to be safely carried out the child should have no

clinical signs of peritonitis. The longer the symptoms have been present, the greater the risk of perforating gangrenous bowel. In adults surgical treatment is invariable as an intussusception is usually caused by a tumour.

Tumours

Polyps

The word 'polyp' means a small mass of tissue, arising from the wall of the bowel projecting into the lumen. Polyps may be sessile or on a stalk, single or multiple. They are best demonstrated with a double contrast barium enema. Most polyps are neoplasms, some the result of inflammatory disease and some due to developmental anomalies.

It is often impossible on radiological grounds to exclude malignancy in a polyp. However, only a tiny minority of polyps less than 1 cm in size and very few less than 2 cm are cancers.

The features that suggest malignancy are: a diameter of more than 2 cm; a short thick stalk; irregular surface; rapid rate of growth as judged by serial barium enema examinations.

The common polyps are:

1. Adenomatous polyp (Fig. 5.68) is a benign neoplasm; there is controversy as to whether such lesions are premalignant. They may be single or multiple and are found most frequently in the rectosigmoid region. In familial polyposis they are numerous and one or more will, in time, undergo malignant change.

Villous adenoma is a benign sessile tumour showing a sponge-like appearance due to barium trapped between the villous strands. They are usually large when first discovered and are frequently mistaken for faeces. The common sites are the rectum and the caecum. There is a high incidence of malignant change.

2. Polypoid adenocarcinoma

3. Juvenile polyps. Almost all isolated polyps in children are benign. They are probably developmental in origin.

4. Inflammatory polyps (*pseudopolyps*) are seen in ulcerative colitis (see p. 151).

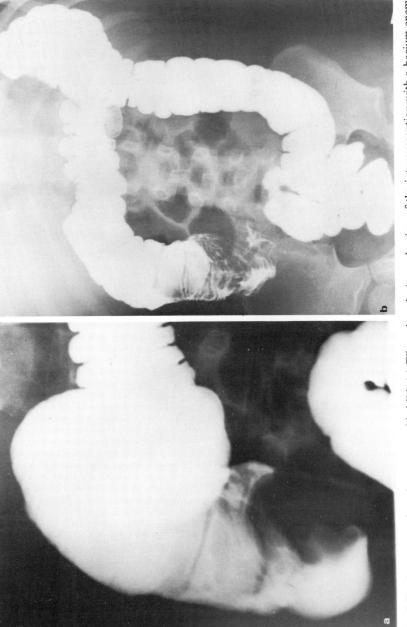

Fig. 5.67 Intussusception. A two-year-old child. (a) Film taken during reduction of the intussusception with a barium enema showing a filling defect in the caecum due to ileum invaginated into the colon; (b) later film showing a 'coiled spring' appearance in the caecum due to stretched mucosal folds of the invaginated ileum.

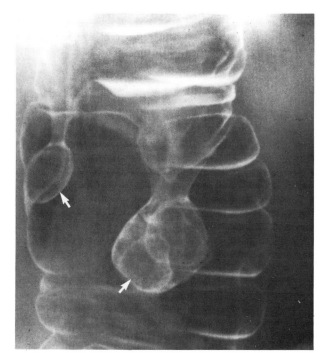

Fig. 5.68 Polyps. Double contrast enema of part of the colon showing two pedunculated adenomatous polyps (arrows).

Carcinoma

Carcinomas may arise anywhere in the colon but they are commonest in the rectosigmoid region and the caecum. The appearance and behaviour of a carcinoma in these two sites is usually quite different. The patient with a rectosigmoid carcinoma often has an annular stricture and presents with alteration in bowel habit and obstruction, whereas with a caecal carcinoma the tumour can become very large without obstructing the bowel, so anaemia and weight loss are the common presenting features.

A barium enema shows the annular carcinoma as an irregular stricture with shouldered edges (Fig. 5.52). Such strictures are rarely more than 6 cm in length. The polypoid or fungating carcinoma (Fig. 5.69) causes an irregular filling defect projecting into the lumen of the bowel.

Multiple primary tumours must be excluded as the patient with one carcinoma of the colon has a higher than normal risk of developing a second colonic cancer. This may be present at the time of the diagnosis or may present after the first tumour has been removed.

Hirschsprung's disease (congenital aganglionosis)

This condition is due to absence of ganglion cells beyond a certain level in the colon, usually in the sigmoid or rectosigmoid region. In time the colon proximal to the aganglionic segment becomes grossly distended, but in those patients who present soon after birth the dilatation may not be obvious.

The aganglionic segment, usually the rectum, is either

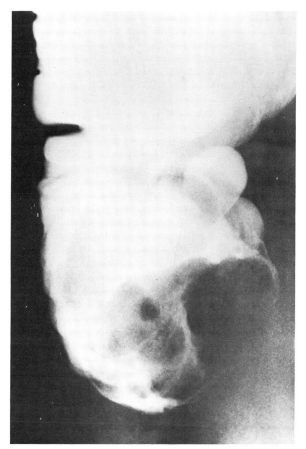

Fig. 5.69 Polypoid carcinoma. A large irregular filling defect is present in the caecum.

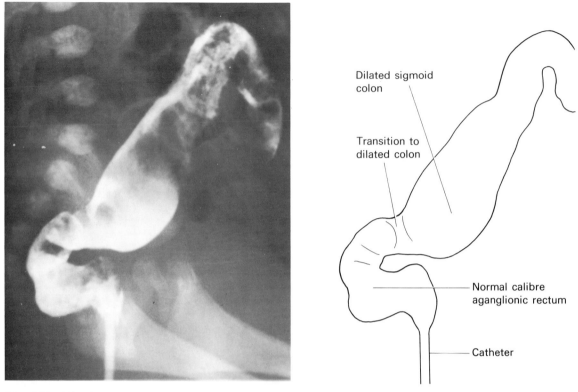

Fig. 5.70 Hirschsprung's disease. Note the transition between the normal calibre aganglionic rectum and the dilated sigmoid colon.

normal or small at barium enema and the diagnosis depends on recognising the transition from the normal or reduced calibre colon to the dilated colon (Fig. 5.70). To prevent the danger of water intoxication from the dilated colon the colon is not washed out before the barium enema. The barium introduced is usually limited to the amount required to show the zone of transition from aganglionic to dilated bowel.

Idiopathic megacolon (functional megacolon)

The cause is believed to be chronic constipation. At barium enema both the rectum and colon are dilated and contain large amount of faeces. The large-sized rectum serves as a differentiating feature from Hirschsprung's disease.

6

Hepatobiliary System and Pancreas

The information obtainable from plain films is limited. Calcification may be seen in the liver, pancreas or biliary system. The size of the liver can be assessed but not accurately, and occasionally it may be possible to diagnose an enlarged gall bladder or pancreatic mass. Because of the limitations of plain films some form of contrast examination is often necessary if diagnostic information is to be obtained. These examinations vary in complexity from the relatively simple, such as cholecystography, to complex procedures such as percutaneous cholangiography, retrograde cholangiopancreatography and selective arteriography. Radioisotopes, ultrasound and computerised tomography are playing an increasingly important part in patient management.

THE BILIARY SYSTEM

The standard examination for the gall bladder is the oral cholecystogram. This is performed over 2 days. On the first day a preliminary film of the right hypochondrium is taken. On the same evening and sometimes again the following morning the patient takes the iodine-containing contrast medium by mouth. No food is taken overnight in order to prevent the gall bladder from emptying. Films of the right hypochondrium are then taken during the morning with the patient both erect and lying flat. A fatty meal or drink is then given to make the gall bladder contract and films of the gall bladder are then repeated.

A cholecystogram will be of no value if the patient is jaundiced with a serum bilirubin above 40 μmol/l (2 mg %) because liver function is too impaired to excrete the contrast in adequate concentration.

The normal cholecystogram

The gall bladder is situated under the right lobe of the liver but may vary considerably in position from being high under the costal margin to overlying the iliac crest. The normal gall bladder shows considerable variation in both the degree of opacification and in its shape, but the outline is always smooth (Fig. 6.1).

Since the gall bladder overlies the colon, superimposed shadowing due to gas and faeces may make it difficult to exclude gall-stones.

The cystic duct leads from the gall bladder to join the common bile duct, which normally has a diameter of less than 1 cm. The common bile duct enters the second part of the duodenum through the papilla of Vater. The cystic duct and common bile duct may be seen at oral cholecystography, particularly after the gall baldder has contracted on the 'after fatty meal' film. The intrahepatic and common hepatic ducts are not normally visualised with a cholecystogram.

The abnormal cholecystogram

Signs on plain films

Calcification. Only gall-stones containing calcium will be seen on the plain films (Fig. 6.2). It should be appreciated that only 20–30% of gall-stones can be so identified. They vary greatly in size and shape and, typically, have a dense outer rim with a more lucent centre. When multiple, the stones are often facetted. Sometimes they appear laminated.

Opaque gall-stones must be distinguished from renal calculi and costal cartilage. Occasionally, a contrast

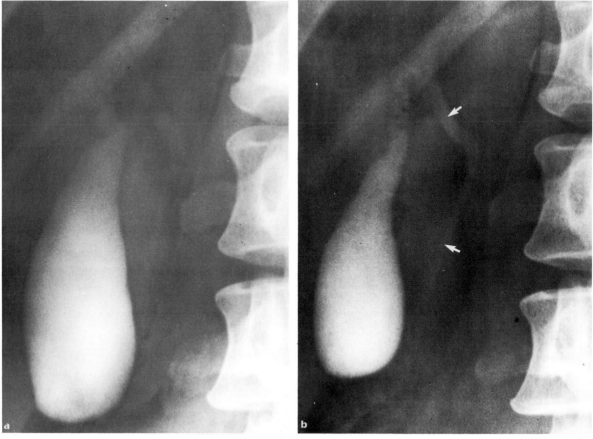

Fig. 6.1   Normal cholecystogram. (a) Note the smooth outline and homogeneous filling of the gall bladder; (b) after a fatty meal: the gall bladder has contracted and the common bile duct is now clearly visible (arrows).

examination of the biliary or urinary systems is needed to do this. Usually the ring-like appearance and the position in relation to the liver edge make the recognition of a gall-stone fairly easy. A gall-stone that overlies the renal outline on the frontal view will be projected clear of the kidney on an oblique view.

Opacification of the gall bladder on a plain film may be due to calcified sludge within the gall bladder known as 'milk of calcium' bile (Fig. 6.3).

Gas in the biliary system (Fig. 6.4)—is seen:

1. Following sphincterotomy or anastomosis of the common bile duct to the bowel.

2. Following a fistula, due to erosion of a gall-stone into the duodenum or colon.
3. With a duodenal ulcer which has penetrated the common bile duct.

Gas may be seen very occasionally in the wall or lumen of the gall bladder in acute cholecystitis with a gas-forming organism.

Signs on films after contrast has been given

Filling defects. By far the commonest filling defects are *gall-stones*. The majority are not radio-opaque and so are not visible on plain films but can be visualised on a

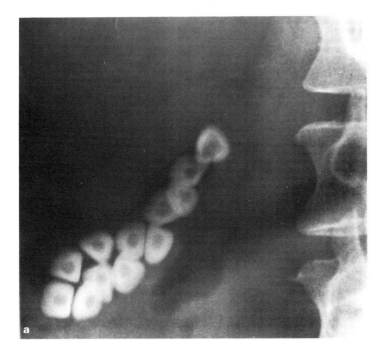

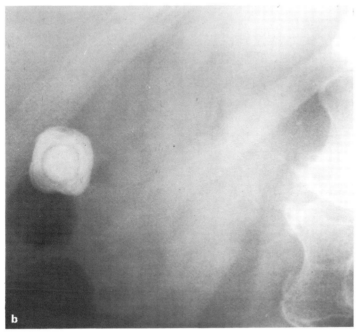

Fig. 6.2 (a) Gall-stones. Multiple facetted stones showing lucent centres; (b) single lamellated gall-stone.

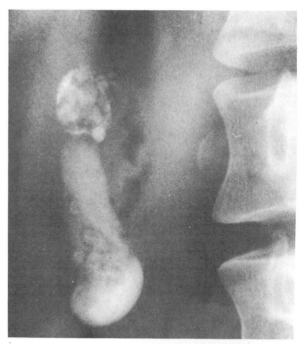

Fig. 6.3 Milk of calcium bile. This plain film shows the gall bladder is full of calcified biliary sludge. This must not be confused with contrast medium in a cholecystogram.

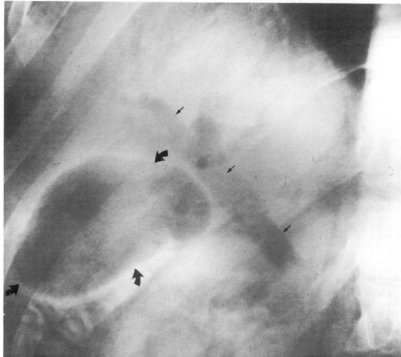

Fig. 6.4 Gas in the biliary tree. The gall bladder (curved arrows) and the duct systems (straight arrows) have been outlined with air. The patient had an anastomosis of the common bile duct to the bowel.

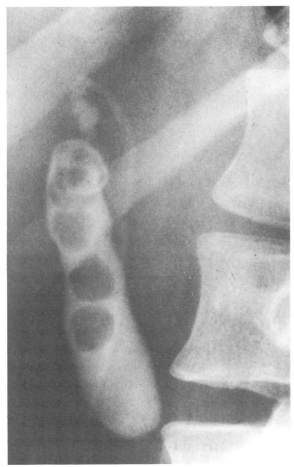

Fig. 6.5 Gall-stones. In this cholecystogram the stones appear as lucent filling defects completely surrounded by contrast.

cholecystogram as lucent filling defects within the gall bladder completely surrounded by contrast (Fig. 6.5).

Occasionally, small stones may be difficult to detect on a film with the patient lying flat, but may become obvious with the patient erect when they sink to the most dependent part of the gall bladder. Sometimes small cholesterol stones can be seen on an erect film floating in a layer across the gall bladder (Fig. 6.6).

Polyps arise from the wall and project into the lumen of the gall bladder. They are usually small and constant in position irrespective of whether the patient is lying or

standing. However, the distinction from non-opaque gall-stones can be difficult.

Outline. Normally, the outline of the gall bladder is smooth but in *adenomyomatosis* there are small projections into the wall which fill with contrast. These are known as Rokitansky–Aschoff sinuses (Fig. 6.7). Other features of adenomyomatosis are shelf-like projections across the gall bladder and a small filling defect at the fundus due to hypertrophied muscle. There is dispute as to whether this condition causes symptoms.

Non-opacification of the gall bladder has many causes. The patient may fail to take the tablets; the contrast may not be absorbed; the liver function may be impaired so that the contrast is not excreted; the cystic duct may be blocked or the gall bladder may be diseased.

If there is no opacification of the gall bladder at cholecystography in a patient who is not jaundiced and if the contrast has definitely been taken (a point that is often confirmed by some remaining unabsorbed in the colon), non-opacification is highly suggestive of a diseased gall bladder, usually chronic cholecystitis. However, a repeat examination is advisable before removing the gall bladder since on occasions a normal gall bladder may not be visualised on one examination but be normal on another. Alternatively an ultrasound examination can be performed.

In acute cholecystitis there is no opacification of the gall bladder, but cholecystography is of little use in the acute phase as it takes too long to obtain a result and the examination gives no positive information.

Ultrasound

The greatest value of ultrasound in diseases of the hepatobiliary system is in distinguishing large duct obstruction from the other causes of jaundice. Ultrasound can reliably demonstrate dilated interhepatic bile ducts. At the present stage of instrument design normal peripheral bile ducts cannot be seen, so identifying intrahepatic bile ducts at ultrasound indicates the presence of dilated ducts (Fig. 6.8). At times even the exact site and nature of the lesion responsible for the obstruction can be

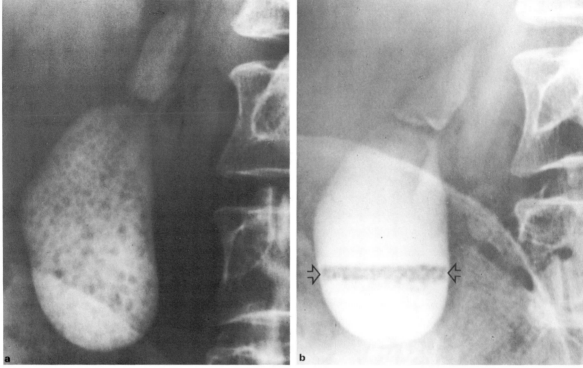

Fig. 6.6 (*above*)   Gall-stones. (a) With the patient lying flat the many small stones are fairly evenly distributed throughout the gall bladder; (b) with the patient erect the stones form a layer across the gall bladder (arrows). Because of the gradation in density from the heavy contrast below to the bile above the stones float in a layer approximating to their own density.

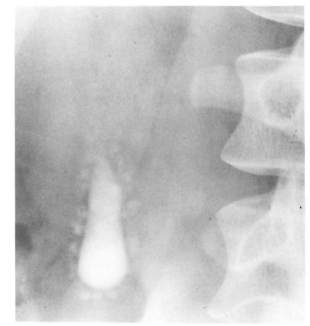

Fig. 6.7   Adenomyomatosis. The gall bladder has an irregular outline due to numerous small projections from the lumen known as Rokitansky–Aschoff sinuses.

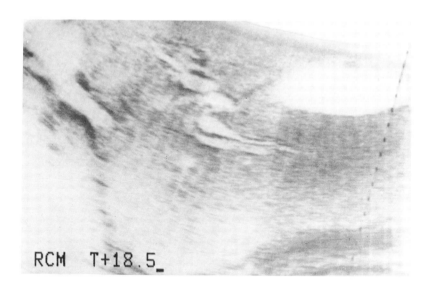

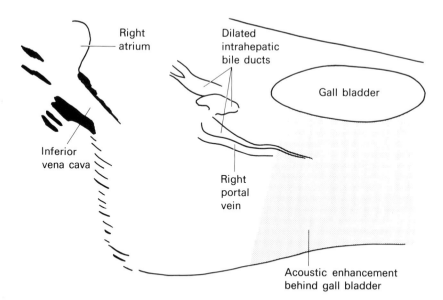

Right atrium

Dilated intrahepatic bile ducts

Gall bladder

Inferior vena cava

Right portal vein

Acoustic enhancement behind gall bladder

Fig. 6.8 Ultrasound scan showing dilated intrahepatic ducts. The patient had obstructive jaundice due to a carcinoma of the head of the pancreas (see also Fig. 6.12b).

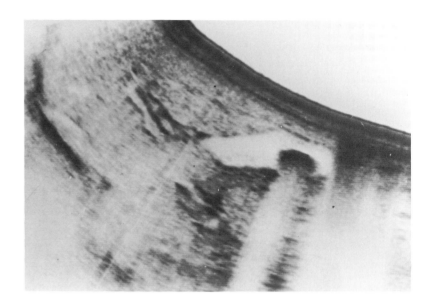

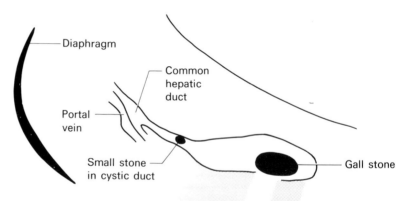

Fig. 6.9 Ultrasound scan showing gall-stones. One large stone is present in the gall bladder and a tiny one in the cystic duct.

identified. If a normal intrahepatic pattern and a normal extrahepatic biliary system is defined, obstructive jaundice can be confidently excluded. However, demonstrating the absence of dilatation of the intrahepatic bile ducts does not exclude obstruction, since in some cases of obstruction there may be selective dilatation of the extrahepatic biliary system.

Gall-stones are readily demonstrated at sonography (Fig. 6.9). The gall bladder is superficially located and stones are seen as highly echogenic foci within this cystic structure. The gall bladder sonogram is optimally performed after an overnight fast. However, unlike the oral cholecystogram it can be performed quickly as an emergency, even in the presence of jaundice or acute pancreatic disease.

Further investigations of the biliary system

Intravenous cholangiography

An iodine-containing contrast medium is injected intravenously; it is then excreted by the liver into the biliary system in sufficient concentration to visualise the ducts and gall bladder directly (Fig. 6.10), without the need for further concentration by the gall bladder. The concentration of contrast achieved is poor, so tomography is often necessary to show the common bile duct. Gall bladder opacification, though adequate is not as good as with an oral cholecystogram. It is rare to achieve diagnostic information if the patient is more than mildly jaundiced.

Adverse reactions due to intravenous biliary contrast media occur several times more frequently than with urographic contrast media, so there must always be a good indication for an intravenous cholangiogram. The complications are very similar to those seen with the urographic agents (p. 177).

The indications for intravenous cholangiography are:

1. Following non-opacification of the gall bladder on an oral cholecystogram in problem cases.
2. To demonstrate the common bile duct in a patient who is mildly jaundiced or one who has recovered from jaundice.
3. Following cholecystectomy.

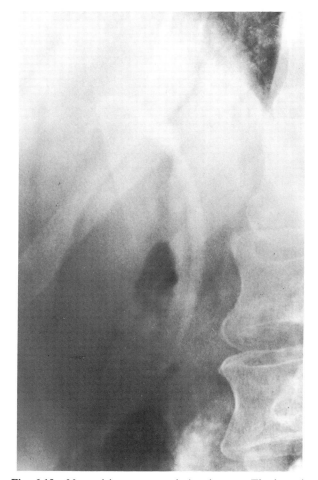

Fig. 6.10 Normal intravenous cholangiogram. The hepatic and common bile duct have been visualised but the gall bladder has not filled.

Stones in the common bile duct are seen as filling defects (Fig. 6.11). Stones which cause obstruction often lead to dilatation of the common bile duct, but sometimes the stones themselves may not be visible at intravenous cholangiography.

Filling of the common bile duct but no opacification of the gall bladder up to 4 hours after injection of contrast medium indicates that the cystic duct is blocked, usually by a stone.

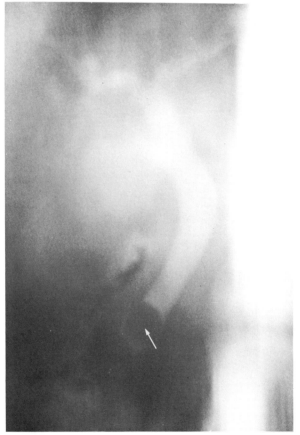

Fig. 6.11 Stone in the common bile duct. The intravenous cholangiogram shows the stone as a round filling defect surrounded by contrast (arrow) in the lower end of a dilated common bile duct.

Percutaneous transhepatic cholangiogram

A fine needle is passed through the abdominal wall under local anaesthetic into the liver in an attempt to inject contrast directly into the bile ducts. Unlike a cholecystogram or intravenous cholangiogram this procedure is only carried out when the patient is jaundiced.

The indications for a percutaneous transhepatic cholangiogram are:

1. To differentiate jaundice due to large duct obstruction from other causes of jaundice in problem cases.

2. In known large-duct obstruction to demonstrate the site of obstruction and to determine its nature.

In patients with jaundice due to large duct obstruction the intrahepatic ducts are dilated so it is easy to enter these ducts with a needle and inject contrast into them. The site of obstruction can be demonstrated and its appearance often suggests the diagnosis. The common causes are: an impacted stone in the common bile duct; a carcinoma of the head of the pancreas; or a carcinoma of the ampulla of Vater (Fig. 6.12). In other forms of jaundice the intrahepatic ducts are not dilated. Such ducts cannot always be successfully entered with a needle.

Haemorrhage is an occasional problem but the main complications are bile leaking into the peritoneum causing biliary peritonitis and septicaemia. In order to prevent these problems, which are both much more likely in an obstructed system, an operation is normally carried out promptly if dilated ducts have been entered. For this reason the percutaneous cholangiogram is usually performed just before the patient is scheduled for operation. Recently, a very fine needle has been designed to puncture the liver and with this the risk of biliary peritonitis is much less.

Endoscopic cannulation of the common bile duct

Contrast can be injected into the common bile duct through a catheter which has been inserted through the papilla of Vater from an endoscope positioned in the duodenum (Fig. 6.13).

The indications for this procedure are:

1. To determine whether jaundice is due to large duct obstruction. In this respect endoscopic cannulation of the common bile duct may be carried out as an alternative to a percutaneous cholangiogram, particularly if normal-sized intrahepatic ducts are shown on ultrasound.

2. The investigation of unexplained abdominal pain thought to be biliary in origin when other investigations have been equivocal. An added advantage is that the pancreatic duct system often fills as well.

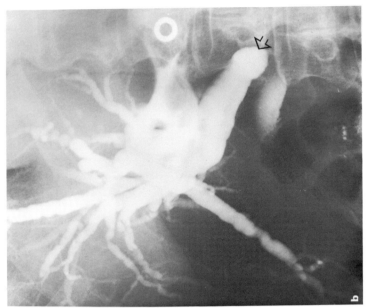

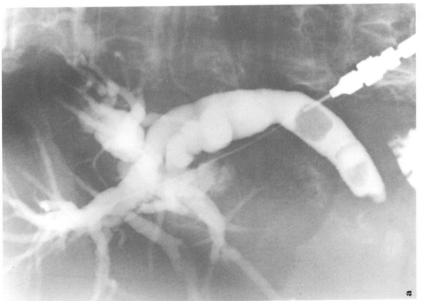

Fig. 6.12 Transhepatic percutaneous cholangiogram. (a) Stones in the common bile duct: contrast has been injected though a cannula into the dilated intrahepatic ducts (normal sized ducts are shown for comparison in (Fig. 6.14); (b) carcinoma of the pancreas: there is a complete obstruction of the common bile duct (arrow). Note the dilated intrahepatic ducts. This is the same patient whose ultrasound scan is shown in Fig. 6.8.

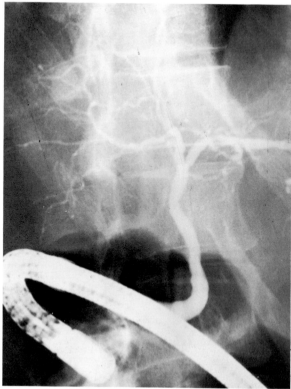

Fig. 6.13 Endoscopic retrograde cholangiography. A normal biliary system has been shown by injecting contrast through a catheter passed from the endoscope into the common bile duct.

Operative cholangiogram

This is performed during an operation by the surgeon who injects contrast directly into the gall bladder, cystic or common bile duct.

The patients are usually undergoing cholecystectomy and the films are needed to show whether there are any stones in the hepatic and common bile ducts and to ensure free passage of contrast into the duodenum. In all cases of cholecystectomy for stones an operative cholangiogram should be performed, as this method is much superior than palpation or exploration for detecting stones in the ducts.

Postoperative T-tube cholangiogram

If the common bile duct is explored at surgery a T-tube is placed in the duct system to provide biliary drainage. Contrast is injected through the T-tube about 10 days postoperatively to ensure that no stones are present in the biliary system, and that the contrast flows freely into the duodenum before the T-tube is pulled out (Fig. 6.14).

PARENCHYMAL LIVER DISEASE

Information regarding parenchymal liver disease can be obtained by a variety of x-ray, ultrasound and radionuclide techniques.

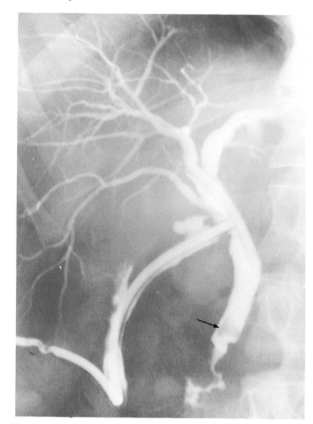

Fig. 6.14 T-tube cholangiogram. There is a stone (arrow) at the lower end of the common bile duct. The intrahepatic ducts are of normal calibre.

The agent most commonly used for radionuclide examinations is ^{99m}Tc-labelled sulphur colloid. After this has been injected intravenously it is taken up by macrophages which in the case of the liver are the Kupfer cells (Fig. 1.9).

The information gained from these tests is largely complementary and the choice of investigation needs to be closely tailored to the clinical requirements. Of the many different problems in which radiology plays an important part only two will be discussed here; namely the investigation of liver masses and portal hypertension.

Liver masses

Metastases are by far the commonest liver tumours. Primary carcinoma of the liver (hepatoma) is uncommon and when seen it often complicates cirrhosis of the liver. Regenerating nodules are another and sometimes a confusing cause of masses in patients with cirrhosis. Other mass lesions of the liver include abscesses, hydatid and other cysts.

The main diagnostic problems in a patient with a suspected liver mass are:

1. Confirmation of the presence of a mass in the liver and determining if it is single or multiple.
2. Determining the nature of the mass.

Radionuclide scans, ultrasound, computerised tomography and angiography are all good methods of deciding that a mass is present. Though they may give some idea of the nature of the mass or masses they rarely provide a definitive answer except in the case of cysts which can be readily diagnosed with ultrasound (Fig. 1.8). However, the combination of the clinical findings together with evidence of one or more masses may be adequate and obviate the need for a tissue diagnosis. For example, in a patient with a known carcinoma of the colon, multiple filling defects on a ^{99m}Tc-sulphur colloid scan provide fairly reliable evidence of metastatic spread of the colonic carcinoma (Fig. 6.15); alternatively, a thick wall fluid-filled space at ultrasound examination in a febrile patient with a leucocytosis is good evidence of a liver abscess.

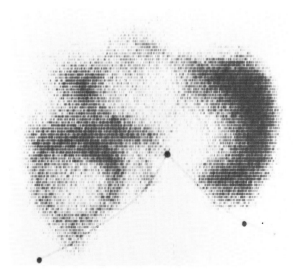

Fig. 6.15 Liver scan. Multiple filling defects are seen in this scan using ^{99m}Tc-sulphur colloid due to metastases in the liver. The dots mark the subcostal margin.

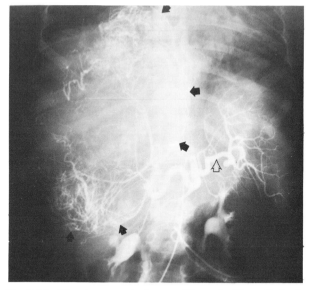

Fig. 6.16 Hepatoma—arteriogram. The end of the catheter is in the coeliac axis. Branches of hepatic arteries are stretched around the large hepatoma (closed arrows). In the centre of the tumour there is an extensive pathological circulation supplying the tumour. The splenic artery (open arrow) and spleen are normal.

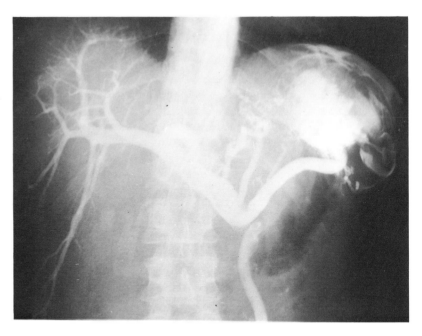

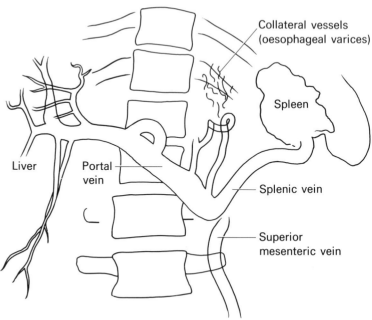

Collateral vessels
(oesophageal varices)

Spleen

Liver

Portal
vein

Splenic vein

Superior
mesenteric vein

Fig. 6.17 Splenoportogram. The injection is made directly into the spleen. Contrast flows to the liver through the splenic and portal veins and it has also refluxed into the superior mesenteric vein. Portosystemic collateral vessels (oesophageal varices) have been shown at the oesophagogastric junction. The patient had cirrhosis.

When the clinical and biochemical features do not point clearly in one direction then some form of biopsy of the mass is required before a final diagnosis can be made.

A coeliac axis or hepatic arteriogram is often useful in the management of a hepatoma because it can help plan any possible operation by demonstrating the vascular anatomy and extent of the tumour (Fig. 6.16).

Portal hypertension

In portal hypertension there is an elevation of the portal venous pressure due to obstruction to the flow of venous blood in the portal vein, liver or hepatic veins. By far the commonest cause of portal hypertension is cirrhosis of the liver. Other causes include thrombosis of the portal vein, particularly following sepsis of the umbilical vein in the neonatal period and following occlusion of the hepatic veins (Budd–Chiari syndrome).

Due to the rise in pressure, blood in the portal venous system flows through anastomatic channels, known as portosystemic anastomoses, to the vena cavae bypassing the liver. They may be found in various sites but the most important, clinically, are varices at the lower end of the oesophagus.

A ^{99m}Tc-sulphur colloid scan will show a large spleen with increased uptake of the isotope and decreased uptake in the liver.

Portal venography may be undertaken to assess the patency of the portal vein for possible surgical portosystemic bypass operations, such as a portocaval shunt. The technique is not used just to show the anastomotic channels; oesophageal varices are more easily demonstrated by a barium swallow or endoscopy. Portal venography can be carried out by two different methods.

Arterial injection (arterioportogram). Contrast is injected into the coeliac axis, splenic artery or superior mesenteric artery. Films taken during the venous phase show the portal venous system.

Direct injection into the spleen (splenoportogram, splenic venography). Contrast is injected through a needle into the splenic pulp from where it passes along the splenic vein into the portal vein (Fig. 6.17). This method carries the risk of bleeding from the spleen, so care must be exercised particularly in patients with cirrhosis in whom the clotting mechanisms may be defective.

THE PANCREAS

The pancreas is an extremely difficult organ to demonstrate by conventional radiographic techniques. The normal pancreas cannot be seen on plain films of the abdomen and no simple method is available for making it opaque. Pancreatic calcification is discussed on p. 113. Radionuclide scanning and arteriography are used but have a limited diagnostic role. Visualisation of the duct system can be achieved by endoscopic cannulation of the pancreatic duct.

Computerised tomography (CT) and ultrasound are proving to be invaluable methods of diagnosis. In those centres where it is available, CT has largely replaced conventional radiology for diagnosing pancreatic disease.

Fig. 6.18 Pancreatic pseudocyst. This has caused extrinsic compression upon the greater curve of the stomach and downward displacement of the duodenojejunal flexure.

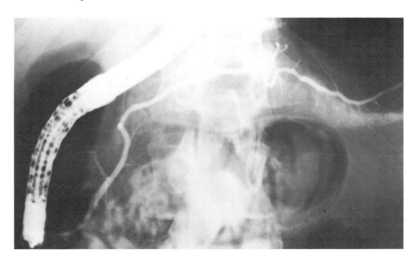

Fig. 6.19 Endoscopic retrograde pancreatography. The pancreatic duct has been cannulated from the endoscope in the duodenum. Contrast has been injected to demonstrate a normal duct system.

Changes in the upper gastrointestinal tract

Both carcinoma and pancreatitis cause enlargement of the head of the pancreas and produce deformity of the duodenal loop, although it is often not possible to distinguish radiologically between these two conditions. These changes in the duodenum seen on a barium meal have already been discussed on p. 136.

Abnormalities of the body and tail of the pancreas are more difficult to detect. A carcinoma may invade the overlying stomach, when it will be difficult to differentiate from a primary neoplasm of the stomach. Extrinsic compression and anterior displacement of the stomach occur with large masses; these are either carcinomas or pseudocysts of the pancreas (Fig. 6.18).

Ultrasound

Ultrasound imaging of the pancreas can be very difficult. The examinations are often unsuccessful due to overlying bowel gas. Despite these problems important information is obtained in a significant number of patients. The pancreas can be imaged directly by ultrasound and its overall size and shape can be assessed. The pattern of echoes from within the pancreas can be used to diagnose cysts (particularly pancreatic pseudocysts), and chronic pancreatitis.

Unfortunately, at the present time, the diagnosis of carcinoma depends on recognising a distortion of pancreatic shape, or an absolute increase in pancreatic size. Therefore, early detection is still beyond the resolution of ultrasound examination.

Ultrasound and computerised tomography of the pancreas may give similar information. Often the information is complementary.

Endoscopic retrograde cholangiopancreatography (ERCP)

The duct system of the pancreas can be visualised by injecting contrast through a catheter passed from an endoscope in the duodenum through the papilla of Vater into the pancreatic duct (Fig. 6.19).

7

Urinary Tract

The standard method of x-ray examination of the kidneys, ureters and bladder is the intravenous urogram (IVU), which involves the intravenous injection of an aqueous solution containing a high concentration of iodine. In most cases a set routine of films is taken but the plain film and each film after the injection of contrast medium is assessed during the course of the examination, so that appropriate extra views or tomograms can be taken.

THE ROUTINE IVU

The contrast medium and its excretion

Urographic media are highly concentrated solutions of organically bound iodine. A large volume, e.g. 50 ml, is injected intravenously and is carried in the blood to the kidneys, where it passes into the glomerular filtrate. Contrast is not absorbed by the tubules, so substantial concentration is achieved in the urine particularly after fluid restriction. The visualisation of the renal substance (the nephrogram) is dependent on the amount of contrast reaching the kidneys; whereas the visualisation of the collecting systems (the pyelogram) depends mainly on the ability of the kidneys to concentrate the urine.

Adverse reactions to intravenous urographic contrast media

Adverse effects may accompany the injection of urographic media. Fortunately, most of these are readily reversible, usually spontaneously.

Most patients experience a feeling of warmth spreading over the body as the contrast medium is injected; a few find this feeling objectionable. Sometimes, particularly with slow injections of the more concentrated solutions, pain occurs in the upper arm and shoulder due to stasis of the contrast medium during injection. If this occurs it is helpful to raise the patient's arm at the end of the injection. If the contrast is inadvertently injected outside a vein it is very painful indeed.

About a fifth of patients experience nausea, vomiting or light-headedness. A few develop an urticarial rash. These phenomena usually subside spontaneously.

In a few, bronchospasm, laryngeal oedema or hypotension develop and these may be so severe as to be life threatening. It is, therefore, essential to be prepared for these dangerous reactions and to have available the equipment and the drugs to cope with them. Approximately one in 40 000 patients dies as a consequence of intravenous injection of urographic agents. This risk, though small, should not be ignored.

Patients with known allergic manifestations, particularly asthma, are more likely to have an adverse reaction to the injection of contrast. Even normal people develop clinically occult bronchospasm, so it is not surprising that asthmatic patients may show an exacerbation of their asthma and on occasion the IVU may precipitate a life-threatening attack. Similarly, patients who have had a previous reaction to contrast agents have a higher than average risk of problems during the IVU. Such patients are usually premedicated with steroids, preferably for at least 18 hours prior to the examination. Antihistamine drugs are also given shortly before the contrast injection.

Other patients with a higher-than-average risk of complications from IVU are:

1. Infants are at risk from a rapid rise in plasma osmolality because of the high osmolality of the injected

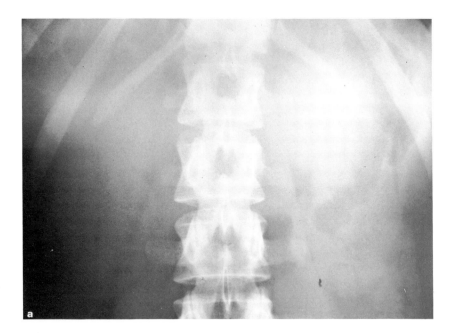

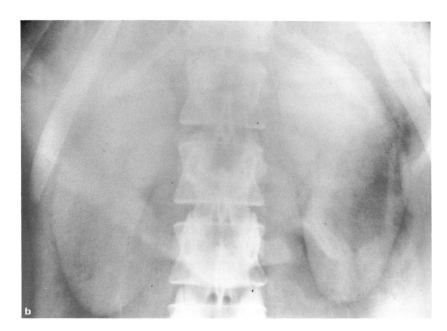

Fig. 7.1   The nephrogram. (a) The
renal areas on a plain film; (b) this film
was exposed immediately after the
intravenous injection of contrast agent.
The renal parenchyma is opacified and
the renal outlines are well shown. Note
that at this early stage the calices are
not seen.

contrast agent. Therefore, the injection rate in infants should be slow—a minimum of 5 minutes.

2. Elderly patients often tolerate the contrast injection poorly, so here too slow injected rates are advised.

3. Those with known heart disease. Arrhythmias are a risk in patients with heart disease. This risk is probably increased with fast injection rates.

4. Those with renal failure, myeloma or severe diabetes are more likely to show a deterioration of renal function due to the contrast medium if they are deprived of fluids prior to the examination.

Preparation of the patient

Fluid restriction is an advantage in patients with normal renal function, as this increases the concentration of the urine and produces a dense pyelogram. The patients are instructed not to drink for at least 8 hours prior to the examination. This does *not* apply to patients in renal failure. They could be harmed by withholding fluids and there would in any event be no advantage, because many are unable to improve the concentration of their urine.

Since colonic shadows can cause confusion when projected over the kidneys aperients or suppositories may be given prior to the examination.

The preliminary film

One or more plain films to show the whole of the urinary tract are taken. The plain film is an integral part of the study and should never be omitted.

Films following the injection of contrast medium

The precise timing of these films will vary in different hospitals. A film is taken 1 minute after the injection, by which time the contrast is in the nephrons producing a nephrogram, but has not yet reached the calices (Fig. 7.1). This film is designed to show the renal parenchyma, particularly the renal outlines. A film of the kidneys is taken at 5 minutes, after which pads are sometimes applied to the lower abdomen to compress the uterers as they cross the pelvic brim, in order to fill out the collecting systems. A further film of the kidneys is taken 5

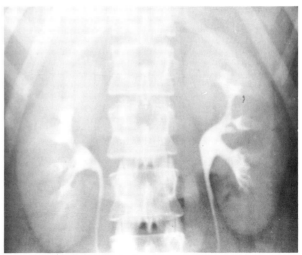

Fig. 7.2 Normal pelvicaliceal system as shown on a film taken 5 minutes after the injection of contrast agent. Note that now the collecting system is well demonstrated.

minutes after the compression has been applied (Fig. 7.2). A full-length abdominal film, exposed immediately after releasing the compression, is the one most likely to show contrast in the ureters. This film, usually taken 15–20 minutes after injection, will probably show the bladder well filled with opacified urine (Fig. 7.3). At the end of the examination the patient is asked to empty his bladder and a film is obtained to assess the residual volume and to provide further details of the bladder.

Extra views

Tomography is very useful when there are overlying bowel shadows in showing the kidneys, particularly to show the renal outlines and renal parenchyma. Some x-ray departments use tomography almost routinely.

Oblique views can be taken before or after contrast. Oblique plain films of the kidneys are valuable in distinguishing calcification within the kidney from calcification overlying it on the AP view (Fig. 7.4). Lateral views are unhelpful in this respect since the kidneys are then projected over the lumbar spine and small calcifications may be lost. Oblique films may be utilised to show

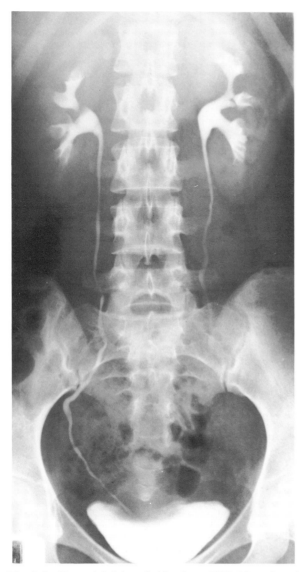

Fig. 7.3 Normal full length 15-minute IVU. Note that the bladder is well opacified. The whole of the right ureter and part of the left ureter are seen. Often only a portion rather than the entire ureters are visualised due to peristalsis emptying certain sections. The bladder outline is reasonably smooth. The roof of the bladder shows a shallow indentation from the uterus.

the relationship of the calcification to the opacified ureters and to provide further details of the bladder.

ROUTINE METHOD OF ASSESSING IVU

The plain film

Identify all calcifications. Decide if they are in a portion of the urinary tract by relating them to the renal outlines (using oblique plain films where necessary), or the expected position of the ureters, bladder, prostate and urethra. Calcification within the urinary tract is discussed on p. 191. Even if the calcification is outside the urinary tract an attempt should be made to decide its nature; it may be the cause of the patient's symptoms.

It is important to realise that calcification can be obscured by contrast medium. Stones would often be missed if no plain film were taken (Fig. 7.5) and renal disorders will be misdiagnosed if calcification is not appreciated.

Look at the other structures on the film, including the bones, just as you would any plain abdominal film. Do not waste time drawing conclusions about the renal outlines on plain films. They are always better seen after contrast has been given. You only need to identify the kidneys if calcification is present and you wish to check its relationship to the kidney.

Films taken after the injection of contrast medium

The nephrogram

Check that the kidneys are in their normal positions (Figs. 7.2 & 7.3) and that their axes are parallel to the outer margins of the psoas muscles. The left kidney is usually higher than the right—the upper margins are opposite T12 or L1 and the lower margins are three vertebrae lower.

There are two basic reasons why the position or axis of a kidney might be abnormal; congenital malposition (see p. 208) or displacement by a retroperitoneal mass (Fig. 7.6).

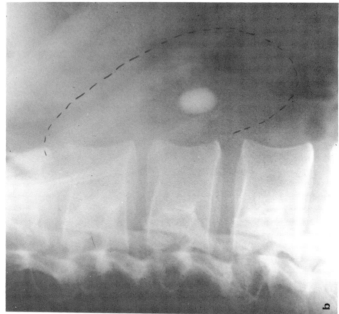

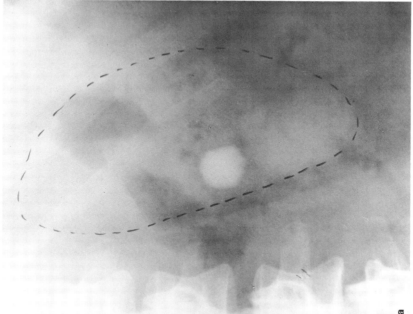

Fig. 7.4 Oblique views in determining whether calcifications are intra- or extrarenal. (a) A rounded calcification is seen overlying the left kidney in the AP plain film; (b) in the oblique plain film, the calcification is in the same position within the renal shadow and is, therefore, a renal calculus;

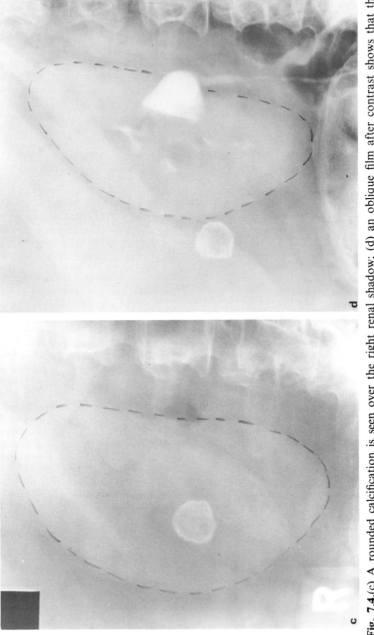

Fig. 7.4.(c) A rounded calcification is seen over the right renal shadow; (d) an oblique film after contrast shows that the calcification lies outside the kidney. It was later confirmed to be a gallstone.

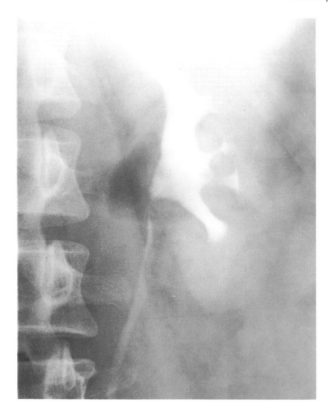

Fig. 7.5 Large calcified calculus in the pelvis of the kidney obscured by contrast medium. This is the same patient as illustrated in Figs 7.4(a) and (b). Since the contrast medium and the calculus have the same radiographic density the calculus is hidden by the contrast medium.

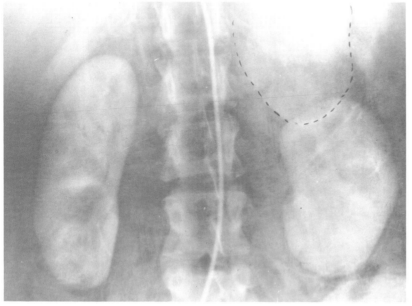

Fig. 7.6 Alteration in the position of the kidney. The left kidney is displaced downward and its upper pole deformed by a large suprarenal mass (the outline of the mass has been drawn in). The axis of the kidney is more vertical than the normal opposite kidney. (This film is the nephrogram phase of an arteriogram.)

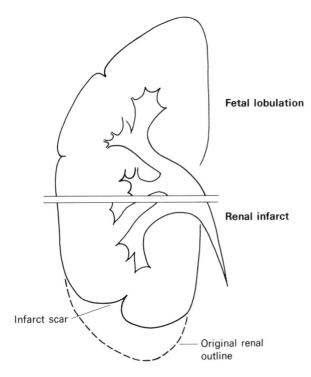

Fetal lobulation

Renal infarct

Infarct scar

Original renal outline

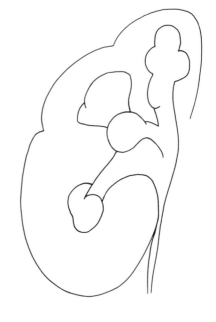

Fig. 7.7 (a) The distinction between fetal lobulation and renal infarction. With fetal lobulation indentations in the renal outline are shallow and correspond to the lobules of the kidney, i.e. the indentations are between calices. With renal infarction the maximal indentation is opposite a calix and there is usually extensive loss of renal parenchyma.

(b) Scars in chronic pyelonephritis (drawing of Fig. 7.11b. The reductions in renal parenchymal width are opposite calices, and these calices are dilated. The overall kidney size is reduced, as is usual. Scars in tuberculosis have much the same appearance but are usually associated with other signs of tuberculosis.

Identify the whole of both renal outlines. They are usually better seen on tomography. If any indentations or bulges are present they must be explained.

1. Local indentations (Fig. 7.7). Minor indentations between normal calices are due to persistent fetal lobations. All other local indentations are scars. Visible scars occur in:

(a) Chronic pyelonephritis, in which the adjacent calices are deformed and usually dilated.

(b) Tuberculosis, in which there is normally other evidence of tuberculosis on the IVU.

(c) Renal infarction, in which the scar is opposite a calix, the calix in question being normal.

2. Local bulges of the renal outline. A bulge of the renal outline usually means a mass. Most masses large enough to deform the renal outline will displace and deform the adjacent calices (Fig. 7.8). An important normal variant causing a bulge of the outline is the so-called 'splenic hump' (Fig. 7.9).

Measure the renal lengths (Fig. 7.10). Although it is possible to measure the other diameters, the length of the kidney is the only measurement in everyday use. The normal adult kidney is between 10 and 16 cm from top to bottom; the length varies with age, being maximal in the young adult. There may be a difference between the two kidneys, but this is normally less than 1·5 cm. A kidney with a bifid collecting system is usually 1–2 cm larger

renal parenchymal width should be uniform and symmetrical—between 1·5 and 2·0 cm, except at the upper and lower poles where there is an extra centimetre of thickness.

The collecting systems

The calices should be evenly distributed and reasonably symmetrical. The term often used to describe the shape of a normal calix is 'cupped' and that to describe the dilated calix 'clubbed' (Fig. 7.11). The normal 'cup' is due to the indentation of the papilla into the calix.

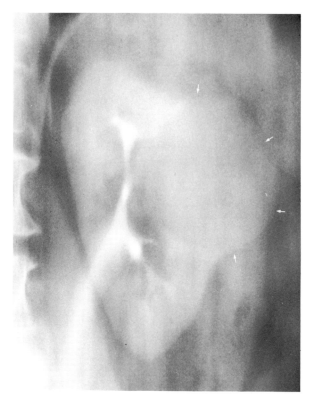

Fig. 7.8 Renal mass. A renal cyst (arrows) has caused a bulge on the lateral aspect of the kidney with splaying of the calices. Note how the cyst is more transradiant than the adjacent normal renal parenchyma.

than the opposite kidney with a single collecting system. Minor changes in size occur in many conditions. Diseases where the increase or decrease in size is sufficient to be recognisable, even without comparison with previous examinations, are listed in Tables 7.1 and 7.2. It will be noted that it is usually possible to diagnose the cause of enlarged kidneys. When only one kidney is small the diagnosis can often be reached by examining the signs on the IVU, but except for bilateral chronic pyelonephritis it is often not possible to distinguish between the various causes of bilaterally small kidneys.

Assess the width of the renal parenchyma (Fig. 7.10). The

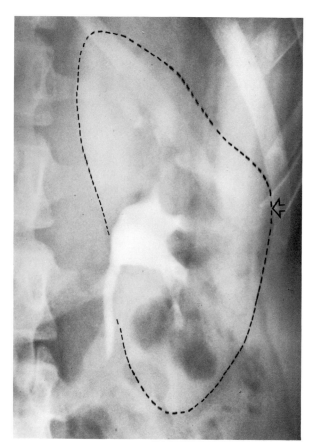

Fig. 7.9 The 'splenic hump'. A bulge is present on the lateral aspect of the left kidney (arrow) but there is no displacement of the calices. This 'splenic hump' is a normal variant.

Table 7.1   Small kidneys

	Diagnosis	*Comments*
Unilateral but may be bilateral	Chronic pyelonephritis	Focal scars and dilated calices (p. 213)
	Tuberculosis	See p. 212
	Following acute infection	Very rare
	Obstructive atrophy	Dilatation of all calices with uniform loss of renal parenchyma
	Renal artery stenosis or occlusion	Outline may be smooth or scarred, but the calices appear normal. In severe unilateral cases the density of the pyelogram on the affected side may be increased. If very severe the pyelogram may be delayed by a minute or two
	Hypoplasia	Very rare; kidneys may be smooth or irregular in outline. Calices may be clubbed
	Radiation nephritis	
Always bilateral	Chronic glomerulonephritis of many types	Usually no distinguishing features. In all these conditions the kidneys may be small with smooth outlines and normal pelvic caliceal systems
	Hypertensive nephropathy	
	Diabetes mellitus	
	Collagen diseases	
	Analgesic nephropathy	Calices often abnormal (p. 213)

Table 7.2   Enlarged kidneys*

	Diagnosis	IVU
Always unilateral	Compensatory hypertrophy	Opposite kidney small or absent
May be unilateral or bilateral	Bifid collecting system (p. 206)	Diagnosis obvious from abnormalities of collecting systems
	Renal mass (p. 203) Hydronephrosis (p. 197) Lymphomatous infiltration	May show obvious masses; however, may be large but otherwise unremarkable
	Renal vein thrombosis	
Always bilateral	Polycystic disease Amyloidosis	Characteristic IVU (p. 210) Non-specific enlargement (rare)
	Acute glomerulonephritis	Slight non-specific enlargement
	Diabetic nephropathy	Slight non-specific enlargement (a rare cause of enlarged kidney)

* Minor degrees of enlargement occur in many conditions. Only those conditions that give rise to easily recognised enlargement are listed here.

Caliceal dilatation has two basic causes:
(a) Obstruction
(b) Destruction of the papilla, the causes of which include
    —chronic pyelonephritis
    —tuberculosis
    —obstructive atrophy
    —papillary necrosis

The first step in sorting out which of these mechanisms is at work is to try and decide whether or not there is obstruction, i.e. dilatation of the collecting system down to a specific point (Fig. 7.11c). If there is no evidence of obstruction the conditions causing papillary destruction will have to be considered.
The pelvis of the kidney is extremely variable in location. It may be almost totally intrarenal or it may be entirely outside the kidney, and yet in both cases be normal. It is also very variable in size and shape. Usually, the inferior border of the pelvis is concave but even normal pelves may show a downward bulge. The pelviureteric junction is normally funnelled but an abrupt change from pelvis to ureter may be normal. True dilatation of the pelvis suggests obstruction. Filling defects within the pelvis are discussed on p. 193.

The ureters

The ureter is usually seen in only part of its length on any one film owing to obliteration of the lumen by peristalsis. No portion of the ureter should be more than 7 mm in diameter, but the normal effects of pregnancy and the contraceptive pill can cause moderate dilatation. Dila-

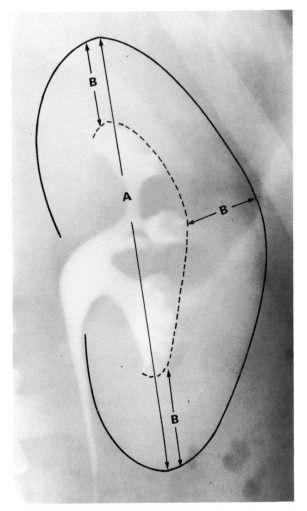

Fig. 7.10 Measurement of renal length and renal parenchymal width. The length of the kidney is obtained by measuring the distance (A) between the upper and lower poles. The distance between the calices and the renal outline (B) represents the renal parenchymal width. Note that this is greater at the poles of the kidney than at the centre.

tation is usually due to obstruction. Occasionally, it is due to vesicoureteric reflux without obstruction.

The course of the ureters is extremely variable and it may be difficult to tell if they are displaced, but:

1. A pronounced local deviation of the ureters suggests displacement by an extrinsic mass. (Fig. 7.12)

2. A ureter lying medial to the pedicle of one of the vertebrae is likely to be displaced by a mass or pulled by retroperitoneal fibrosis.

The bladder

The bladder is a centrally located structure which should have a smooth outline. It often shows normal indentations from above due to the uterus and the sigmoid colon, and from below by muscles of the pelvic floor. After micturition the bladder should be empty apart from a little contrast trapped in the folded mucosa.

SPECIAL TECHNIQUES

Retrograde pyelogram

A catheter is placed at cystoscopy into the ureteric orifice and can then be passed a variable distance along the ureter. Alternatively, a bulb-tipped catheter can be inserted snugly into the ureteric orifice, but not passed up the ureters. A standard urographic contrast medium is then injected via the catheter to outline the collecting system. This is best done under fluoroscopic control, so that the optimum volume can be injected and areas of interest can be filmed repeatedly, to determine the constancy of the findings.

The indications for retrograde pyelography are limited to those situations where the information cannot be achieved by intravenous urography: notably the 'non-functioning kidney' and cases of hydronephrosis where further information about the nature of the obstruction is required.

Retrograde pyelography has many disadvantages. It carries the risk of introducing infection with possible septicaemia, of producing obstruction secondary to oedema of the ureteric orifice and of extravasation of contrast and urine due to overdistension at the time of injection. It also provides limited information, since there is no opacification of the renal parenchyma; and in cases with obstruction it may only be possible to see the lower end of the responsible lesion.

With the proper use of adequate doses of intravenous

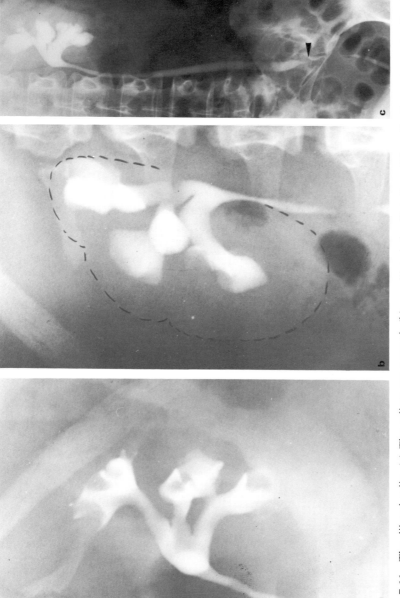

Fig. 7.11 The dilated calix. (a) The calices are normal; (b) many of the calices are clubbed. There is scarring of the parenchyma of the upper half of the kidney indicating that the diagnosis is chronic pyelonephritis; (c) all the calices are dilated, the dilatation of the collecting system extending down to the point of obstruction (arrow), in this case due to a malignant retroperitoneal lymph node.

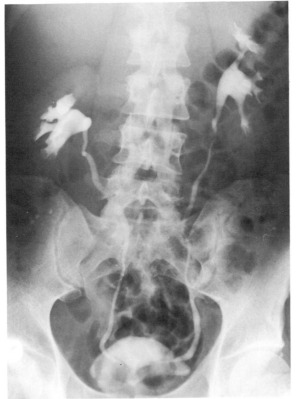

Fig. 7.12 Medial deviation of the mid portion of the right ureter by a retroperitoneal mass of lymph nodes (Hodgkin's disease). The course of the left ureter is normal.

contrast media and delayed films it is rarely necessary to do a retrograde pyelogram.

Percutaneous antegrade pyelography and percutaneous nephrostomy

This procedure is an alternative, often a preferable one, to retrograde pyelography. The commonest indication is in the diagnosis of an obstructed kidney. The examination is preceded by an intravenous injection of a large dose of urographic contrast medium in order to try and define the position of the kidney. Alternatively, the kidney may be localised using ultrasound and the needle introduced under ultrasonic guidance. A needle with a

closely fitting teflon sheath is passed through the patient's back directly into the pelvicaliceal system. Urine specimens are obtained and water soluble contrast medium injected. The procedure is easily accomplished under local anaesthesia.

The pelvicaliceal system is demonstrated and detailed films of any obstructing lesion are taken. If indicated, the teflon sheath can be left in position for drainage purposes. Drainage is sometimes the prime indication for performing percutaneous puncture of an obstructed pelvicaliceal system ('percutaneous nephrostomy').

Cyst puncture

When a cyst is diagnosed on IVU and ultrasound it is possible to confirm the diagnosis by percutaneous needle puncture. Cyst puncture is readily performed under local anaesthesia on out-patients.

When performed under fluoroscopic control an intravenous injection of urographic contrast medium is given in order to outline the pelvicaliceal systems and demonstrate the position of the cyst. When ultrasound guidance is used no preliminary contrast injection is required. A fine needle with a closely fitting teflon sheath is passed through the back directly into the cyst. Cyst fluid is aspirated, and analysed for abnormal cells. The fluid in a benign cyst is a clear urine-like liquid. Contrast medium is then injected into the cyst and films are taken.

Micturating cystogram

Contrast medium is run into the empty bladder through a catheter. The bladder is filled to capacity and films are taken during micturition. The entire process is observed fluoroscopically so that any vesicoureteric reflux can be observed and bladder contractility assessed.

The major indications are:

1. To identify and quantify vesicoureteric reflux.
2. To demonstrate the emptying of the bladder and the control of micturition.
3. To investigate the anatomy of the bladder neck and of the urethra, particularly to show obstructions such as strictures or urethral valves.

4. Occasionally, to assess bladder tumours and diverticula.

Urethrography (Fig. 7.13)

Urethrography can be performed as part of the investigation of micturating cystography or by a retrograde injection technique. In the retrograde injection technique contrast medium, either an iodine-containing jelly or a conventional water soluble medium, is injected under fluoroscopic control via a cannula which fits tightly into the meatus of the penile urethra (the retrograde technique is rarely performed in females).

The usual indications for the examination are the demonstration of strictures and extravasation following trauma.

Renal arteriography

Renal arteriography is performed via a catheter intro-

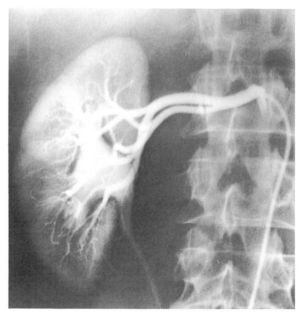

Fig. 7.14 Normal selective right renal arteriogram. Note that not only are the arteries well shown but there is also an excellent nephrogram. The renal pelvis and ureter are opacified because of a previous injection of contrast.

duced into the femoral artery by the Seldinger technique (see p. 5). The usual technique involves injecting contrast into the aorta just above the origin of the renal arteries as well as selective injections into one or both renal arteries (Fig. 7.14). The usual indications are to diagnose the nature and extent of a renal mass suspected to be a tumour, and the assessment of renal artery stenosis.

CALCIFICATION IN THE URINARY TRACT

Calcification in the urinary tract has numerous causes, calculi being by far the commonest. There are five main categories of calcification:

1. Calculi.

2. Nephrocalcinosis.

3. Localised renal parenchymal calcification.

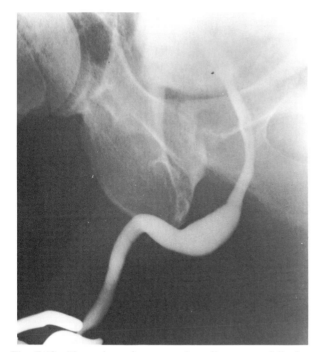

Fig. 7.13 Normal urethrogram. A radio-opaque jelly has been injected retrogradely to distend and opacify the urethra.

4. Bladder wall calcification.
5. Prostatic calcification.

Calculi. Nearly all urinary calculi are calcified and appear partly or totally opaque. Only pure uric acid and xanthine stones are radiolucent. Many are uniformly calcified but some, particularly bladder stones, are laminated (Fig. 7.15). Small renal calculi are often round or oval; the larger ones frequently assume the shape of the pelvicaliceal system in which they lie and are known as stag-horn calculi (Fig. 7.16).

It is essential to examine the plain film carefully for calcified calculi, because even large calculi can be completely hidden within the opacified collecting system once contrast has been given (Fig. 7.5, p. 183).

Stones may cause obstruction to the collecting system and the IVU is the best method of determining that obstruction exists and showing its precise site. The IVU is also important in identifying any mechanical obstruction or deranged anatomy responsible for stasis, that could predispose to stone formation.

Nephrocalcinosis is the term used to describe numerous irregular spots of calcium in the parenchyma of both kidneys (Fig. 7.17). The causes fall into two main categories:

1. Nephrocalcinosis associated with hypercalcaemia and/or hypercalcuria: namely hyperparathyroidism, renal tubular acidosis and sarcoidosis.
2. Nephrocalcinosis *not* associated with disordered calcium metabolism: namely medullary sponge kidney (p. 211); and widespread papillary necrosis (p. 213).

Localised renal parenchymal calcification occurs in:

1. Renal tuberculosis (Fig. 7.18). In these cases: there will be other evidence to support the diagnosis on the IVU (see p. 212).
2. Tumours. A small proportion of renal carcinomas calcify (Fig. 7.19). When this happens it is always possible to appreciate that the calcification is in a renal mass. (Calcification in the wall of a benign cyst is exceedingly uncommon.)

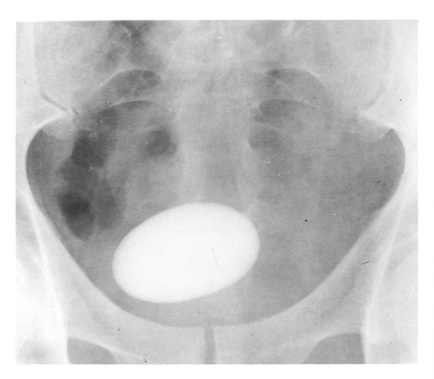

Fig. 7.15 Bladder calculus.

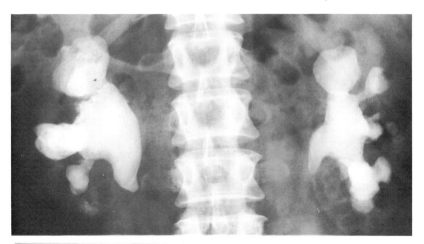

Fig. 7.16 Plain film showing a calcified staghorn calculus in each kidney.

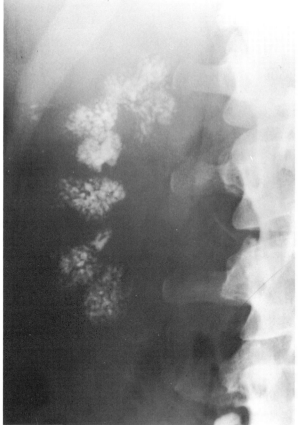

Fig. 7.17 Nephrocalcinosis. There are numerous calcifications in the pyramids of both kidneys (the left kidney is not illustrated).

Bladder wall calcification. Calcification in the wall of the bladder is rare. When it is seen it is either due to schistosomiasis or to calcium deposited on the surface of a bladder tumour. With the latter, the filling defect of the tumour, once the bladder is opacified at IVU, is usually much more obvious than the calcification on the plain film.

Prostatic calcification is due to numerous prostatic calculi. It is so common that it can be regarded as a normal finding in older men. It shows no correlation with the symptoms of prostatic hypertrophy nor any relation to prostatic carcinoma. It always appears as numerous flecks of calcification of varying size, approximately symmetrical about the midline just beneath the bladder shadow (Fig. 7.20).

FILLING DEFECTS WITHIN THE COLLECTING SYSTEMS

The two common causes of a filling defect in the opacified urinary tract are calculi and tumours. Blood clot and pyeloureteritis cystica are rarer causes.

Calculi

Nearly all urinary stones contain visible calcification

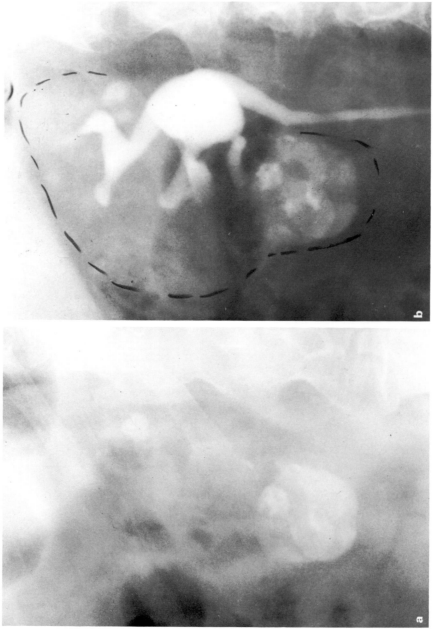

Fig. 7.18 Renal parenchymal calcification due to tuberculosis. (a) Plain film. (b) After contrast.

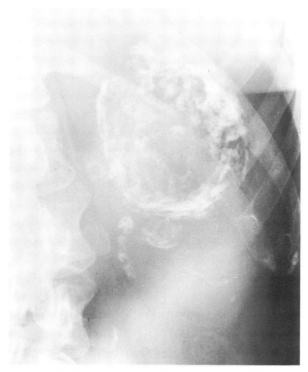

Fig. 7.19 Renal cell carcinoma. This is partially calcified.

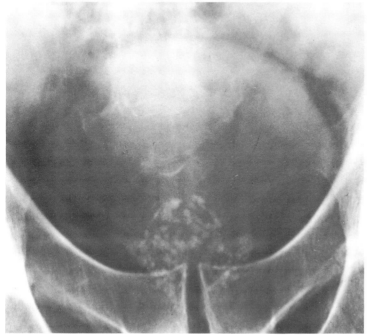

Fig. 7.20 Prostatic calcification. Numerous calculi just above the pubic symphysis are present in the prostate.

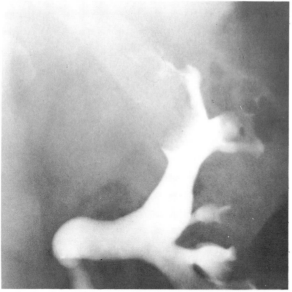 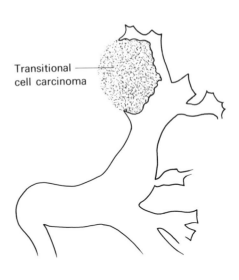

Fig. 7.21   Filling defect in an upper calix due to transitional cell carcinoma.

and virtually all calcified filling defects are stones. Therefore, provided the preliminary film is reviewed, filling defects due to stones very rarely present a diagnostic problem.

Tumours

Virtually all the tumours that grow within the collecting systems of the kidneys are *transitional cell carcinomas*. In the pelvicaliceal system they are seen as lobulated or, very occasionally, as fronded filling defects either projecting into the lumen or completely surrounded by contrast (Fig. 7.21). It is easy to confuse such tumours with overlying gas shadows. However, gas shadows often vary during the course of the examination and usually can be traced beyond the lumen of the urinary tract. Tomography may be necessary to solve the problem.

Urothelial tumours may obstruct the ureter in which they are growing, in which case it is usually only possible to determine the site of obstruction. Further information can be obtained at retrograde or antegrade pyelography, which allow one to see the true extent of the tumour and to look for further tumours, since these tumours often occur in multiple sites. Arteriography plays little part in diagnosis as these tumours are usually small and relatively avascular.

Blood clot

Blood clot from any cause of profuse bleeding will result in a filling defect. The diagnosis in such cases rests on knowing that the patient has severe haematuria and noting the smooth outline of the filling defect (Fig. 7.22). Sometimes the distinction between tumour and clot is difficult. If clot is suspected it is important to look for the cause of bleeding elsewhere in the urinary tract. Repeat examination after some days will usually show change or clearing if the filling defect is due to blood clot.

Pyeloureteritis cystica

In this condition there are numerous translucent fluid-containing cysts in the submucosa of the collecting systems and bladder. The cause is unknown but there is a

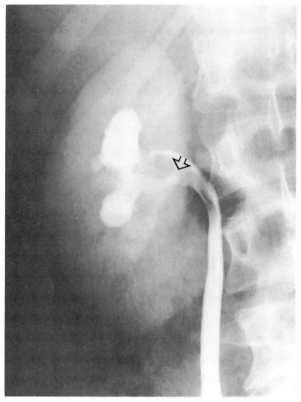

Fig. 7.22 Filling defect due to blood clot in the pelvis and upper ureter (arrow).

definite correlation with urinary tract infection (Fig. 7.23).

URINARY TRACT OBSTRUCTION

The principal feature of obstruction is dilatation of the collecting system. All the affected calices are dilated to approximately the same degree, the degree depending on the chronicity as well as the severity of the obstruction. The obstructed collecting system is dilated down to the level of the obstructing pathology (Fig. 7.24); demonstrating this level is a prime objective of radiological investigation.

Opacification of the urine in an obstructed system following an intravenous injection of contrast is usually delayed. Delayed films are, therefore, an essential part of any IVU where the level of obstruction is not shown on the routine films. Even if the flow of urine has stopped completely, providing glomerular filtration is still occurring, the collecting system will opacify in an IVU, but it may take many hours. In acute obstruction—usually due to a stone in the ureters—contrast accumulates and is concentrated in the tubules producing a very dense nephrogram. The pyelogram phase is greatly delayed, but in time the collecting systems and the level of obstruction can be demonstrated (Fig. 7.25).

Prolonged obstruction causes atrophy of the kidney substance which is recognised by observing the reduction in renal parenchymal width (Fig. 7.26).

In certain situations obstruction is intermittent. Such patients usually have renal colic and the IVU may be normal between attacks. If, however, the urogram is performed during an attack of colic (the so-called 'emergency pyelogram') the level of obstruction is nearly always demonstrated.

Clearly, if there is no concentration of contrast by the obstructed kidney, or if the concentration is poor, it may not be possible to diagnose the level of obstruction with certainty. In these cases a retrograde or antegrade pyelogram is undertaken.

Once the site of obstruction is established:

1. The plain film must be reviewed to confirm or exclude an opaque calculus responsible for the obstruction.
2. Occasionally, it will be possible to detect a filling defect in the contrast and so make a diagnosis of tumour or blood clot.
3. If the ureters are deviated at the site of obstruction the cause must lie outside the ureters, and retroperitoneal disorders will have to be considered.

Often none of these diagnostic clues are present and the possibilities will then depend on site alone.

Obstruction to the ureters and pelvicaliceal systems

There are many causes of obstruction to the ureters and pelvicaliceal systems. Calculi, tumours and strictures

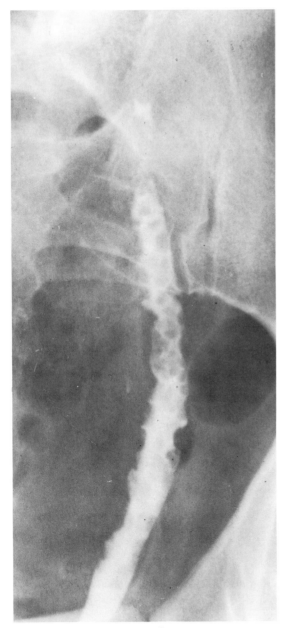

Fig. 7.23 Pyeloureteritis cystica showing small smooth spherical indentations projecting from the wall of the ureter.

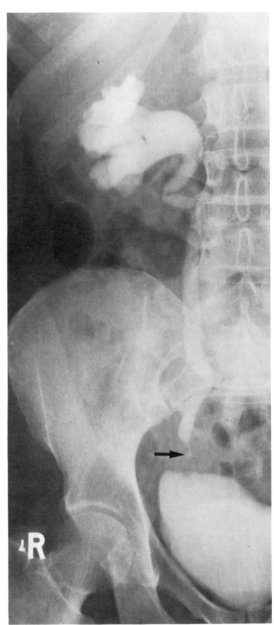

Fig. 7.24 Ureteric obstruction. The pelvicaliceal system and ureter are dilated down to the level of the obstructing pathology (arrow)—in this instance a small calculus.

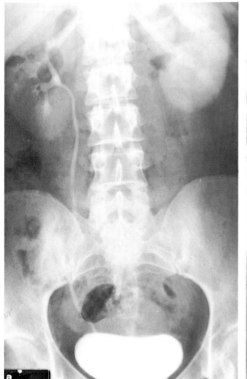
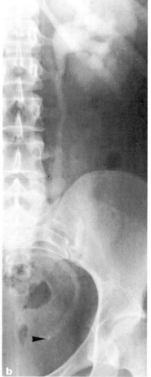

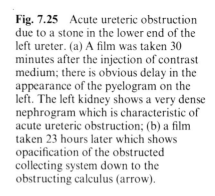

Fig. 7.25 Acute ureteric obstruction due to a stone in the lower end of the left ureter. (a) A film was taken 30 minutes after the injection of contrast medium; there is obvious delay in the appearance of the pyelogram on the left. The left kidney shows a very dense nephrogram which is characteristic of acute ureteric obstruction; (b) a film taken 23 hours later which shows opacification of the obstructed collecting system down to the obstructing calculus (arrow).

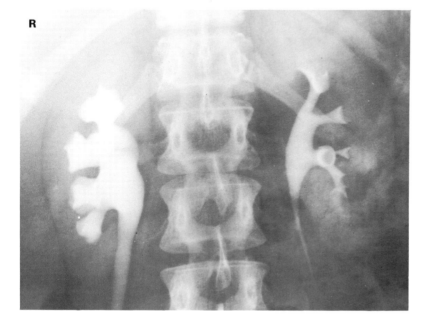

Fig. 7.26 Obstructive atrophy. The right kidney is small with a smooth outline due to reduction in renal parenchyma and shows dilatation of the collecting system. This loss of renal tissue was due to obstruction by a stone in the ureter.

may be responsible for obstruction at any level. Obstruction at the pelviureteric junction is usually due to the congenital disorder 'intrinsic pelviureteric junction obstruction'. Aberrant renal artery, retroperitoneal fibrosis and tumours are further causes of ureteric obstruction.

Calculi

Calculi are by far the commonest cause of obstruction to the urinary tract. A calcified opacity will usually be visible on the plain film, but since parts of the ureter overlie the transverse processes of the vertebrae and the wings of the sacrum, the calculus may be impossible to see.

Tumours

1. Transitional cell carcinomas of the calices or pelvis usually produce a recognisable filling defect. But if the tumour is in the ureter it may be impossible to diagnose the cause of the obstruction from the IVU. This requires retrograde or antegrade pyelograms. Carcinoma of the bladder causing ureteric obstruction can usually be easily identified as a filling defect in the bladder.
2. Renal parenchymal masses may narrow and obstruct a calix or the pelvis but the explanation is usually obvious because they cause other displacements as well.

Infective strictures

These are mostly due to tuberculosis or schistosomiasis. In the case of tuberculosis there is usually other evidence on the IVU to suggest the diagnosis (p. 212).

Congenital intrinsic pelviureteric junction obstruction

In this disorder peristalsis is not transmitted across the pelviureteric junction, i.e. a functional obstruction exists, but there is no naked-eye evidence of the cause. The disease may present at any age but it is usually discovered in children or young adults. The radiological diagnosis depends on identifying dilatation of all the calices and of the pelvis, with an abrupt change in calibre

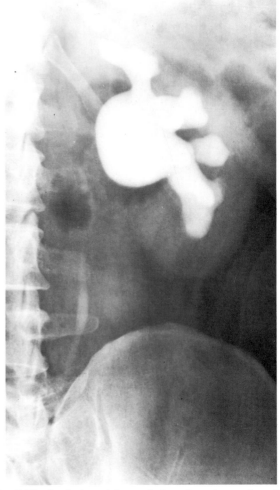

Fig. 7.27 Intrinsic pelviureteric junction obstruction. The pelvicaliceal system is considerably dilated, but the ureter from the pelviureteric junction onward is normal in calibre.

at the pelviureteric junction (Fig. 7.27). Often the ureter cannot be identified at all on the IVU. If it is seen, it will be either narrow or normal in size. The differential diagnosis includes the other processes described in this section which cause obstruction of the pelviureteric junction.

The branches of the renal artery

These branches pass very close to the pelvis and the necks of the calices, occasionally causing obstruction.

Retroperitoneal disease

1. Tumours. Carcinoma of the cervix and rectosigmoid junction and malignant lymph node enlargement are frequent causes in this category. The ureters may be visibly deviated by such tumours but frequently the ureteric course is normal. Because some of these tumours originate in the midline or are bilateral, both ureters may be obstructed.

2. Retroperitoneal fibrosis. In most cases no cause can be found for this fibrosis which encases the ureters and causes obstruction. When first seen only one side may be obstructed but eventually the condition becomes bilateral. The affected ureters may be displaced medially. The obstruction is usually at the L.4/5 level.

An important diagnostic feature is the ease with which the ureteric catheter passes up the affected ureters at retrograde pyelography.

Obstruction to the outflow tract of the bladder

Such obstruction may be caused by benign or malignant enlargement of the prostate, urethral strictures, bladder tumours and, in little boys, by posterior urethral valves. Patients with neurological deficit may have neurogenic obstruction to bladder emptying.

All these conditions may give rise to bilateral hydronephrosis and sometimes to vesicoureteric reflux as well. The indication that the basic abnormality lies at the level of the bladder outflow will be dilatation of the ureters right down to the bladder with thickening and trabeculation of the bladder wall with a significant post-micturition residue. The precise findings will depend on the cause.

Ureteric dilatation in pregnancy

Pronounced unilateral or bilateral ureteric dilatation with dilatation of the pelvicaliceal systems is often seen during the second half of pregnancy, or in women taking the contraceptive pill. It may take 3 months or more to return to normal; the mechanism is uncertain. It is therefore wise to wait, if possible, for 4 months before performing an IVU after pregnancy.

THE UNILATERAL 'NON-FUNCTIONING' KIDNEY

The phrase unilateral 'non-functioning' kidney implies non-visualisation of one kidney, not as the name would suggest a total absence of function by that kidney. The phrase was originally introduced at a time when contrast dosage was much lower than it is today and before the introduction of routine tomography. Because of its convenience the term has stuck despite its inaccuracy. Total non-visualisation is rare; a nephrogram is often seen and may provide valuable diagnostic information. Delayed films should be part of the routine in those patients where the early films do not show a pyelogram on one side. They may demonstrate hydronephrosis and show the level of obstruction.

The common causes of a 'non-functioning' kidney are surgical removal of the kidney and ureteric obstruction. Other causes are renal agenesis, renal artery occlusion and renal vein thrombosis.

Ureteric obstruction causing unilateral absence of a pyelogram may be due to any of the causes discussed on p. 196. In many such cases it is possible to see a nephrogram. The dilated calices which contain unopacified urine may be seen as rounded defects in the nephrogram. If this so-called 'negative pyelogram' is seen, the diagnosis of hydronephrosis and, therefore, obstruction is certain (Fig. 7.28).

It is worth noting that renal parenchymal tumours (renal cell carcinoma and Wilm's tumour) are very rare causes of 'non-function' because they virtually never destroy sufficient parenchyma. The usual mechanism by which they produce 'non-function' is to grow into and obstruct the ureter or to produce ureteric obstruction due to blood clot from bleeding by the tumour. On the other hand transitional cell carcinomas of the pelvis, ureter and bladder are common causes of 'non-function', since they so readily obstruct the drainage of a kidney.

Most cases of unilateral non-visualisation at IVU require further investigation. These procedures usually centre around confirming or excluding obstruction to the collecting system. Radioisotope studies are most

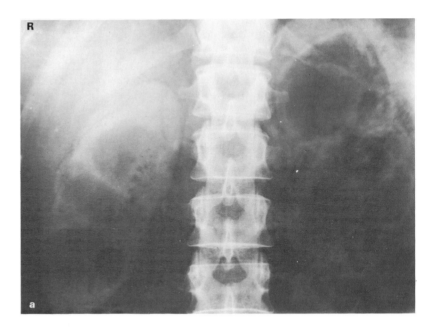

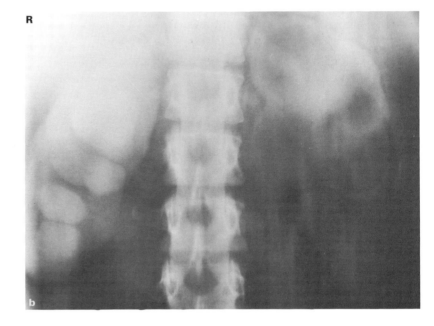

Fig. 7.28 Hydronephrosis causing a 'negative pyelogram'. There is bilateral hydronephrosis due to ureteric obstruction. (a) The dilated calices are seen as translucencies within the nephrogram (film taken 5 minutes after injection of contrast); (b) tomogram taken 10 minutes later, showing opacification of the right hydronephrosis. The left hydronephrosis is now seen as a 'negative pyelogram'. Note the severe obstructive atrophy of the renal parenchyma.

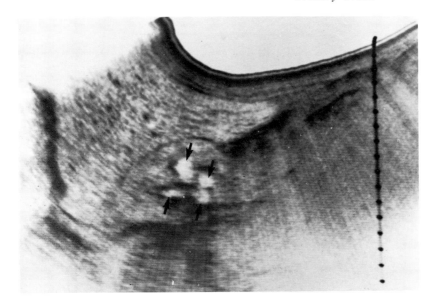

Fig. 7.29 Hydronephrosis. Ultrasound examination showing hydronephrosis of the right kidney. Longitudinal scan through right upper quadrant. The arrows point to dilated calices.

helpful in excluding renal artery occlusion or renal agenesis by demonstrating blood flow to the affected kidney and in showing the presence (though not the cause) of obstruction to the collecting system. Ultrasound examination is very helpful in establishing the presence of a hydronephrosis (Fig. 7.29). Antegrade or retrograde pyelography is often necessary to show the level of obstruction and sometimes may indicate the precise cause.

In renal agenesis the opposite kidney will show compensatory hypertrophy, providing it is normal. There is complete absence of blood flow and function on the affected side on radioisotope studies, and no renal tissue can be identified with ultrasound examination. Further confirmation of the diagnosis can be obtained, where indicated, by cystoscopy when the trigone of the bladder will be seen to be deficient and the ureteric orifice absent on the relevant side. In a few cases it may prove necessary to do an arteriogram which will demonstrate absence of the kidney and its artery.

With renal artery occlusion the kidney may be seen on plain film but no nephrogram is seen after contrast is given.

Acute renal vein thrombosis is a rare cause of non-visualisation. With time collateral venous channels open up and a visible pyelogram usually returns.

RENAL MASSES

Almost all solitary renal masses are either *malignant tumours* or *simple cysts*. In adults the malignant tumour is almost certain to be a renal cell carcinoma, whereas in young children the common neoplasm is the Wilm's tumour. Because of the close anatomical relationship of the adrenal gland to the kidney, neuroblastoma can be confused with Wilm's tumour in children.

Occasionally, invagination of normal cortical tissue into the renal medulla (sometimes called a 'renal pseudotumour') may produce the signs of a localised mass. Here the distinguishing feature is uniform opacification of the 'mass' on high dose nephrotomography and on arteriography.

Unusual causes of a renal mass include: renal abscess, hydatid cyst, benign tumours and metastasis. Multiple renal masses usually indicate: multiple simple cysts, polycystic disease (see p. 210) or malignant lymphoma.

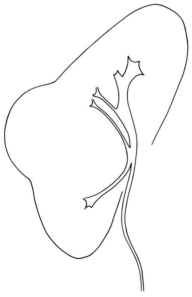

Fig. 7.30 Signs of a renal mass. The renal outline is bulged and there is a corresponding displacement of the calices. (This diagram is a drawing of the IVU on the patient whose cyst puncture is illustrated in Fig. 7.31).

The basic signs of a renal mass on an IVU are (Fig. 7.30):

1. Bulging of the renal outline. Sometimes the outline is indistinct so that the bulge cannot be appreciated.
2. Displacement of major and minor calices.
3. Increase in renal size, particularly if the mass is at the upper or lower pole where its bulk adds to the renal length.

Having diagnosed the presence of a mass an attempt may be made to diagnose its nature:

Signs suggesting a cyst

1. No opacification of the mass.
2. Clear-cut border between the mass and the adjacent parenchyma.
3. A claw of parenchyma pulled out by the mass.

Signs suggesting a malignant tumour

1. Opacification of any part of the mass.
2. Indistinct border between the mass and the adjacent

nephrogram on a good quality nephrotomogram.
3. Any solitary mass in a young child.
4. Calcification in the mass—particularly if the calcification is more than just a thin line at the periphery (Fig. 7.19, p. 195).

With good quality nephrotomography it is possible to be 85–90% accurate in diagnosing the nature of a renal mass; however, the main object of the IVU is to detect renal masses since ultrasound can improve the accuracy to 95% or more.

Ultrasound in renal masses

A cyst shows clear echoes from the front and back wall with no intervening echoes and enhanced echoes behind the cyst (Fig. 7.31a). A solid lesion also shows echoes from front and back wall but there are also numerous echoes arising within the mass (Fig. 7.31b).

Cyst puncture

If a mass is thought to be a cyst, puncture can be performed to confirm the diagnosis and exclude tumour. Cyst fluid can then be obtained for cytology and contrast medium injected to check that the walls are smooth and that the size of the cyst corresponds to the size of the mass diagnosed on the IVU (Fig. 7.32).

Renal arteriography in renal masses

If on clinical, urographic and ultrasound evidence a renal mass is thought to be a malignant tumour, arteriography is the next investigation. This involves a free flush aortogram which, in addition to showing the mass, shows the anatomy of the renal arteries, the state of the opposite kidney and may detect metastases outside the kidney. The free flush aortogram is followed by a selective renal arteriogram which provides details of the renal mass and may demonstrate the patency of the renal veins.

Arteriographic signs of tumours versus cyst

There will be displacement of blood vessels round the

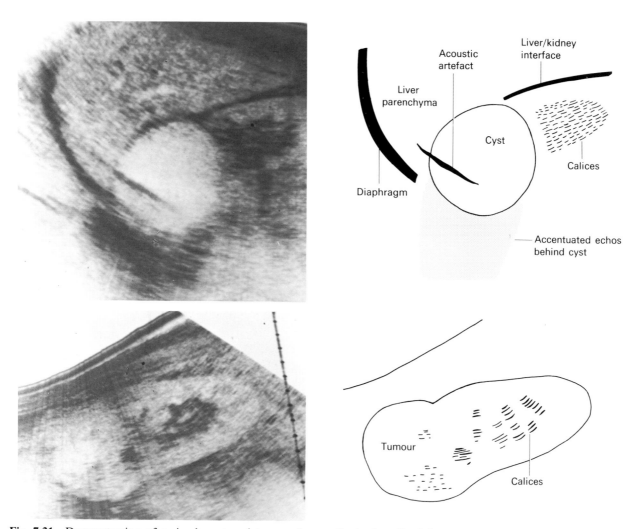

Fig. 7.31 Demonstration of a simple cyst and tumour by ultrasound. (*above*) Simple renal cyst (anterior scan); (*below*) renal carcinoma (posterior scan) shown by ultrasound. The renal cyst is trans-sonic with enhancement of the echoes from the back wall of the cyst. The renal tumour shows echoes arising from within the mass and there is no accentuation of the echoes from the posterior wall of the tumour.

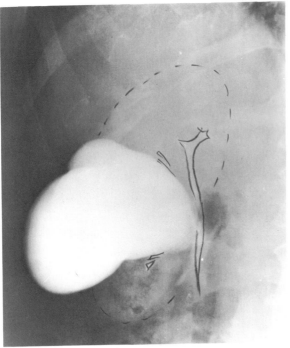

Fig. 7.32 Benign cyst diagnosed with percutaneous cyst puncture and contrast opacification. The cyst contained clear fluid and shows a smooth outline completely accounting for the mass seen in the IVU (Fig. 7.30).

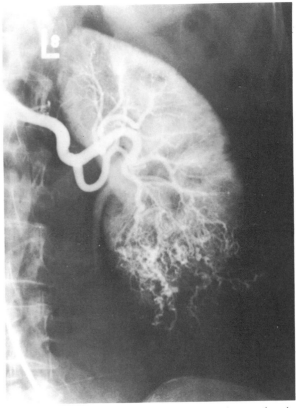

Fig. 7.33 Renal arteriogram in renal cell carcinoma showing a lower pole tumour with numerous abnormal vessels within the mass. The remainder of the kidney is normal.

mass whatever its nature. Malignant tumours usually show a so called 'pathological circulation', namely new blood vessels within the tumour mass. Contrast often pools within these abnormal vessels (Fig. 7.33). Sometimes no discrete vessels are seen but parts of the tumour opacify. Cysts are always radiolucent and never contain vessels within them, though normal vessels may be superimposed. Tumours show an ill-defined border between the mass and the adjacent nephrogram whereas with a cyst a clear-cut border is seen.

Ninety-five per cent of malignant tumours in the kidney can be easily diagnosed on renal arteriography. The remainder are relatively avascular and resemble cysts. Such avascular masses should be punctured for final diagnosis.

CONGENITAL VARIATIONS

Congenital variations in the anatomy of the urinary tract are very common. They are of many types, too numerous to discuss in detail. Only the commoner anomalies are discussed here (congenital pelviureteric junction obstruction is discussed on p. 200, and renal agenesis is discussed on p. 203).

Bifid collecting systems (Fig. 7.34)

Bifid collecting systems are the commonest congenital variations. The condition may be unilateral or bilateral. The pelvicaliceal systems and ureters may be divided for

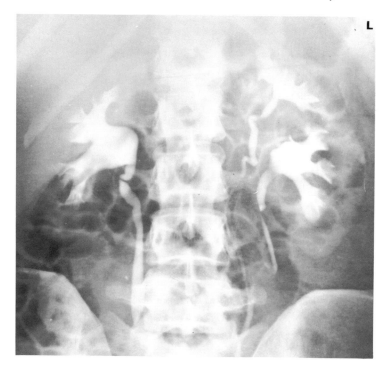

Fig. 7.34 Bifid collecting systems. There is a bifid collecting system on the left with the two ureters joining a the levels of the transverse process of L5. Note how the left kidney is larger than the right.

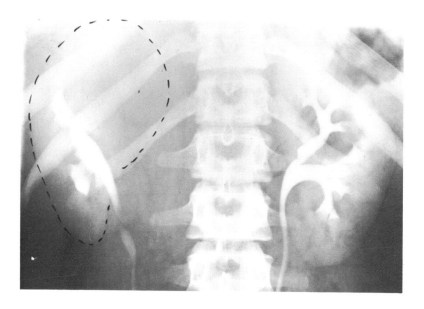

Fig. 7.35 Obstructed ectopic ureterocele. There is a bifid collecting system on the right. The upper moiety is obstructed and dilated causing deformity of the lower moiety. The obstructed moiety does not opacify.

a variable distance. Sometimes just the pelvis is bifid; an anomaly of no importance. At the other extreme the two ureters may be separate throughout their length and have separate openings into the bladder, the ureter draining the upper moiety crossing the ureter draining the lower moiety to insert lower in the bladder closer to the bladder neck. The ureter draining the upper moiety may open into the bladder in an abnormal site or may drain outside the bladder, e.g. into the vagina or urethra, producing incontinence if the opening is beyond the urethral sphincter. Such ureters, known as ectopic ureters, are frequently obstructed (Fig. 7.35) and lead to dilatation of the entire moiety. The lower end of the dilated ureter often causes a smooth filling defect in the bladder.

Kidneys with bifid collecting systems are up to 2 cm longer than the normal opposite side.

Ureterocele

A ureterocele is a dilatation of the lower ureter due to a narrow ureteric orifice. There are two forms:

1. The 'ectopic ureterocele' is always associated with substantial obstruction. It occurs in patients with complete ureteric duplication, the ureter draining the upper moiety being the one with an ectopic insertion. The dilated distal ureter may bulge into the bladder (Fig. 7.35).

2. In 'simple ureteroceles' the ureter opens into the bladder. It is dilated, often only at its lower end, and it prolapses into the bladder lumen (Fig. 7.36). Most simple ureteroceles are of little or no significance to the patient. They are usually first discovered in adults.

Ectopic kidney (Fig. 7.37)

During fetal development the kidneys ascend within the abdomen. An ectopic kidney results if this ascent is halted. Such kidneys are usually in the lower abdomen and rotated so that the pelvis of the kidney points forward. The ureter is short and travels directly to the bladder. Chronic pyelonephritis, hydronephrosis and calculi are all more common in ectopic kidneys, but

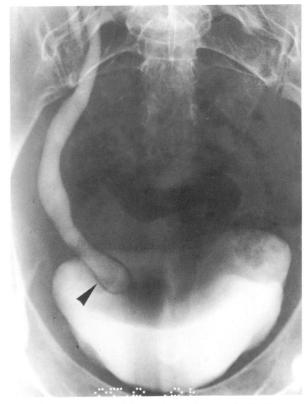

Fig. 7.36 Ureterocele in a patint with ureteric duplication. One of the duplicated ureters is dilated at its lower end and has prolapsed into the bladder. The wall of the prolapsed ureter (arrow) is seen as a lucent line between contrast in the bladder and contrast within the lumen of the ureter.

ectopic kidneys are often incidental findings of no consequence to the patient, except as a cause of diagnostic confusion. There are cases on record where laparotomies have been performed for a lower abdominal mass, where the surgeon discovers to his surprise that the patient just has an innocent ectopic kidney. Such mistakes are avoided if an IVU is performed preoperatively in such patients.

The rotated kidney

One or both kidneys may point forward instead of medially. The pelvis and calices are then seen more on

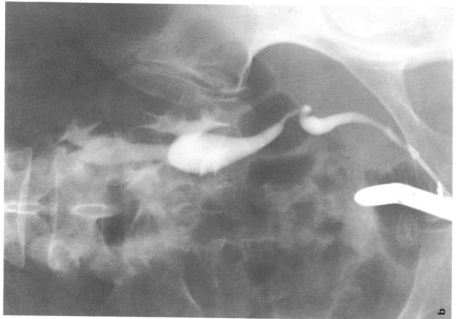

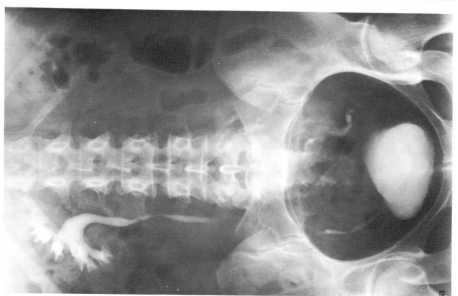

Fig. 7.37 Ectopic kidney. This congenitally ectopic kidney lies within the pelvis. (a) On the IVU: the left kidney is not seen in its expected position; it can, however, be identified in the pelvis. As is often the case the ectopic kidney could easily be overlooked; (b) a retrograde examination demonstrates a short ureter, the length being appropriate to the position of the kidney, not redundant as it would be were it displaced from its normal position by a mass.

end than usual. This anomaly is of no significance but it may be confused with displacement by a retroperitoneal mass.

Horseshoe kidney

The common form of fusion is fusion of the lower poles giving rise to a horseshoe kidney. The radiological features are shown in Fig. 7.38.

The condition may be an incidental finding and of no significance, but obstruction to the collecting systems and calculi formation are fairly common.

Sometimes the kidneys remain fused and both lie on the same side of the abdomen. This is known as *crossed fused ectopia*.

Cystic disease of the kidneys

There are many varieties of cystic renal disease varying from simple cysts (see p. 203), which may be single or multiple, to complex renal dysplasias. The commonest complex dysplasia encountered in clinical practice is polycystic disease; a familial disorder which though con-

genital in origin usually presents between the ages of 35 to 55 years with features of hypertension, renal failure or haematuria, or by the discovery of bilaterally enlarged kidneys. The reason for the late presentation is that the cysts are initially small and do not cause trouble for a long time. As the patient gets older the cysts enlarge causing compression of the adjacent renal substance, rendering it ischaemic, often resulting in hypertension and interfering with function causing renal failure.

The radiological signs of polycystic disease are (Fig. 7.39):

1. Both kidneys are very enlarged with poorly visualised outlines. Since the kidney is now largely composed of cysts the renal outline can no longer be identified. The recognition of the increased size depends largely on noting the increased distance between the calices.

2. The calices are grossly stretched and distorted by the cysts.

3. On high dose nephrotomography the nephrogram may show multiple lucencies due to the cysts.

The diagnosis can be readily confirmed by ultrasound examination

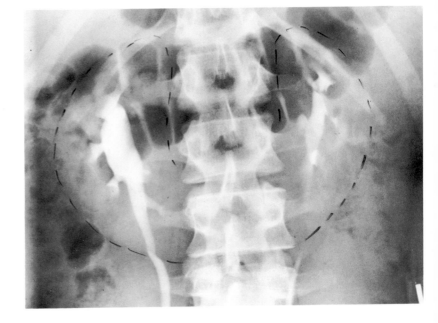

Fig. 7.38. Horseshoe kidneys. The two kidneys are fused at their lower poles. The striking feature is the alteration in the axis of the kidneys; the lower calices are closer to the spine than the upper calices. The kidneys are rotated so that their pelves point forward and the lower calices point medially. The medial aspects of the lower poles cannot be identified.

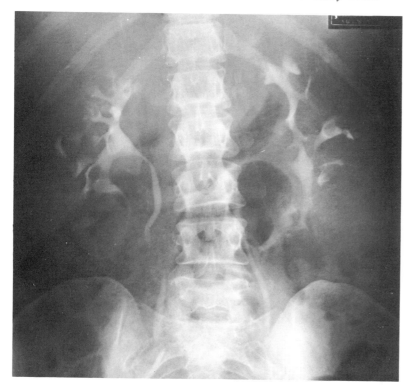

Fig. 7.39 Polycystic disease. The kidneys are greatly enlarged and the renal outlines cannot be identified. The pelvicaliceal systems are stretched and deformed by innumerable cysts.

Medullary sponge kidney

Medullary sponge kidney is the name given to a condition with cystic dilatation of the collecting tubules in the pyramids of the kidneys. The pyramids are enlarged and the contrast pooling in the dilated tubules looks like a collection of dots, rarely more than 3 mm in diameter, within the papillae. Calculi may form within the dilated portions of the tubules and on plain film the appearances are then indistinguishable from other causes of nephrocalcinosis.

Unless calculi are present medullary sponge kidney is asymptomatic and of no consequence to the patient. If calculi have formed the symptoms and problems become those of calculous disease.

ACUTE PYELONEPHRITIS

Acute pyelonephritis is usually due to bacterial infection from organisms which enter the urinary system via the urethra. Anatomical abnormalities such as stones, duplex systems and obstructive lesions, and conditions such as diabetes mellitus, predispose to infection.

In most cases the IVU is normal even during the acute attack. In very severe cases with suppuration in the kidney the caliceal system is compressed by the swelling of the renal substance and the concentration of contrast on the affected side is reduced. In a few cases no pyelogram may be seen at all, but a retrograde pyelogram shows a normal collecting system.

Renal and perinephric abscesses are usually difficult to diagnose. Abscesses are recognised by the presence of a mass, the clinical setting suggesting the nature of the mass but in many cases there is pre-existing hydronephrosis or calculous disease leading to poor visualisation of the kidney at urography. Ultrasound examination is a much better method than the IVU for diagnosing these abscesses.

TUBERCULOSIS

Urinary tuberculosis follows blood-borne spread of *Mycobacterium tuberculosis*, usually from a focus of infection in the lung. A normal chest film, however, does not exclude the diagnosis, since the tuberculous lesion in the lung is often too small to be identified.

The tubercle bacilli infect the cortex of both kidneys; most foci heal, but one or more infected sites may enlarge and coalesce to form abscesses in the cortex. These ulcerate into the renal pelvis and the infection may then spread to involve other portions of the urinary and genital systems. Multiple sites of involvement are commonly seen radiologically; an observation of great diagnostic value, since tuberculosis is by far the commonest disease to show radiological findings in the kidney, ureter and bladder simultaneously. Inflammatory oedema may progress to stricture formation, the common sites of which are: the necks of the calices, the pelviureteric junction and the lower ureter. If the obstruction is severe, total destruction of the kidney may take place leading to the so-called 'autonephrectomy' (Fig. 7.40).

In the early stages of the disease, even though tubercle bacilli can be recovered from the urine, the IVU may be normal. Signs that develop in the later stages of urinary tuberculosis are:

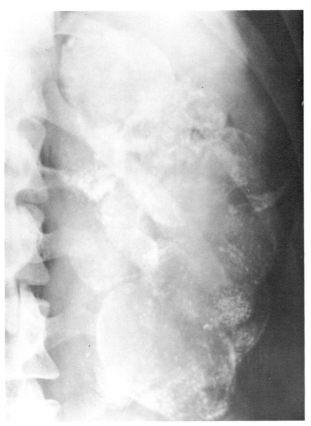

Fig. 7.40 Tuberculous autonephrectomy. Plain film showing calcification of a hydronephrotic kidney. The shape of the grossly dilated calices can be made out.

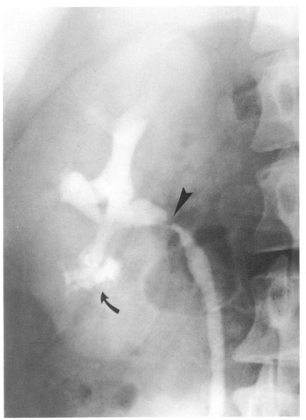

Fig. 7.41 Tuberculosis showing irregularity of the calices (curved arrow) and stricture formation of the pelvis (arrow head).

Plain film

Calcification is common (Fig. 7.18). Usually, there are one or more patches of irregular calcification, but in advanced cases with long-standing tuberculous pyonephrosis and 'autonephrectomy' the majority of the kidney and hydronephrotic collecting system are calcified. Calcification implies healing but does not mean that the disease is inactive. Indeed, the opposite is true; active infection is almost invariably present when untreated calcified tuberculous lesions are first seen.

After contrast

The earliest change is irregularity of a calix. Later a definite contrast-filled cavity is seen adjacent to the calix (Fig. 7.41). With progression of the disease the cavities enlarge and the renal substance is destroyed producing loss of parenchymal width.

Strictures of any portion of the pelvicaliceal systems or ureters may occur, producing dilatation of one or more calices. If the obstruction is severe non visualisation of the obstructed calices results. When localised, the dilated calices may then mimic a renal mass.

If the bladder is involved the wall is irregular due to inflammatory oedema; advanced disease causes fibrosis resulting in a thick-walled small volume bladder. Multiple strictures may be seen in the urethra.

CHRONIC PYELONEPHRITIS

Chronic pyelonephritis is due to reflux of infected urine from the bladder into the pyramids of the kidneys, leading to destruction and scarring of the adjacent renal substance with consequent dilatation of the calices. The upper and lower calices are the most susceptible to damage from reflux. The condition is often bilateral. Most damage is believed to occur in infancy and reflux may not be detectable on micturating cystography later in life.

The usual cause of reflux is a congenital abnormality of the ureterovesical junction, with a shorter than normal intramural section of ureter. A similar arrangement is found in the ureter draining the lower moiety of a kidney with a bifid collecting system.

The radiological signs are (Fig. 7.42):

1. Local reduction in renal parenchymal width (scar formation). The distance between the calix and the adjacent renal outline is usually substantially reduced and may be as little as 1 or 2 mm. Since there is little function in these scarred areas the renal outline is often difficult to see. It may, therefore, be easy to miss the fact that extensive scarring exists. The only way to avoid this error is to trace the entire renal outline in every IVU.
2. Dilatation of the calix in the scarred areas is the result of destruction of the pyramid.
3. Overall reduction in renal size is partly due to local loss of renal substance and partly because the scarred areas do not grow.
4. Dilatation of the affected collecting system may be seen due to reflux.
5. Vesicoureteric reflux may be demonstrated at micturating cystography. It is believed that scarring only occurs with severe reflux forcing infected urine into the kidney substance via the collecting ducts. The severity of the reflux diminishes as the child gets older and may have ceased by the time the diagnosis is made on IVU.

PAPILLARY NECROSIS

In papillary necrosis part or all of the papilla sloughs and may be passed with the urine. There are a number of conditions with strong associations with papillary necrosis. The commonest are:

1. High analgesic intake—particularly phenacetin but also aspirin.
2. Diabetes mellitus.
3. Sickle cell disease—related to the propensity for this disease to cause vascular obstruction.
4. Infection—usually only with very severe infections.
5. Obstruction is a rare cause.

Radiological signs (Fig. 7.43)

In most cases the renal outlines are smooth or slightly

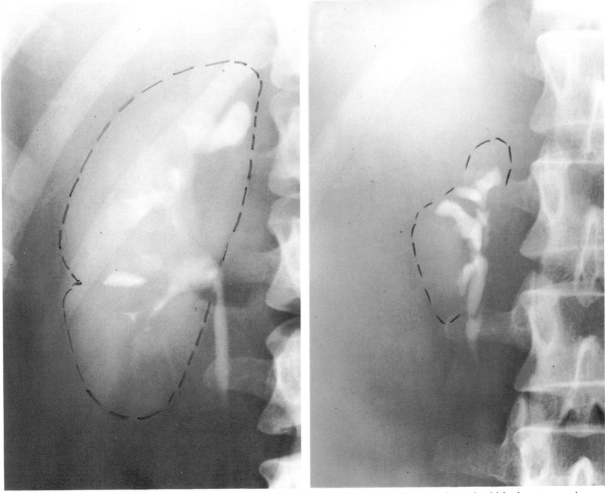

Fig. 7.42 Chronic pyelonephritis. Two examples: (a) Showing reduction in renal parenchymal width due to scarring and dilatation (clubbing) of the adjacent calices; (b) a severly shrunken kidney showing multiple scars and clubbed calices.

dimpled over the affected calices. The renal length may be reduced. The disease is usually patchy in its distribution and severity. The pattern of destruction of the papilla takes many forms. If the papilla is only partially separated contrast can be seen tracking into it, but if the papilla is totally sloughed the calix appears spherical having lost its papillary indentation. The papilla may be seen as a filling defect in a spherical calix, or may pass down the ureter.

The necrotic papilla can calcify; this occurs prior to sloughing so calcification may be seen in the site of the normal papilla, or in the sloughed papilla in the collecting system.

TRAUMA TO THE URINARY TRACT

Renal trauma

Injuries to the kidney are usually the result of a direct blow to the loin. Loin pain and haematuria are the

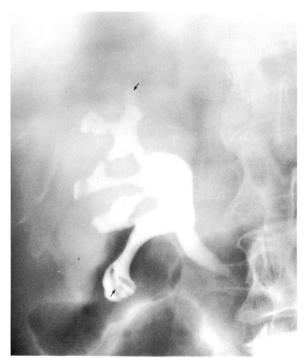

Fig. 7.43 Papillary necrosis showing dilated calices due to loss of the papillae. Some of the papillae have sloughed and appear as filling defects with the calices (lower arrow). The upper arrow points to a contrast-filled cavity within a papilla.

major presenting features. The IVU is indicated to assess damage and to ensure the normality of the opposite kidney; a point of particular importance where surgery to the injured kidney is being contemplated.

Plain films may show fractures of the vertebrae or ribs.

The appearances after contrast depend on the extent of injury. Minor injury produces swelling of the parenchyma which compresses the calices. If the kidney substance is torn the renal outline is irregular and the calices separated at the site of rupture. Extravasation of contrast may be seen (Fig. 7.44). Retroperitoneal haemorrhage may displace the kidney and may cause ureteric obstruction. If thrombosis or rupture of the renal artery occurs the kidney will be 'non-functioning'.

There is controversy over the role of renal angiography in the management of renal trauma. Arteriography is very accurate in predicting the type and severity of damage, but it is probably only necessary to carry it out if there is a good clinical or urographic indication for surgery.

Trauma to the bladder and urethra

Rupture of the bladder may result from a direct blow to the distended bladder or may be part of extensive injury such as occurs with fractures of the pelvis. If the rupture is intraperitoneal, contrast introduced into the bladder will leak out into the peritoneal cavity.

A common site of rupture is at the bladder base, in which case cystography will show elevation and compression of the bladder by extravasated urine and

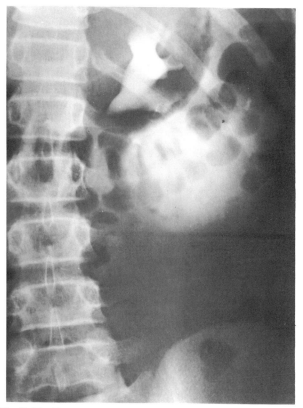

Fig. 7.44 Traumatic rupture of the kidney. The lower pole of the kidney has been ruptured and a pool of extravasated contrast can be seen.

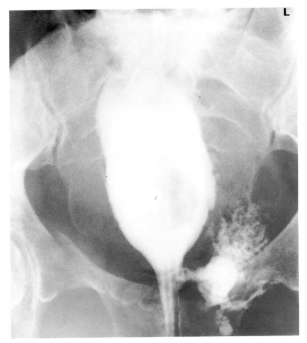

Fig. 7.45 Rupture of the base of the bladder. Cystogram showing extravasation of contrast into the extraperitoneal space on the left, and deformity of the bladder due to surrounding hematoma and urine. There is a fracture of the right pubic bone.

haematoma (Fig. 7.45). The site of extravasation is frequently demonstrated.

Damage to the urethra is a serious complication of pelvic fractures. An ascending urethrogram using a water-soluble contrast medium may show rupture of the urethra with extravasation of contrast into the adjacent tissues.

HYPERTENSION IN RENAL DISEASE

There are a large number of diseases, both renal and non-renal, that are associated with hypertension. The renal conditions include chronic glomerulonephritis, chronic pyelonephritis, renal artery stenosis, polycystic disease, polyarteritis nodosa, Kimmelstiel–Wilson kid-ney and occasionally renal masses. The feature common to all the renal conditions is a reduction in blood supply to all or part of the kidney.

In the cases of glomerulonephritis, polyarteritis nodosa and the Kimmelstiel–Wilson kidney the changes are usually bilateral uniform reduction in renal size without specific features. Essential hypertension may cause identical changes at IVU and the decision whether the small kidneys are the cause or the result of hypertension cannot be made radiologically.

Chronic pyelonephritis (see p. 213) polycystic disease (see p. 210) and renal artery stenosis all give rise to specific radiologic abnormalities.

Renal artery stenosis

Renal artery stenosis is sometimes a cause of hypertension. It is, however, found fairly commonly at post mortem or angiographically in normotensive patients. Abnormalities are seen at IVU only in those cases where the stenosis is haemodynamically significant, and even with significant renal artery stenosis the IVU may be normal.

The common causes are atheroma which usually involves the origin or proximal third of the renal artery, and fibromuscular hyperplasia (Fig. 7.46), a disease of younger people which usually leaves the proximal third of the renal artery free and causes one or more stenoses in the distal two-thirds and branches of the renal artery.

The signs of renal artery stenosis on IVU are:

1. Reduction in the size of the kidney with either a smooth or scarred outline, the calices being structurally normal.
2. Delayed visualisation of the pyelogram on the affected side.
3. Small volume pelvicaliceal system containing highly concentrated contrast. Because of the slow circulation in the renal tubules, water reabsorption is greater and, therefore, better concentration of the urine occurs compared with the normal side.
4. Notching of the ureter may be seen due to enlarged ureteric arteries acting as collaterals.

Arteriography is the best method of demonstrating

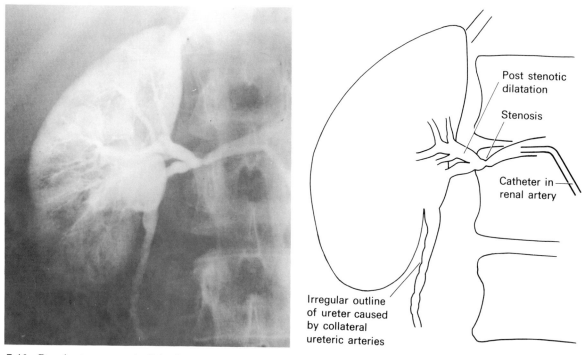

Post stenotic
dilatation

Stenosis

Catheter in
renal artery

Irregular outline
of ureter caused
by collateral
ureteric arteries

Fig. 7.46 Renal artery stenosis. Selective renal arteriogram showing a stenosis (fibromuscular hyperplasia) of the right renal artery in a 16-year-old girl. This kidney was 3 cm shorter than the opposite one.

renal artery stenosis, but even with arteriography it is not always possible to predict the haemodynamic significance of a stenosis from its angiographic appearance.

The role of IVU and arteriography in managing patients with hypertension

At one time most patients with significant hypertension had an IVU in the hope of detecting unilateral renal disease amenable to surgical treatment. However, surveys have now shown that even though routine IVU examination in adult hypertensive patients will reveal renal abnormalities in approximately 20% of cases, but in only 1% will this lead to a change in management. This low figure is mainly due to the effectiveness of modern drug therapy for hypertension.

IVU and arteriography is nowadays largely limited to children with severe hypertension and those patients whose hypertension is inadequately controlled by drugs.

UROGRAPHY IN RENAL FAILURE

Until the mid 1960s intravenous urography played little or no part in the management of patients in real failure. It was even believed by some that the contrast medium was dangerous if the patient was in renal failure. This is no longer believed to be the case. Nowadays renal failure is regarded as an important indication for IVU and, since the kidneys cannot concentrate well, large doses of contrast agent are used. The contrast medium is excreted by the liver when the patient's kidneys fail.

The technique of performing the IVU is modified. Tomography is used extensively, since the concentration of contrast is very poor; and films may be taken for many hours after injecting contrast as the passage of contrast through the collecting systems can be very slow in the patient with ureteric obstruction.

Providing prerenal renal failure is excluded on general

grounds intravenous urography permits one to draw the following conclusions:

1. Diagnose acute tubular necrosis with confidence.
2. Distinguish chronic renal failure due to intrinsic renal disease ('end-stage kidney') from that due to obstruction. This is important as surgery may be indicated to relieve the obstruction. The demonstration of a normal pelvicaliceal system rules out obstructive uropathy. Since the patient is in renal failure the visualisation of the pelvicaliceal system is bound to be poor. But even if only a few calices are seen, and these are normal in shape and size, obstruction as the cause of the renal failure can be excluded.

The ability to exclude obstruction, and by implication diagnose renal parenchymal disease, is just as important as being able to make a diagnosis of obstruction.

Acute tubular necrosis (Fig. 7.47)

In acute tubular necrosis, from whatever cause, early films during the IVU show a clear nephrogram—this is often denser than in the average normal patient. The diagnostic feature is that the nephrogram persists for up to 24 hours, without visible caliceal filling. It is important to continue for this length of time in order to exclude delayed filling of obstructed calices.

Renal failure due to obstructive uropathy (Fig. 7.48).

In obstructive uropathy the calices are dilated. The concentration of the contrast medium in the calices is usually faint, so tomograms are frequently required for adequate visualisation. The obstructed dilated calices fill late and it may be necessary to continue taking films at intervals for many hours in order to see them properly. Further films are taken to demonstrate, if possible, the level of obstruction.

Retrograde pyelography may be required in the few cases where the details obtainable from the IVU are inadequate. However, retrograde pyelography should preferably be avoided, since infection may be introduced and patients in renal failure are particularly vulnerable to urinary tract infection.

Intrinsic renal disease

The radiological features vary with the cause. Most cases have small kidneys due to reduced renal parenchyma, smooth outlines and normal calices. There are many causes for such an appearance (see Table 7.1, p. 186). Sometimes the kidneys appear structurally normal, as in chronic glomerulonephritis and sometimes the kidneys show irregular outlines as in chronic pyeloneph-

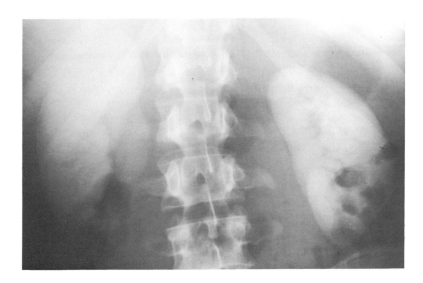

Fig. 7.47 Acute tubular necrosis. This film, taken 9 hours after injection of contrast, shows a dense persistent nephrogram with no opacification of the collecting systems.

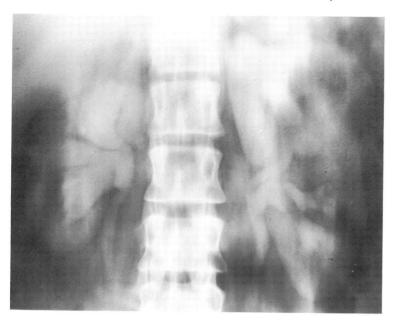

Fig. 7.48 Renal failure due to obstructive uropathy. The collecting systems of both kidneys are dilated. They are poorly seen and tomography was necessary. (In this instance the bilateral hydronephroses were due to a pelvic tumour obstructing both ureters.)

ritis. Multiple renal infarcts and papillary necrosis may be demonstrated.

THE BLADDER

Tumours

The bladder is the commonest site for neoplasms in the urinary tract. Almost all are transitional cell tumours, of varying degrees of malignancy. They vary in shape; some are delicate fronded papillary lesions, some sessile irregular masses, and others form flat plaque-like growths that infiltrate widely.

Tumours produce appropriate filling defects in the bladder shadow at IVU or cystography (Fig. 7.49). Very rarely, calcium is seen deposited on the surface of the tumour. It may be difficult to distinguish overlying gas and faeces in the sigmoid colon or rectum from a bladder neoplasm. Usually, it is possible to show on oblique views that the colonic contents lie partly outside the bladder; whereas the filling defect of a bladder tumour is of necessity always within the bladder shadow.

The nature and extent of a tumour in the bladder is

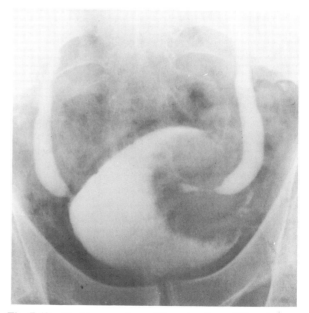

Fig. 7.49 Bladder neoplasm. There is a large filling defect in the left side of the bladder due to a transitional cell carcinoma.

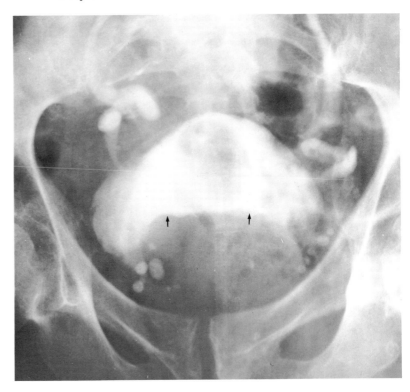

Fig. 7.50 Prostatic enlargement. The bladder base is lifted up and shows an impression from the enlarged prostate. (arrows). The ureters are tortuous and enter the bladder horizontally. A balloon catheter is in the bladder.

best observed at cystoscopy, so the main value of the IVU is in demonstrating ureteric obstruction. Small lesions readily identifiable at cystoscopy may be invisible on IVU or cystogram.

Prostatic enlargement (Fig. 7.50)

Enlargement of the prostate can be due to benign hypertrophy or to carcinoma. The distinction between these two is not possible on IVU unless metastases are visible in the bones. The diagnosis is made on clinical grounds and confirmed by biopsy. The role of radiology is in the assessment of the obstructive effects of prostatic enlargement on the remainder of the urinary tract. The enlarged prostate may be visible as a filling defect in the bladder base.

The signs of bladder outflow obstruction are:

1. Increased trabeculation and thickness of the bladder wall, often with diverticula formation.
2. Residual urine in the bladder after micturition.
3. Dilatation of the collecting systems.

Bladder diverticula (Fig. 7.51)

Bladder diverticula may be either the consequence of chronic obstruction to the bladder outflow, or of congenital origin. Because of urinary stasis they predispose to infection and stone formation and tumours may arise within them. Diverticula are well demonstrated at micturating cystography and on the after micturition film during IVU, because bladder contraction is often necessary before they fill with contrast medium.

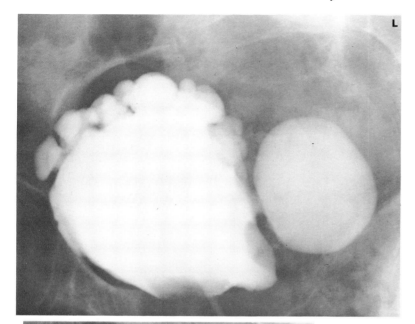

Fig. 7.51 Bladder diverticula. Cystogram showing numerous out-pouchings from the bladder with a very large diverticulum projecting to the left.

Neurogenic bladder

There are two basic types of neurogenic bladder (attempts to correlate these types with specific neurological lesions have not been satisfactory):

1. The large atonic smooth-walled bladder with poor or absent contractions and a large residual volume.

2. Small volume thick-walled bladder with gross trabeculation and sacculation (Fig. 7.52). The bladder neck may be narrow or the whole of the posterior urethra may be widened.

URETHRA

Posterior urethral valves (Fig. 7.53)

Congenital valves may occur in the posterior urethra at the level of the verumontanum. They cause significant obstruction and are almost invariably discovered during infancy or childhood. The IVU in such children will show hydronephrosis, hydroureter and a large poorly emptying bladder. The urethral valves cannot be demonstrated by retrograde urethrography since there is no obstruction to retrograde flow. They are easily

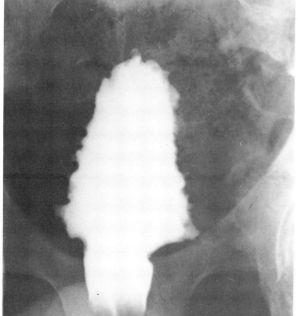

Fig. 7.52 Neurogenic bladder. The outline of the bladder is very irregular due to trabeculation of the bladder wall. The bladder has a small volume with an elongated shape. This appearance has been described as a 'fir tree badder'. There is a balloon catheter in the dilated posterior urethra.

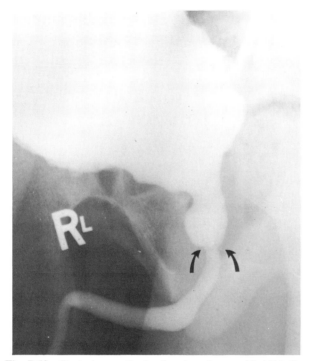

Fig. 7.53 Posterior urethral valves in a 6-year-old child. On this micturating cystogram the site of the valves is arrowed. Their presence is recognised by dilatation of the posterior urethra. Note the irregular outline of the thick-walled bladder due to chronic obstruction.

demonstrated at micturating cystourethrography where substantial dilatation of the posterior urethra is seen which terminates abruptly in a convex border formed by the valves. The distal urethra is usually collapsed but otherwise normal.

Urethral stricture (Fig. 7.54)

The majority of urethral strictures are due to previous trauma or infections. Some are neoplastic or congenital in origin. Posttraumatic strictures are usually in the proximal penile urethra; the most vulnerable portion of the urethra to external trauma. Such strictures are usually smooth in outline and relatively short. Inflammatory strictures, which are usually gonococcal or occasionally tuberculous in origin, may be seen in any portion of the urethra, but are usually found in the anterior urethra. They may be single or multiple and may be associated with fistulae to the skin. The best technique for demonstrating urethral strictures is retrograde urethrography. Visualising the urethra at micturating cystography is an alternative method.

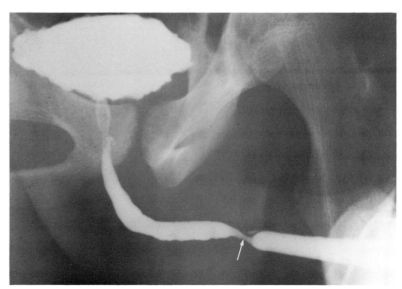

Fig. 7.54 Urethral stricture. A stricture in the penile urethra is arrowed. The patient had gonorrhoea.

8

Ultrasound and Radiology in Obstetrics

X-ray examination now has a very limited role in obstetrics because of the ease and safety with which it is possible to determine gestational age, multiparity, placenta praevia, fetal death and trophoblastic disease by ultrasound examination. Ultrasound examination also allows one to diagnose hydramnios and its causes as well as uterine or ovarian tumours. X-ray examination is still, however, used in pelvimetry and in the evaluation of suspected skeletal deformities of both fetus and mother.

Unlike x-ray examination (see p. 10 for discussion of radiation hazards), no biological damage has as yet been attributed to ultrasound as used in current obstetric practice. Over the past 25 years clinical studies of pregnant women and their babies exposed to ultrasound have demonstrated no adverse effects or increased incidence of fetal anomalies.

Pelvimetry

Pelvimetry is a method used to determine the relative pelvic and fetal head dimensions. An erect lateral film is always taken and in some centres a supine AP film is added. Anteroposterior and side-to-side pelvic diameters are measured from bone surface to bone surface and corrections are made for magnification. These measurements take no account of the soft tissues which may be all important in determining the effective diameters of the pelvis available to the fetus during delivery. For this reason the use of pelvimetry varies greatly from centre to centre.

ULTRASOUND IN OBSTETRICS
Technique

It is important to realise, when viewing the ultrasonic images in this chapter, that the images represent sections 1 cm thick obtained in a variety of planes. In obstetrics all images are obtained with the patient in the supine position. Longitudinal sections are oriented as if viewed from the patient's right side; transverse sections are oriented as if viewed from the patient's feet. Each section is identified on a coordinate system; the umbilicus is used as the standard reference point for labelling the sections.*

Obstetric ultrasound examinations are always undertaken with a full urinary bladder because the bladder acts as an acoustic window to the bulk of the uterus in early pregnancy and to the lower uterine segment in later pregnancy.

The developing fetus

Approximately 4 weeks after the last menstrual period (all 'gestational dates' given in this chapter are calculated from the time since the last menstrual period), a focus of high intensity echoes can be identified within the uterus (Fig. 8.1). These echoes represent the first ultrasonic evidence of conception. At this point one cannot be certain whether one is identifying an early

* Sections which pass through the umbilicus are labelled TO or LO depending on whether the sweeps are transverse (T) or longitudinal (L). All longitudinal sections obtained to the right of the midline sagittal plane are identified as L+, the number following the '+' being the distance in centimetres from the midline. Similarly, sections to the left of the midline are identified as L−. A similar system applies to the transverse sections where sections above the umbilicus are referred to as T+ sections and those below the umbilicus as T− sections with a figure indicating the distance from the umbilicus in centimetres.

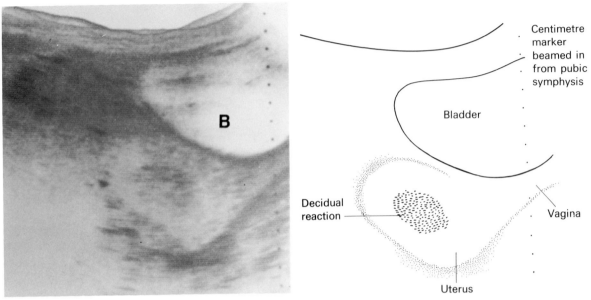

Fig. 8.1 Earliest sonographic sign of pregnancy: the decidual reaction within the uterus (longitudinal scan).

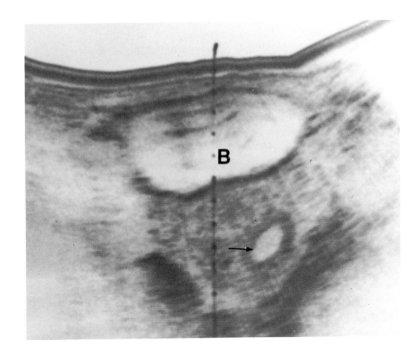

Fig. 8.2 The gestational sac. Transverse section 12 cm below the midline demonstrates an enlarged uterus with a well-defined gestational sac within it (arrow). The period of amenorrhea was 5 weeks and the volume of the sac is less than 2 ml. (B = bladder.)

developing gestation or an involuting missed abortion. A subsequent scan 2 weeks later should clarify this problem. If there is a viable pregnancy, a gestational sac should develop. The normal gestational sac is seen as a small cystic structure appearing between 5 and 6 weeks (Fig. 8.2).

As the normal pregnancy progresses internal echoes within the gestational sac will become apparent. These echoes represent the developing fetus. One should be able to see these echoes by 7 weeks. The volume of the gestational sac can be assessed; these volumes are of great value in the early diagnosis of an anembryonic pregnancy (a blighted ovum). A blighted ovum is diagnosed when a pregnancy has a sac volume of 2·5 ml or greater in which no fetal structures can be identified at sonography. It is noteworthy that gestational sacs which demonstrate a markedly irregular contour or a low implantation often herald pregnancy failure.

As the fetal echoes become visible a whole new method of dating the pregnancy becomes available. The crown-rump length (the longest demonstrable length of the echoes within the gestational sac) can be measured and compared with a normogram from which the gestational age of the fetus can be read directly (Fig. 8.3). This is a highly reliable method of dating a pregnancy from 7 to 12 weeks. After 12 weeks the widest biparietal diameter of the fetal head can be measured to assess gestational age. The biparietal diameter is used because it is

an easy diameter to identify. Note how in Fig. 8.4 there is a well-defined midline echo within the fetal head. This is the falx cerebri. At the level of the falx you will notice that the head is oval shaped rather than circular. It is also possible to identify linear echoes paralleling the falx which represent the side walls of the lateral ventricles. When both these features are seen one can reliably assume that the measurement obtained is the *widest* biparietal diameter possible.

From 12 weeks to term the fetal biparietal diameter is used to assess gestational age. The diameter obtained is compared to a normogram. The figure quoted for fetal age is always given as the date from the last menstrual period.

As the fetal skull becomes defined, sonographically, other fetal structures can be identified.

The placenta and placenta praevia

By 9 weeks the placenta is seen as a well-defined intra-uterine structure of variable thickness lining the inner wall of a portion of the uterine cavity. The placenta demonstrates medium to low level echoes with a well-defined interface between it and the amniotic fluid. The fetal surface of the placenta, the chorionic plate, can be recognised by a thin line of increased echoes (Fig. 8.5). In placenta praevia the placenta encroaches on the lower uterine segment. (Identifying the position of the

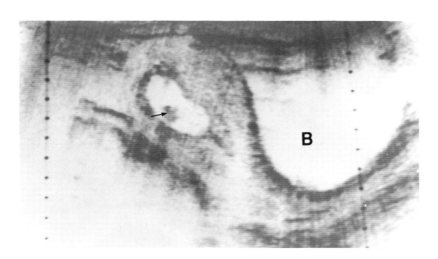

Fig. 8.3 Crown rump length. Longitudinal section 1 cm to the left of the midline. The solid structure within the gestational sac represents the developing fetus (arrow). The crown rump length in this patient corresponds to a 7-week period of amenorrhea. (B=bladder.)

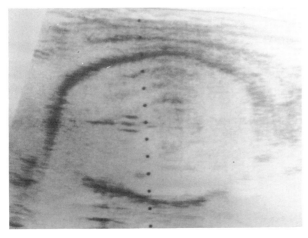

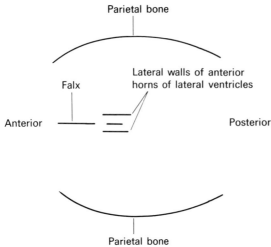

Parietal bone

Falx   Lateral walls of anterior
horns of lateral ventricles

Anterior   Posterior

Parietal bone

Fig. 8.4   Fetal head biparietal diameter.

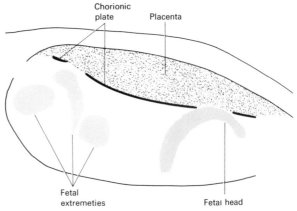

Chorionic
plate   Placenta

Fetal
extremeties   Fetal head

Fig. 8.5   The chorionic plate. Longitudinal scan.

placenta is of particular importance in identifying placenta praevia in patients who present with third-trimester vaginal bleeding.)

The lower uterine segment is defined, sonographically, as that portion of the uterus lying below a line connecting the pubic symphysis with the sacral promontory. The internal cervical os cannot be identified with certainty by sonography, but one would expect to find it within the lower uterine segment (behind the bladder) within 1 cm of the midline. When placental tissue is seen covering or extending to the expected position of the internal cervical os placenta praevia can be diagnosed. A word of caution is needed here. Often in the early stages of pregnancy the placenta occupies a relatively greater portion of the uterus than it does in the third trimester. As the pregnancy progresses and differential uterine growth occurs a low-lying placenta will be 'carried up' away from the lower uterine segment. Therefore, an apparent placenta praevia in an asymptomatic patient examined within the second trimester has a totally different significance than placenta praevia seen in the third trimester.

The 'large for dates' uterus

A common problem in clinical obstetrics is the patient

whose uterus is larger than expected. Often the cause is a false estimation of the gestational age usually due to a mistake in calculating the date of conception. By using the method outlined earlier in this chapter it is possible to determine the fetal age with considerable accuracy.

Less common causes for a uterus larger than expected for the gestational age include:

1. Multiparity.
2. Trophoblastic disease.
3. Hydramnios.
4. Associated ovarian or uterine tumours, either of which may be mistaken for generalised enlargement of the uterus.

Multiparity is readily diagnosed by sonography or conventional x-ray films. Sonographically, the diagnosis is established when two or more fetal heads are demonstrated on a single scan. Care must be taken by the examiner to ensure that the same head is not being identified in two different positions as the fetus moves.

Trophoblastic disease is a spectrum of pathology ranging from the relatively benign hydatidiform mole to the malignant choriocarcinoma. Plain radiographs are not helpful in making the diagnosis. Sonographically, multiple vesicular (cystic) structures are demonstrated within the tumour. These cysts are identified as multiple echo-free structures of varying sizes (Fig. 8.6). All varieties of trophoblastic disease, whether benign or malignant, are indistinguishable from one another on ultrasound examination. In most cases no fetal parts will be identified, but trophoblastic disease may, very rarely, coexist with a pregnancy.

Uterine tumours. The most common uterine tumour associated with pregnancy is the fibromyoma (fibroid). These are present in 1% of all pregnancies and can increase in size under the hormonal stimulation of pregnancy. Usually, they arise in the body or fundus of the uterus but when situated low in the uterus they can obstruct labour. They are readily demonstrated by ultrasound (Fig. 8.7), but cannot usually be diagnosed on plain film radiographs.

Ovarian tumours may be confused with uterine enlargement. Those associated with pregnancy are usually cysts, particularly corpus luteum cysts. Being cysts they are readily diagnosed by ultrasound. Most corpus luteum cysts will involute as the pregnancy progresses.

Hydramnios, like oligohydramnios, remains a judgment decision on ultrasound as there are no reliable quantitative estimations. When the fetus is seen to be entirely surrounded by amniotic fluid hydramnios is assumed to be present. Hydramnios can be associated with a multitude of fetal and maternal abnormalities including

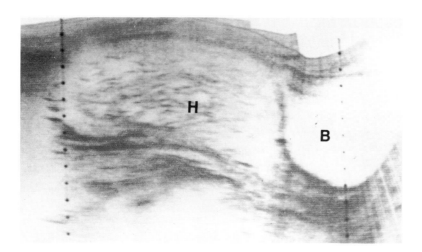

Fig. 8.6 Hydatidiform mole. Longitudinal section illustrating multiple irregular vesicular structures within an enlarged uterus. (H = hydatidiform mole; B = bladder.)

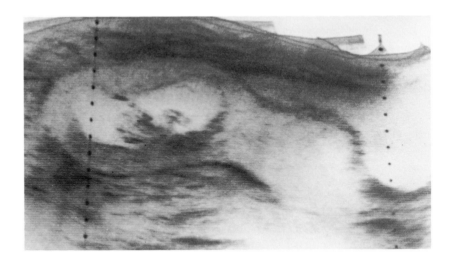

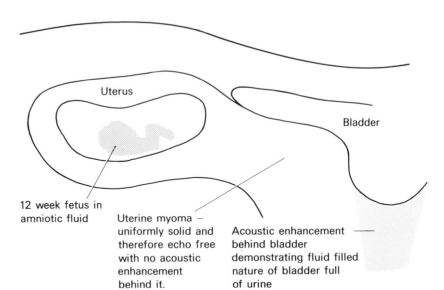

Uterus

Bladder

12 week fetus in
amniotic fluid

Uterine myoma –
uniformly solid and
therefore echo free
with no acoustic
enhancement
behind it.

Acoustic enhancement
behind bladder
demonstrating fluid filled
nature of bladder full
of urine

Fig. 8.7   Lower uterine segment
leiomyoma. Longitudinal scan; the
centimetre markers are at the
umbilicus and at the pubic symphysis.

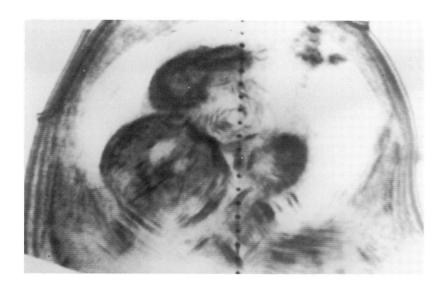

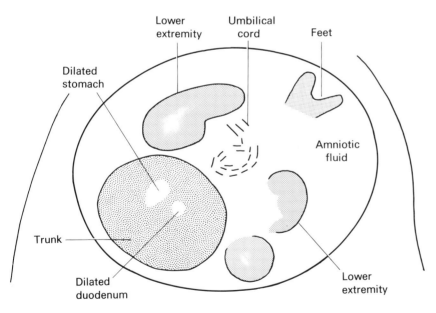

Fig. 8.8 Hydramnios. Transverse scan. In this case the hydramnios was due to duodenal atresia. Duodenal obstruction was diagnosed on this scan from the presence of the dilated fluid-filled stomach and duodenum.

maternal diabetes, erythroblastosis fetalis, anencephaly and duodenal atresia. A number of these abnormalities can be demonstrated by ultrasonic examination of the fetus (Fig. 8.8).

Ectopic pregnancy

Ectopic pregnancies generally present with sudden pain due to rupture of the ectopically placed gestational sac. When seen sonographically they are identified as adnexal, i.e. broad ligament, complex masses having more solid than cystic characteristics.

Fetal death

Fetal death can be inferred from both the static sonogram and the radiograph. Similar anatomic findings are present on both examinations. Overlapping of the bones of the non-engaged fetal skull and sharp angulation of the fetal spine are demonstrated with both techniques. Gas in the fetal circulatory system provides unequivocal evidence of fetal death and to date has only been seen on radiographic examination. However, all of these findings are seen relatively late. Therefore, the 'Doppler probe' examination of the fetal heart, a structure that is readily identified during sonographic scanning, is the method of choice to establish that the fetus is alive. Here the sound beam is reflected by the beating fetal heart and the change in frequency of the reflected sound can be converted to an *audible* signal. Use of this device makes the radiographic assessment of fetal death redundant.

Scanners with the ability to demonstrate motion as a two-dimensional cross-sectional image are now in use. These are referred to as 'real time scanners'. They can easily demonstrate fetal cardiac, limb and trunk movement quickly and accurately.

Other abdominal problems in pregnancy

The established safety of ultrasound compared with x-ray examination, permits evaluation of abdominal abnormalities which may coexist with pregnancy. Diseases of the biliary tract and of the urinary tract may be readily diagnosed by ultrasound and radiological examination of these structures with its attendant dangers of radiation can in some instances be completely avoided.

9

Bone Disease

The anatomy of a normal long bone is shown in Fig. 9.1.

BASIC SIGNS

The radiological responses of bone to any pathological process are limited, thus similar x-ray signs occur in widely different conditions.

Decrease in bone density which may be focal, when it is usually referred to as a 'lytic area' or an area of 'bone destruction'; or it may be generalised. When generalised, decrease in bone density is best referred to as 'osteopenia' until a specific diagnosis such as osteomalacia or osteoporosis can be made.

Increase in bone density (sclerosis) which may also be focal, or generalised. Sometimes lesions show a mixed lytic and sclerotic pattern.

Periosteal reaction. The periosteum is not normally visible on a radiograph. The term 'periosteal reaction' refers to excess bone produced by the periosteum, which occurs in response to such conditions as neoplasm, inflammation or trauma. A periosteal reaction may take several forms (Fig. 9.2) but the individual pattern is not specific enough to allow us to state the precise cause. At the edge of a very active periosteal reaction there may be a cuff of new bone known as a Codman's triangle (Fig. 9.2d). Although often seen in highly malignant primary bone tumours, e.g. osteosarcoma, a Codman's triangle is also found in other aggressive conditions.

Cortical thickening also involves the laying down of new bone by the periosteum (Fig. 9.3), but here the process is very slow. The result is that the new bone, though it may be thick and irregular, shows the same homogeneous density as does the normal cortex. There are no separate lines or spicules of calcification as is seen in a periosteal reaction. The causes are many including chronic osteomyelitis, healed trauma, response to chronic stress or benign neoplasm. The feature common to all is that the process is either very slow or has healed.

Alteration in trabecular pattern is a complex response usually involving a reduction in the number of trabeculae with an alteration in those that remain. In Paget's disease, for example, the trabeculae that remain are thickened and trabeculation is seen in the normally compact bone of the cortex (Fig. 9.4).

Alteration in the shape of a bone may be congenital in origin as in a bone dysplasia or may be acquired, as in acromegaly. This subject is discussed further on p. 262.

Alteration in bone age. The time of appearance of calcification and fusion of the various epiphyseal centres depends on the age of the child. Sets of standard films have been published which provide an indication of skeletal maturation. For the measurement of 'bone age' it is usually most convenient to take a film of the hand and wrist, but in the neonatal period films of the knee provide the most accurate assessment.

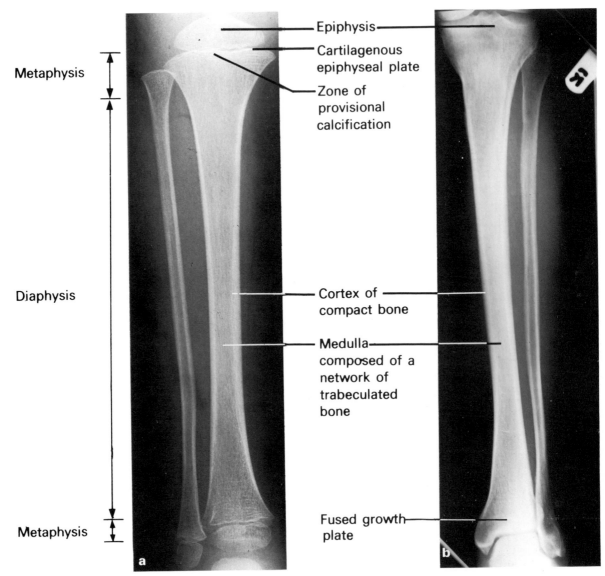

Metaphysis

Diaphysis

Metaphysis

Epiphysis

Cartilagenous epiphyseal plate

Zone of provisional calcification

Cortex of compact bone

Medulla composed of a network of trabeculated bone

Fused growth plate

Fig. 9.1 Normal long bones. (a) a child; (b) adult. Increase in length takes place at the epiphyseal plate. In the growing child calcification of cartilage occurs at the interface between the radiolucent growing cartilage and the bone to give the zone of provisional calcification which is seen as a dense white line forming the ends of the shaft and surrounding the bony epi-physes. This calcified cartilage becomes converted to bone. (If there is temporary cessation of growth then the zone of provisional calcification may persist as a thin white line, known as a 'growth line', extending across the shaft of the bone.) As the child grows older the epiphyseal plate becomes thinner until, eventually, there is bony fusion of the epiphysis with the shaft.

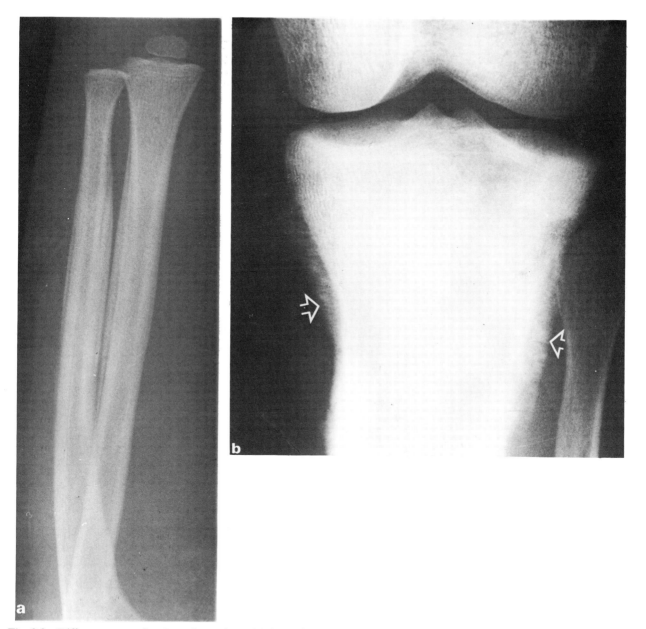

Fig. 9.2 Different types of periosteal reactions. (a) Smooth lamellar periosteal reaction on the radius and ulna in a case of a battered baby. (b) Spiculated (sunray) periosteal reaction in a case of osteogenic sarcoma (arrows).

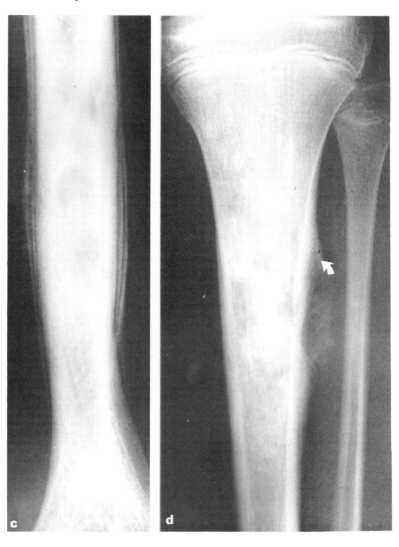

Fig. 9.2(c)   Onion skin periosteal reaction in a case of Ewing's sarcoma (arrows). Here the periosteal new bone consists of several distinct layers. (d) Codman's triangle in a case of osteogenic sarcoma. At the edge of the lesion the periosteal new bone is lifted up to form a cuff (arrow).

It takes time for all the above signs to develop, e.g. in adults it takes several weeks for a periosteal reaction to be visible after trauma and in a child with osteomyelitis the clinical features are present 7 to 10 days before the first sign is visible on the radiograph. In general, the signs take longer to develop in adults than they do in children.

For diagnostic purposes bone diseases can be divided into
—solitary lesions
—multiple focal lesions, i.e. several discrete lesions in one or more bones
—generalised lesions, where all the bones are diffusely affected

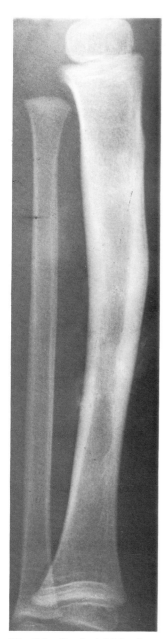

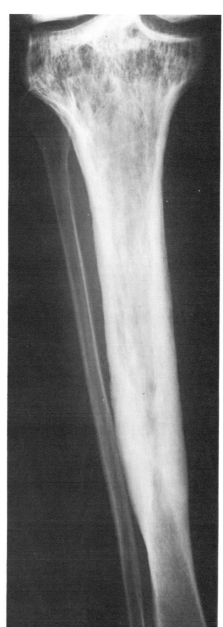

Fig. 9.3 Cortical thickening. Note the thickened cortex in the midshaft of the tibia due to old, healed osteomyelitis. Same cases as in Fig. 9.12 taken 1 year later.

Fig. 9.4 Alteration of trabecular pattern. Paget's disease involving the upper part of the tibia leaving the lowest part of the tibia and the fibula unaffected. Note the coarse trabeculae. The other features of Paget's disease—thickened cortex and bone expansion—are also present.

SOLITARY LESIONS (for fractures see Chapter 12)

The common solitary lesions are localised areas of lysis or sclerosis, or a combination of the two. They may be accompanied by a periosteal reaction or a fracture and are usually one of the following:

1. Bone tumours.
   (a) Benign.
   (b) Malignant (primary or secondary).
2. Osteomyelitis.
3. Bone cysts, fibrous dysplasia or other non-neoplastic defects of bone.
4. Histiocytosis X and osteoid osteoma.

   The radiological diagnosis of a localised bone lesion can be a problem. Some conditions are readily diagnosed, but in others even establishing which broad category of disease is present can be difficult. The initial radiological decision is usually to try and decide whether the lesion is benign, i.e. stationary or very slow growing, or whether it is aggressive, e.g. a malignant tumour or an infection. This decision is easier in primary bone lesions than in metastatic disease because some metastatic cancers and myeloma have certain features in common with benign lesions. In these cases the presence of a known primary tumour or multiple lesions on films or isotope bone scans will usually prevent the erroneous diagnosis of a benign lesion.

   It is always important to know the age of the patient since certain lesions tend to occur in a specific age range.

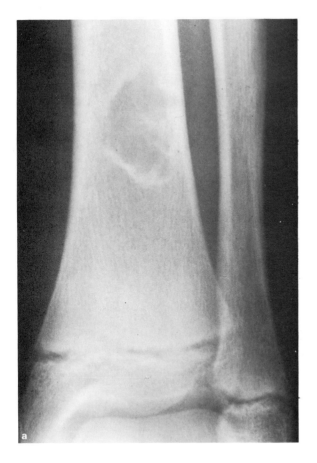

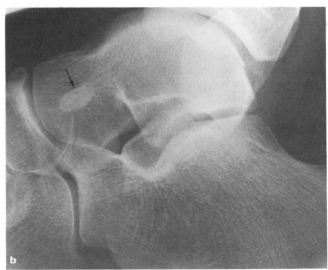

Fig. 9.5 The localised lesion. (a) With a well-defined sclerotic edge indicating a benign lesion—a fibrous cortical defect; (b) bone island: there is a small, well-defined area of compact bone in the talus (arrow). This common finding is without significance.

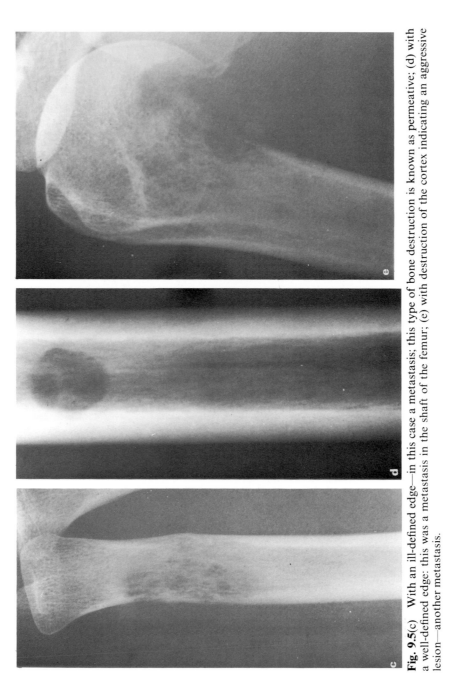

Fig. 9.5(c) With an ill-defined edge—in this case a metastasis; this type of bone destruction is known as permeative; (d) with a well-defined edge: this was a metastasis in the shaft of the femur; (e) with destruction of the cortex indicating an aggressive lesion—another metastasis.

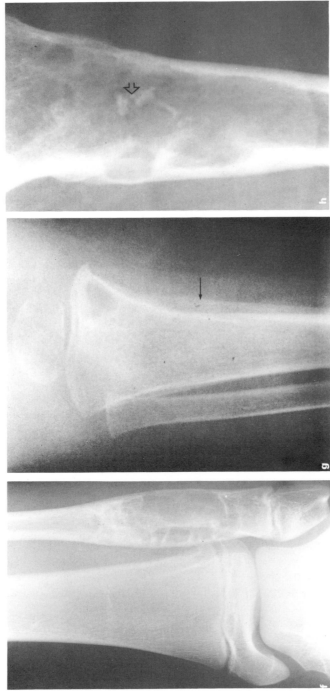

Fig. 9.5(f)   With expansion of the cortex—fibrous dysplasia; (g) with periosteal reaction (arrow)—osteomyelitis; (h) containing calcium (arrow)—a cartilage tumour; in this case a chondrosarcoma.

The signs to look for when trying to decide the nature of a localised bone lesion are:

1. The edge. Look carefully at the edge of any lytic or sclerotic lesion to see whether it is well demarcated or whether there is a wide zone of transition between the normal and the abnormal bone. There are two extremes. A lesion with a well-defined sclerotic edge is almost certainly benign, e.g. a fibrous cortical defect (Fig. 9.5a) or a bone island (Fig. 9.5b), whereas a lytic or sclerotic area with an ill-defined edge is likely to be aggressive (Fig. 9.5c). In the middle of this spectrum lies the lytic area with no sclerotic rim, which may be a benign lesion, a metastasis or myeloma (Fig. 9.5d).

2. The adjacent cortex. Any destruction of the cortex indicates an aggressive lesion (Fig. 9.5e).

3. Expansion. Bone expansion with an intact well-formed cortex usually indicates a slow growing lesion, e.g. enchondroma, fibrous dysplasia (Fig. 9.5f).

4. Periosteal reaction. The presence of an active periosteal reaction in the absence of trauma usually indicates an aggressive lesion (Fig. 9.5g). The causes of localised periosteal reactions are:

    (a) Osteomyelitis.

    (b) Malignant bone tumour, particularly Ewing's sarcoma and osteosarcoma.

    (c) Occasionally metastasis, particularly neuroblastoma.

    (d) Histiocytosis X.

    (e) Trauma—a periosteal reaction following trauma may lead to an erroneous impression of an aggressive process.

5. Consistency. Look for any densities within the lesion, e.g. patchy calcification usually indicates a cartilage tumour (Fig. 9.5h).

6. Soft tissue swelling. Look for an adjacent soft tissue mass. The presence of a soft tissue mass suggests an aggressive lesion—the better defined it is the more likely it is that the lesion is a neoplasm. Ill-defined soft tissue swelling often indicates oedema due to infection. Sometimes a tumour arising primarily in the soft tissues may be responsible for the bone lesion by pressure erosion or direct invasion.

7. Site. The site of the lesion is important, since certain lesions tend to occur at certain sites.

Bone tumours

The precise diagnosis of a bone tumour is often notoriously difficult both for the radiologist and the pathologist. Bone tumours may be primary or secondary. Metastatic malignant tumours are by far the commonest outnumbering many times primary malignant tumours. They are discussed on p. 244.

There is no entirely satisfactory classification of primary bone tumours, but the easiest one to comprehend is based on the tissue of origin (Table 9.1). Some conditions, e.g. fibrous dysplasia are included in this classification although they are not true tumours.

Table 9.1 Tumours and tumour-like conditions.

Cell type	Benign	Malignant
Osteoid	Osteoma	Osteosarcoma
Cartilage	Enchrondroma	Chondrosarcoma
	Osteochondroma	
Fibrous	Fibrous dysplasia	Fibrosarcoma
	Fibrous cortical defect	
Miscellaneous	Bone cyst	Giant cell tumour
	Osteoid osteoma	Ewing's sarcoma
	Aneurysmal bone cyst	
	Histiocytosis X	

Malignant tumours

Primary malignant tumours usually have poorly defined margins, often with a wide zone of transition between the normal and abnormal bone. The lesion may destroy the cortex of the bone. A periosteal reaction is often present and an additional feature is that a soft tissue mass may be seen.

Osteosarcoma (osteogenic sarcoma) occurs mainly in the 5 to 20-year-old age group, but is also seen in the elderly from malignant change in Paget's disease. The tumour usually arises in the metaphysis, most commonly around the knee. There is usually bone destruction with new bone formation and typically a florid periosteal reaction is present producing spiculation

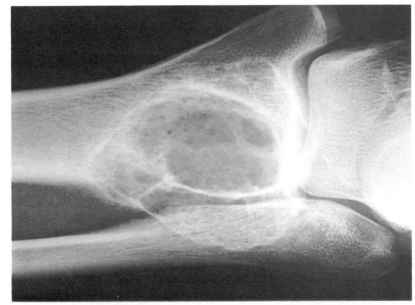

Fig. 9.8 Giant cell tumour. Eccentric expanding lytic lesion crossed by strands of bone which has thinned the cortex. The subarticular position is characteristic of this tumour.

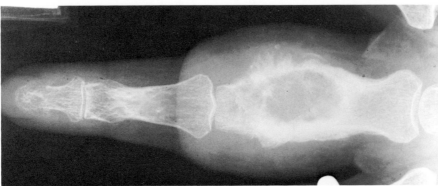

Fig. 9.7 Chondrosarcoma showing a lucent area in the finger which has expanded the bone and destroyed the cortex—there is a florid periosteal reaction.

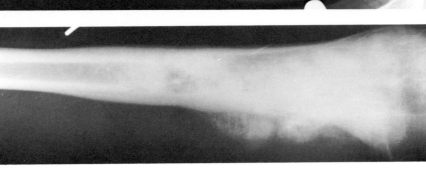

Fig. 9.6 Osteogenic sarcoma showing patchy bone destruction with marked sclerosis and spiculated periosteal reaction.

('sun ray appearance'; Fig. 9.6). The tumour may elevate the periosteum to form a Codman's triangle.

Chondrosarcoma occurs mainly in the 30 to 50-year-old age group most commonly in the pelvic bones, scapulae, humeri and femora. A chondrosarcoma produces a lytic expanding lesion containing flecks of calcium, a sign that indicates its origin from cartilage cells. It can be difficult to distinguish from its benign counterpart the enchondroma, but a chondrosarcoma is usually less well defined in at least one portion of its outline and it may show a periosteal reaction (Fig. 9.7). A chondrosarcoma may arise from malignant degeneration of a benign cartilagenous tumour.

Ewing's tumour is a highly malignant tumour, commonest in children, usually arising in the shafts of long bones. It produces ill-defined bone destruction with periosteal reaction that is typically 'onion skin' in type.

Metastases and multiple myeloma

These are discussed on p. 244.

Giant cell tumour

A giant cell tumour (Fig. 9.8) has features of both malignant and benign tumours. It is locally invasive but rarely metastasises. It occurs most commonly around the knee and at the wrist after the epiphyses have fused. It is an expanding destructive lesion which is subarticular in position. The margin is fairly well defined but the cortex is thin and may in places be completely destroyed.

Benign tumours and tumour-like conditions

Under this heading are included benign tumours such as enchondroma, certain conditions not regarded as tumours such as fibrous dysplasia and some abnormalities which are difficult to classify such as osteoid osteoma and histiocytosis X. Benign lesions usually have an edge which is well demarcated from the normal bone by a sclerotic rim. They cause expansion but rarely breach the cortex. There is no soft tissue mass and a

Fig. 9.9 Enchondromas in the metacarpal, proximal and middle phalanges showing lytic areas that expand but do not breach the cortex.

periosteal reaction is invariably absent unless there has been a fracture through the lesion.

An enchondroma is seen as a lytic expanding lesion, most commonly in the bones of the hand (Fig. 9.9). It often contains a few flecks of calcium.

Fibrous cortical defects (non-ossifying fibromas) are common chance findings in children and young adults. They are well-defined lucent areas in the cortex of long bones (Fig. 9.5a, p. 236).

Fibrous dysplasia may affect one or several bones. It occurs most commonly in the long bones and ribs as a lucent area with a well-defined edge and may expand the bone (Fig. 9.5f, p. 238). There may be a sclerotic rim around the lesion.

A bone cyst has a wall of fibrous tissue and is filled with fluid. It occurs in children, most commonly in the humerus and femur. It appears as a lucency across the width of the shaft of the bone with a well-defined edge. The cortex may be thin and the bone expanded (Fig. 9.10). It is often a pathological fracture that draws attention to it.

An osteoid osteoma is a painful condition occurring most commonly in the femur and tibia in young adults with a characteristic radiological appearance. There is a small lucency known as a nidus surrounded by dense sclerotic rim and a periosteal reaction may be present as well (Fig.

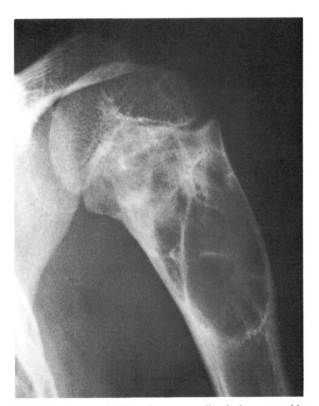

Fig. 9.10 Bone cyst. There is an expanding lesion crossed by strands of bone in the upper end of the humerus in a child. The lesion extends to but does not cross the epiphyseal plate.

9.11). The nidus may contain a small speck of calcification.

Histiocytosis X occurs in children and young adults and produces lytic lesions which may be single or multiple, most frequently in the skull, pelvis, femur and ribs. In the skull the lesions may be extensive to give the so-called 'geographical skull' (Fig. 13.4). In the long bones there is bone destruction which may be well defined or ill defined and may have a sclerotic rim. A periosteal reaction is sometimes seen. Histiocytosis X of the spine is discussed on p. 285.

Osteomyelitis

Osteomyelitis is most often caused by *Staphylococcus aureus* and usually affects infants and children. The initial radiographs are normal as bone changes are not visible until 7 to 14 days after the onset of the infection. The earliest change is bone destruction in the metaphysis with a periosteal reaction, which eventually may become very extensive and surround the bone to form an involucrum (Fig. 9.12). A part of the original bone may die and form a separate dense fragment known as a sequestrum (Fig. 9.13). In chronic osteomyelitis the bone becomes thickened and sclerotic with loss of differentiation between cortex and medulla and within the bone there may be sequestra and areas of destruction. Occasionally an abscess, known as Brodie's abscess, is seen as a lucency surrounded by an area of sclerosis (Fig. 9.14).

Distinction of neoplasm from osteomyelitis

It is not always possible, radiologically, to distinguish osteomyelitis from a bone tumour. The clinical history is clearly important. With malignant bone tumours the radiographs are usually abnormal when the patient first presents, compared with osteomyelitis where the initial films are often normal. But if early films are not available difficulties may arise in distinguishing an acute osteomyelitis from a highly malignant tumour such as a Ewing's tumour or an osteosarcoma. Chronic osteomyelitis may simulate a benign bone tumour radiogra-

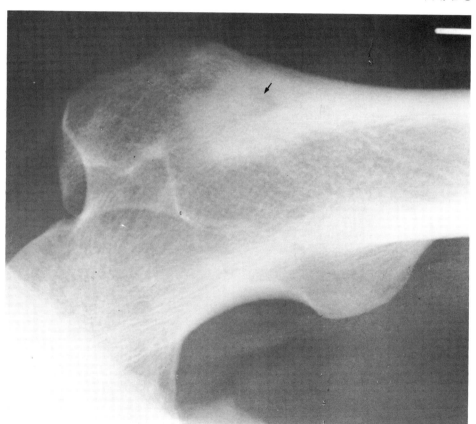

Fig. 9.11 Osteoid osteoma. There is an area of sclerosis in the upper end of the femur containing a central lucency (arrow) known as a nidus.

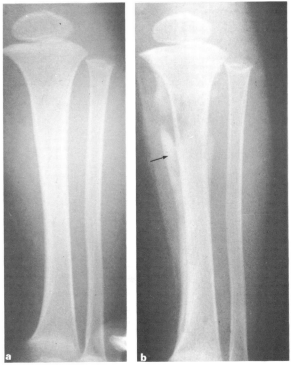

Fig. 9.12 Osteomyelitis. (a) Initial films reveal no abnormality; (b) films taken 3 weeks later show some destruction of the upper end of the tibia and an extensive periosteal reaction along the tibia, particularly the medial side (arrow).

phically, but the presence of fever and sometimes of discharging sinuses usually helps to diagnose an infective lesion.

MULTIPLE FOCAL LESIONS

Metastases

Metastases are by far the commonest bone tumour outnumbering many times primary malignant bone tumours. Metastases may be sclerotic, lytic or a mixture of lysis and sclerosis. Those bones containing red marrow are the ones mainly affected—spine, skull, ribs, pelvis, humeri and femora.

Many tumours may metastasise to bone but lytic metastases in adults most commonly arise from a car-

cinoma of the bronchus, breast, kidney and thyroid and in children from neuroblastoma and leukaemia. Lytic metastases give rise to well-defined or ill-defined areas of bone destruction without a sclerotic rim. The lesions may vary from small holes to large areas of bone destruction (Fig. 9.15). In the long bones metastases usually arise in the medulla and as they grow they enlarge and may destroy the cortex. Metastases and myeloma are virtually the only cause of multiple lytic lesions. Expansion of the bone is uncommon with metastases but when present suggests secondary deposits from a thyroid or renal carcinoma.

Sclerotic metastases appear as ill-defined areas of in-

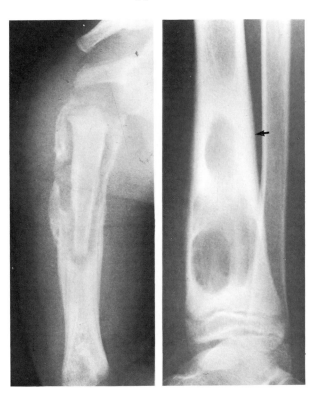

Fig. 9.13 (*left*) Osteomyelitis. The upper part of the humerus has separated to form a sequestrum. It is surrounded by an extensive periosteal reaction to form an involucrum.

Fig. 9.14 (*right*) Chronic osteomyelitis (Brodie's abscess) showing a lucency surrounded by substantial sclerosis. A faint periosteal reaction is present (arrow).

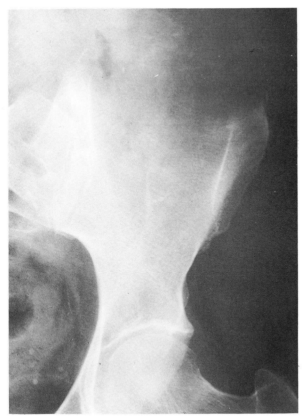

Fig. 9.15 Metastases from a carcinoma of the colon. There is a large area of bone destruction with an ill-defined edge in the iliac crest.

creased density of varying size with ill-defined margins. In men they are most commonly due to metastases from carcinoma of the prostate (Fig. 9.16), and in women from carcinoma of the breast.

Mixed lytic and sclerotic metastases are not uncommon. They are often seen with carcinoma of the breast.

A periosteal reaction is uncommon with metastases except in neuroblastoma.

Multiple myeloma

Although myeloma deposits may be found in any bone,

they are most frequently seen in bones with active haemopoiesis. The bone lesions in multiple myeloma may resemble metastases in every way with widely scattered lytic lesions (Fig. 9.17). They often tend to be better defined and may cause expansion of the bone. Diffuse marrow involvement may give rise to generalised loss of bone density producing a picture similar to that of osteoporosis (Fig. 9.18).

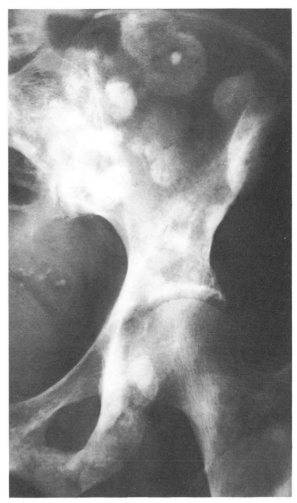

Fig. 9.16 Sclerotic metastases showing scattered areas of increased density.

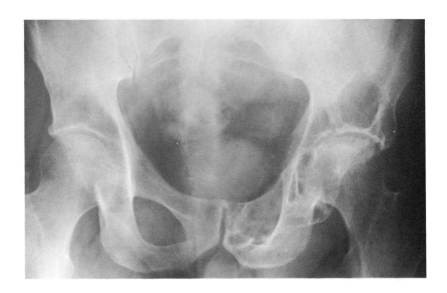

Fig. 9.17 Localised myeloma deposits causing lysis and expansion of bone.

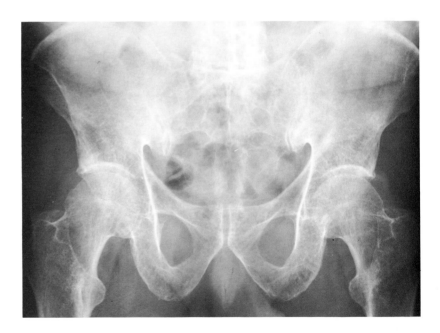

Fig. 9.18 Myeloma. Diffuse involvement of the pelvis in myeloma causing a generalised reduction in bone density resembling osteoporosis.

Hodgkin's disease

This disease may give rise to lytic or sclerotic areas closely resembling metastases.

Heavy metal poisoning

In children a dense line is seen across the metaphysis due to disordered bone formation following incorporation of metal, notably lead, into the growing bone (Fig. 9.19). The ends of the bone may have an abnormal shape because of interruption of normal modelling.

Bone infarction

As this usually occurs in the intra-articular parts of bones, it is described in the chapter on joint disease (p. 275). However infarcts in the medulla can occur in caisson disease, following radiation therapy or sickle cell disease. Sometimes they are found, incidentally, in older people with no known cause. These healed infarcts appear as irregular calcification in the medulla of the long bones (Fig. 9.20).

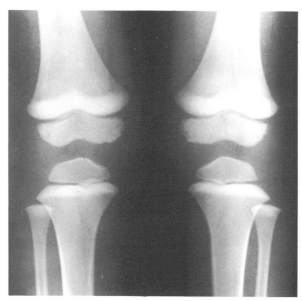

Fig. 9.19 Heavy metal poisoning. In this case due to lead, showing dense bands across the metaphyses.

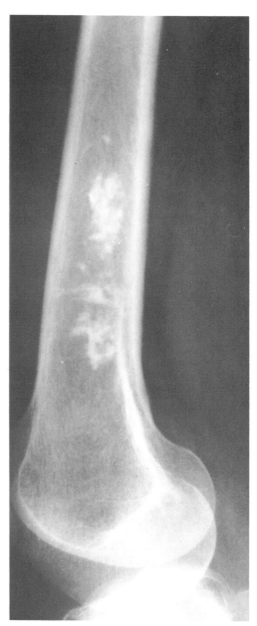

Fig. 9.20 Bone infarct. There is calcification in the medulla of the lower end of the femur.

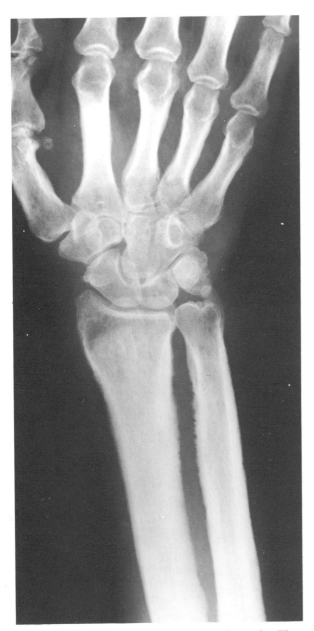

Fig. 9.21 Hypertrophic pulmonary osteoarthropathy. There is a periosteal reaction which was present bilaterally along the shafts of the radius and ulna and the metacarpals. In this case it was associated with a bronchial carcinoma.

Multiple periosteal reactions

These are seen in conjunction with other signs in:

1. Battered-baby syndrome (see p. 312).
2. Scurvy (see p. 250).
3. Venous stasis and ulceration of the legs; where there is a low-grade periosteal reaction and cortical thickening of the tibia and fibula.
4. In hypertrophic pulmonary osteoarthropathy there is widespread periosteal reaction around the bones of the forearms and lower legs which, when severe, extends to involve the hands and feet (Fig. 9.21). Finger clubbing is invariably present as well. Hypertrophic pulmonary osteoarthropathy is seen in a number of conditions, mostly intrathoracic, of which carcinoma of the bronchus is by far the commonest. The bone changes may occur before the bronchial carcinoma is apparent.

GENERALISED BONE CHANGES

Generalised decrease in bone density (osteopenia)

The radiographic density of bone is dependent on the amount of calcium present in the bone. This may be reduced due to a disorder of calcium metabolism as in osteomalacia or hyperparathyroidism, or to a reduction in the protein matrix as in osteoporosis. The radiological diagnosis of decreased bone density is often difficult especially as the appearances of the bones are markedly affected by radiographic exposure factors. Before deciding that the bone density is reduced assess the exposure by looking at the soft tissues.

The main causes of generalised decrease in bone density are

—osteoporosis
—osteomalacia
—hyperparathyroidism
—multiple myeloma, which may cause generalised loss of bone density (Fig. 9.18), but there may also be patchy bone destruction.

Each of these conditions may have other radiological features which enable the diagnosis to be made but when they are lacking, as they frequently are in osteoporosis and osteomalacia, it becomes very difficult to distinguish between them radiologically.

Osteoporosis

Osteoporosis is due to a deficiency of the protein matrix (osteoid). Because the matrix is reduced in quantity there is necessarily a reduction in calcium content.

The causes of osteoporosis are:

1. Idiopathic—often subdivided according to age of onset, e.g. juvenile, postmenopausal, senile. Senile osteoporosis is the commonest form.
2. Cushing's syndrome and steroid therapy.
3. Osteogenesis imperfecta.
4. Disuse.
5. Scurvy.

A radiological diagnosis of osteoporosis is only made after other diseases have been excluded. Therefore, whenever there is generalised decreased bone density look for features of bone destruction, particularly the pedicles of the vertebral bodies which would indicate malignant disease, and look for signs of hyperparathyroidism and osteomalacia since these conditions can closely resemble osteoporosis.

The changes of osteoporosis are best seen in the spine (Fig. 9.22). Although there is an overall decrease in bone density the cortex stands out clearly as if pencilled in. An important feature is collapse of the vertebral bodies. These are compression fractures which result in the vertebral bodies appearing wedged or biconcave and several vertebrae may be involved. The disc spaces often appear widened.

The long bones show a thin cortex. Many of the trabeculae are resorbed but those that remain stand out clearly.

In Cushing's disease and steroid therapy the appearances are similar to idiopathic osteoporosis but may show the additional feature of florid callus around any fractures that may be present.

Osteogenesis imperfecta is a dysplasia of bone which may be of varying degrees of severity. There is always generalised osteoporosis and the bones are so fragile

that they fracture easily and bend. Bowing of the bones does not usually occur in osteoporosis except in osteogenesis imperfecta. There can be great deformity of the skeleton (Fig. 9.23). An important diagnostic feature is

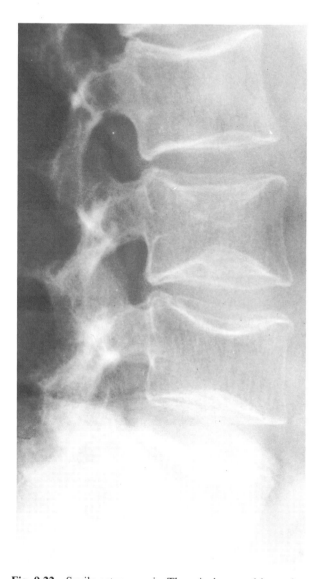

Fig. 9.22 Senile osteoporosis. There is decreased bone density but the edge of the vertebral bodies are well demarcated. Note the partial collapse of several of the vertebral bodies and the widening of the disc spaces.

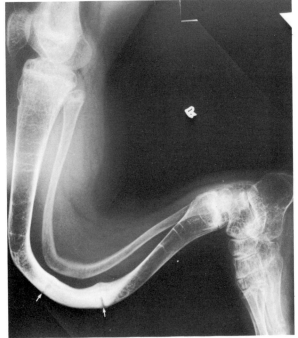

Fig. 9.23 Osteogenesis imperfecta. There is decreased bone density with a gross deformity of the leg. Old fractures can be seen in the tibia (arrows).

the presence of a large number of Wormian bones in the skull.

Disuse osteoporosis. Osteoporosis can occur with disuse, the commonest causes being pain and immobilisation in plaster (Fig. 9.24). Besides a reduction in density and thinning of the cortex the bone may have a spotty appearance.

Sudeck's atrophy is a mysterious condition where there is severe osteoporosis and oedema of the soft tissues following a fracture. The degree of osteoporosis is disproportionate to the trauma and the degree of disuse.

Scurvy is due to lack of vitamin C which results in a deficiency of the protein matrix in the bone. Calcification of cartilage is not affected and therefore the zone of provisional calcification is seen as a relatively dense white line. Similarly, the outline of the epiphyseal

centres show a dense ring of calcification. Transformation of calcified cartilage to bone is impaired resulting in a radiolucent zone. Due to the weakened bone fractures may occur, usually through the metaphyses (Fig. 9.25).

In the active stage of the disease large subperiosteal haemorrhages may form which calcify during the healing stage.

Rickets and osteomalacia

These are conditions in which there is lack of calcium in

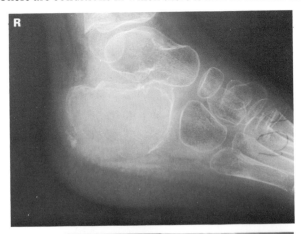

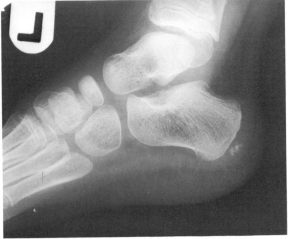

Fig. 9.24 Disuse osteoporosis due to osteomyelitis of the right calcaneum. The calcaneum is partly destroyed by infection. The remaining bones of the right foot show a marked reduction in bone density with well-defined cortex. Compare these bones with those in the normal left foot.

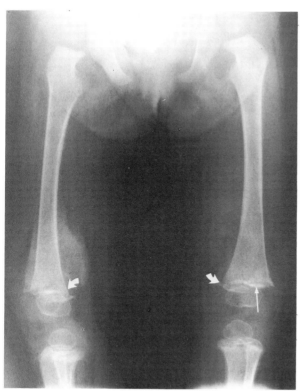

Fig. 9.25 Scurvy. Note the dense white line at the metaphysis (arrow), metaphyseal fractures (curved arrows) and calcifying subperiosteal haematoma. Reduction in bone density is not marked in this case.

the body tissues with poor mineralisation of osteoid. If this occurs before epiphyseal closure the condition is known as rickets—in adults it is known as osteomalacia.

The important causes of rickets and osteomalacia are:

1. Dietary deficiency of vitamin D or lack of exposure to sunlight resulting in decreased production of endogeneous vitamin D.
2. Malabsorption resulting in impaired absorption of calcium or vitamin D.
3. Renal disease, where rickets develops despite normal amounts of vitamin D in the diet, hence the term vitamin-D-resistant rickets.

   (a) Tubular defects—hypophosphataemia, Fanconi syndrome and renal tubular acidosis.

   (b) Chronic renal failure—impaired ability to activate vitamin D.

Regardless of the cause of the osteomalacia or rickets the bone changes are similar, but if it is due to chronic renal failure, or in some cases of malabsorption the changes of hyperparathyroidism may also be present.

Rickets. The changes are maximal where bone growth is

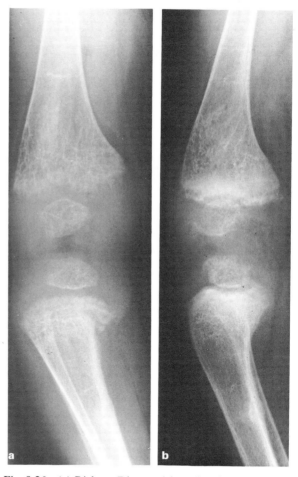

Fig. 9.26 (a) Rickets. Dietary rickets showing widening and irregular mineralisation of the metaphyses which have a frayed appearance. There is reduced bone density and bowing of the limbs; (b) after commencement of vitamin-D treatment, mineralisation of the metaphysis has occurred.

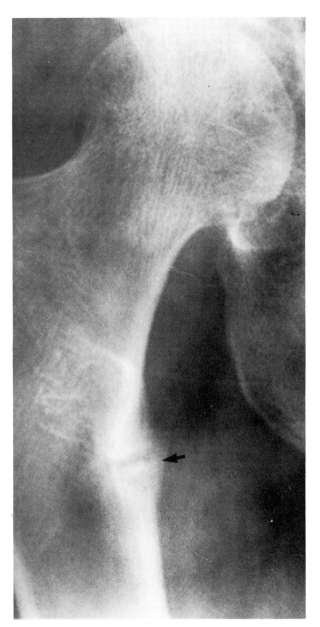

Fig. 9.27 Osteomalacia: Looser's zone showing the horizontal lucent band with sclerotic margins running through the cortex of the medial side of the upper femur (arrow).

occurring so they are best seen at the knees, wrists and ankles. The zone of provisional calcification is deficient and the metaphyses are irregularly mineralised, widened and cupped (Fig. 9.26). This results in an increased distance between the epiphysis and the shaft of the bone. The generalised decrease in bone density, however, may not be very obvious. There is retardation of growth. Deformities of the bones occur because the under-mineralised bone is soft. Greenstick fractures are common.

Osteomalacia. The characteristic features of osteomalacia are loss of bone density, thinning of the trabeculae and the cortex and Looser's zones (pseudofractures) (Fig. 9.27). These are short lucent bands running through the cortex at right-angles and usually going only part way across the bone. They may have a sclerotic margin making them more obvious. They are commonest in the scapulae, medial aspects of the femoral necks and shafts and the pubic rami.

Bone deformity consequent upon bone softening is an important feature. In the spine the vertebral bodies are biconcave, the femora may be bowed and in severe cases the side walls of the pelvis may bend inwards giving the so-called 'triradiate pelvis'.

Hyperparathyroidism

Excess parathyroid hormone secretion mobilises calcium from the bones, resulting in a decrease in bone density.

Hyperparathyroidism may be primary due to hyperplasia or a tumour of the parathyroid glands, or secondary to chronic renal failure.

Many patients with primary hyperparathyroidism present with renal stones and only a small minority have bone changes radiologically.

The signs of hyperparathyroidism in the bones are:

1. A generalised loss of bone density occurs with loss of the differentiation between cortex and medulla. The trabecular pattern may have a fine lacework appearance.

2. The hallmark of hyperparathyroidism is subperios-

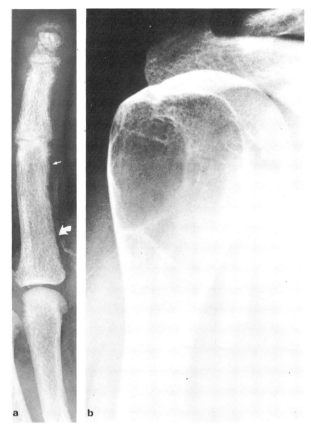

Fig. 9.28 Hyperparathyroidism. (a) Note the characteristic features of subperiosteal bone resorption (straight arrow), absorption of the tip of the terminal phalanx and the altered bone architecture. Arterial calcification is also present (curved arrow). (b) Brown tumour. There is a lytic area in the upper end of the humerus with a well-defined edge.

teal bone resorption (Fig. 9.28a). This occurs particularly in the hands on the radial side of the middle phalanges and the tips of the terminal phalanges. There may also be absorption of the outer ends of the clavicles.
3. Soft tissue calcification, vascular calcification and chondrocalanosis may occur. With advanced disease there may be marked deformity of the skeleton.
4. 'Brown tumours' are lytic lesions which may be single or multiple. They are of varying size and may expand the bone. They occur most commonly in the mandible and pelvis but any bone may be involved (Fig. 9.28b).

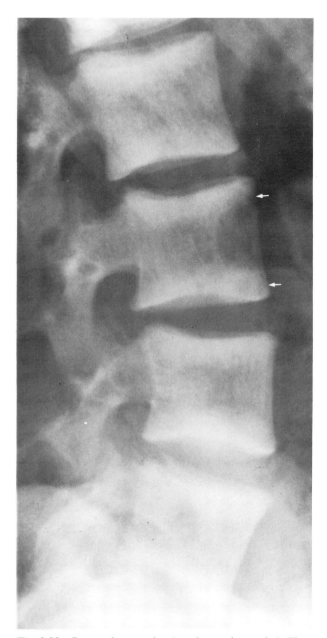

Fig. 9.29 Rugger jersey spine (renal osteodystrophy). There are sclerotic bands running across the upper and lower ends of the vertebral bodies of the lumbar spine (arrows).

The bone changes in primary and secondary hyperparathyroidism are similar except that 'brown tumours' are much rarer and vascular calcification commoner in secondary hyperparathyroidism.

Renal osteodystrophy

Three distinct bone lesions can occur, often together, in patients with chronic renal failure.

1. Osteomalacia in adults; rickets in children.
2. Hyperparathyroidism.
3. Sclerosis is an infrequent sign. It may be seen in the spine as bands across the upper and lower ends of the vertebral bodies giving the so-called 'rugger jersey' spine (Fig. 9.29) or at the metaphyses of the long bones.

Generalised increase in bone density

Sclerotic metastases particularly from prostatic or breast carcinoma may diffusely affect the skeleton (Fig. 9.30).

Osteopetrosis (marble bone disease). In this congenital disorder of bone formation the bones are densely sclerotic because most of the bone is in the form of compact bone rather than trabecular bone and, therefore, there is no differentiation between cortex and medulla (Fig. 9.31). The bones are brittle and may fracture readily but if fractured they heal easily.

Myelosclerosis is a form of myelofibrosis in which in addition to the replacement of bone marrow by fibrous tissue the process extends to lay down extra trabecular bone usually of a rather patchy nature (Fig. 9.32). The spleen is invariably enlarged due to it becoming the site of haemopoiesis. It may reach a very large size and forms an important sign on an x-ray of the abdomen.

Fluorosis is caused by drinking water containing a high fluoride concentration. The bones become uniformly sclerotic and in addition there is calcification of ligaments and muscle insertions to the bones.

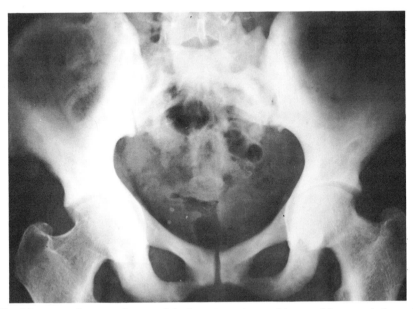

Fig. 9.30 Metastases from carcinoma of the breast causing a widespread increase in bone density.

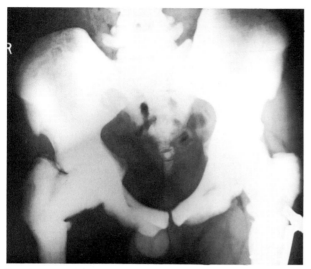

Fig. 9.31 Osteopetrosis. This is a marked generalised increased bone density affecting all bones. There are multiple healed fractures with a pin and plate in the left femur.

ALTERATION OF TRABECULAR PATTERN

Paget's disease

This is a common disorder in the elderly often encountered as a chance finding. One or more bones may be affected. It is most common in the pelvis, spine, skull and long bones.

Although there is a rare lytic form of Paget's disease, e.g. osteoporosis circumscripta of the skull, the cardinal features are thickening of the trabeculae and of the cortex leading to loss of corticomedullary differentiation and an increased bone density together with enlargement of the affected bone (Fig. 9.33a).

In the skull there are many circumscribed areas of sclerosis scattered in the skull vault giving a mottled appearance which has been likened to cotton wool. An increased thickness of the calvarium is a particularly obvious feature (Fig. 13.15, p. 324).

These changes of sclerosis, cortical thickening, coarse trabeculae and particularly increase in the size of the bone distinguish Paget's disease from metastases due to prostatic or breast carcinoma which are both common in the elderly.

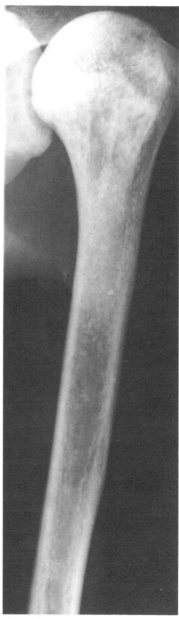

Fig. 9.32 Myelosclerosis. Patchy increase in bone density in the humerus is seen. In this condition the bone marrow becomes replaced with bone.

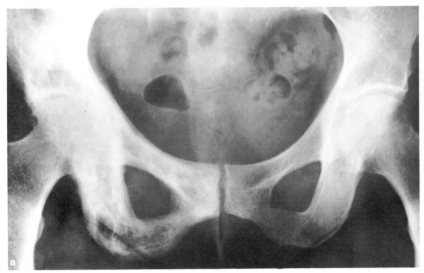

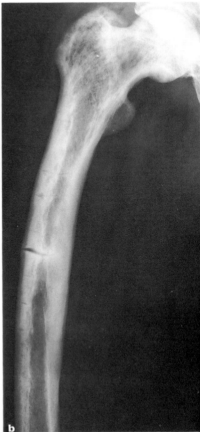

Fig. 9.33 (a) Paget's disease. There is sclerosis with coarse trabeculae in the right pubic and ischial rami. Note that the width of the affected bones is increased; (b) Paget's disease showing incomplete fractures known as infractions on the lateral aspect of the femur. Note the marked thickening of the cortex and bowing of the femur.

Bone softening causes bowing and deformity of the bones and basilar invagination in the skull (p. 323). Pathological fractures often seen as incomplete or complete transverse fractures across the bone may occur (Fig. 9.33b).

Malignant degeneration with development of an osteosarcoma in abnormal bone is an occasional occurrence (Fig. 9.34).

Haemolytic anaemia

There are several types of haemolytic anaemia but radiological changes are important in two main types, namely thalassaemia and sickle-cell disease. Both show changes of marrow hyperplasia but sickle-cell anaemia may also show evidence of bone infarction and infection.

Marrow hyperplasia. Overactivity and expansion of the bone marrow causes thinning of the cortex and resorption of some of the trabeculae so that those that remain are thickened and stand out more clearly. Loss of bone density may be apparent. In the skull there is widening of the diploë and there may be perpendicular striations

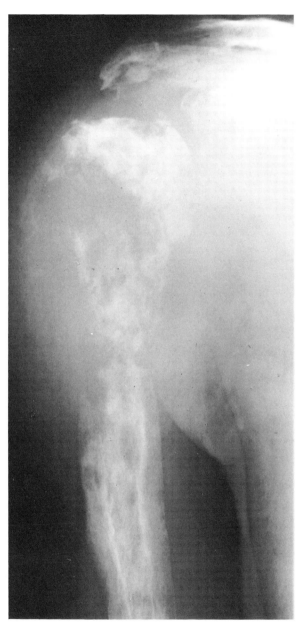

Fig. 9.34 Sarcoma in Paget's disease. There is extensive bone destruction in the humeral head and shaft. Evidence of the underlying Paget's disease can be seen.

giving an appearance known as 'hair-on-end' (Fig. 9.35a). The ribs may enlarge and the phalanges become rectangular (Fig. 9.35b).

Infarction and infection. Infarction at bone ends gives flattening and sclerosis of the humeral and femoral heads.

Areas of bone destruction with periosteal new bone formation or just a periosteal reaction may be seen in the shafts of the bones (Fig. 9.36). These changes may sometimes be extensive leading to the appearances of a bone within a bone. These signs are due to bone infarction. It is not possible to determine from the radiographs whether or not these infarcts are infected. In sickle-cell anaemia the vertebral bodies may have a biconcave appearance due to infraction of their upper and lower surface.

Sarcoidosis

This occasionally involve the bones. The phalanges of the hands and feet are virtually the only bones affected. The signs are either small cysts with a well-defined edge or areas of bone destruction showing a lace-like pattern (Fig. 9.37). If the bones are involved there is invariably evidence of sarcoidosis in the chest.

CHANGES IN BONE SHAPE

Bone dysplasias

Bone dysplasias are congenital disorders resulting in abnormalities in the size and shape of the bones. There are a large number of different dysplasias; many of them are hereditary and all of them are rare. Only two of the commoner examples will be described here: achondroplasia and diaphyseal aclasia. (Osteogenesis imperfecta has been described on p. 249 and osteopetrosis on p. 254.)

In *achrondroplasia* there is defective ossification of the bones formed in cartilage. It results in dwarfism characterised by shortening of the shafts of the long bones (Fig. 9.38a). There is reduction in the growth of the skull base, as this is formed in cartilage, and the vault is dispropor-

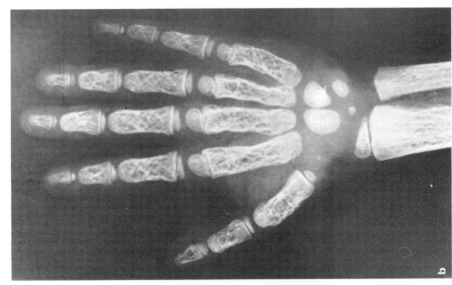

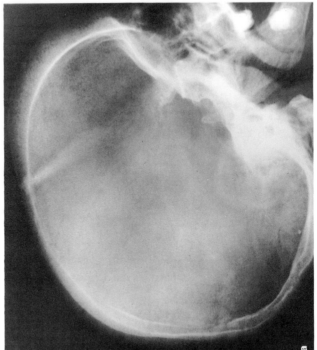

Fig. 9.35   Haemolytic anaemia—thalassaemia. (a) Skull showing thickened diploe; (b) hand: due to marrow expansion the bones are expanded and those trabeculae that remain are very thickened.

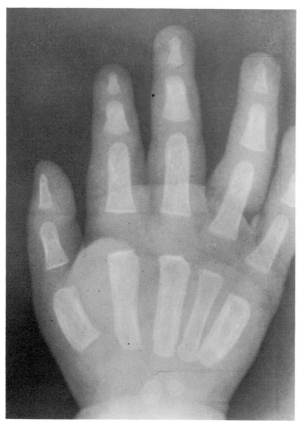

Fig. 9.36 Haemolytic anaemia—sickle-cell disease. Periosteal reactions due to bone infarction are seen along the shafts of the metacarpals.

tionally large. An important characteristic feature is that the distance between the pedicles narrows going down the lumbar spine (Fig. 9.38b). This is the reverse of what happens in the normal. The pelvis is contracted and sacrum is set low within the pelvis.

In *diaphyseal aclasia (multiple exostoses)* there are bony projections from the shaft of the bone known as osteochondromas or exostoses. They have a cartilagenous cap which may have calcification within it. When osteochondromas occur on the long bones they are near the metaphyses and are directed away from the joint. There is associated deformity and lack of modelling of

Fig. 9.37 Sarcoid: showing the characteristic lace-like trabecular pattern in the middle phalanx.

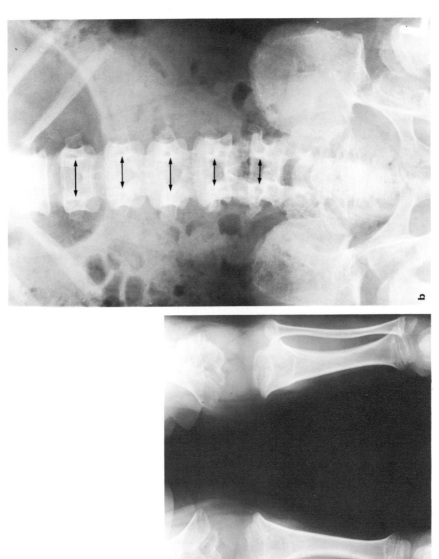

Fig. 9.38 (a) Achondroplasia: this child shows shortening of the bones with expansion of the metaphyses; (b) the lumbar spine shows narrowing of the interpedicular distances (arrows). These distances decrease from L1 to L5 which is the reverse of normal. Note the horizontal acetabular roofs.

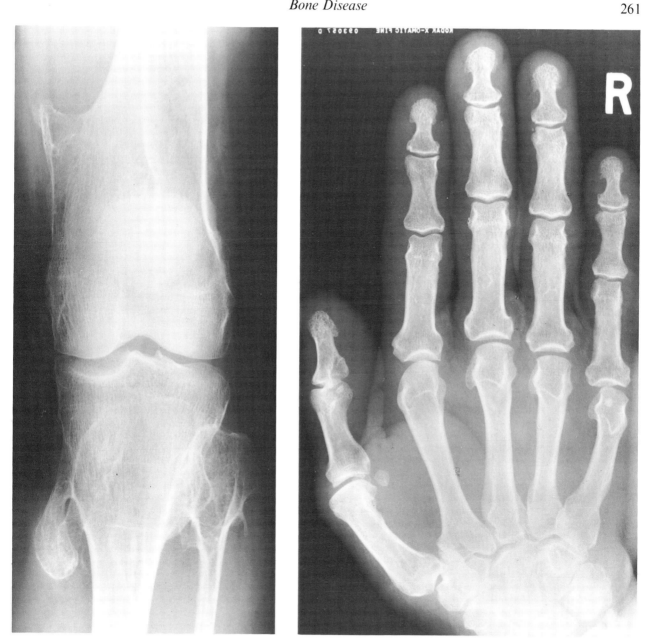

Fig. 9.39 (*left*) Diaphyseal aclasia. Several bony projections (exostoses) are arising around the knee and directed away from the joint. The opposite knee was similarly affected.

Fig. 9.40 (*right*) Acromegaly. The hand is large with prominent tufts to the terminal phalanges. There is widening of metacarpophalangeal joint spaces due to overgrowth of articular cartilage.

the shafts and metaphyses of the bones (Fig. 9.39). Osteochondromas may be single.

Occasionally, a chondrosarcoma may develop in the cartilage cap. This should be suspected if there is either rapid growth, an ill-defined edge to the bone or extensive calcification extending into the soft tissues.

Gigantism and acromegaly

Both these conditions are due to overproduction of growth hormone by the anterior pituitary. When this occurs before normal growth has ceased bony overgrowth results in gigantism; after epiphyseal closure acromegaly will develop. The bone and soft tissue overgrowth is maximal in the hands, feet and face. The tufts of the terminal phalanges enlarge and overgrowth of the articular cartilage in the hands and feet results in widened joint spaces. The outline of the bones becomes irregular (Fig. 9.40). The changes, however, may be

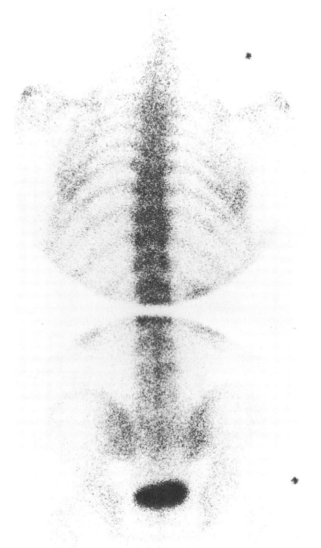

Fig. 9.42 Normal bone scan: posterior scans of thorax and pelvis. Note the isotope in the bladder.

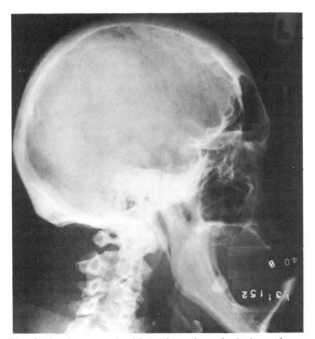

Fig. 9.41 Acromegaly. Note the enlarged pituitary fossa, thickened skull vault, large frontal sinuses and prognathous jaw.

difficult to differentiate from a normal hand in a manual worker.

In the skull the pituitary fossa is frequently enlarged by the eosinophil adenoma responsible for the excess growth hormone. The skull vault may be thickened and the sinuses and mastoid air cells enlarged (Fig. 9.41). A

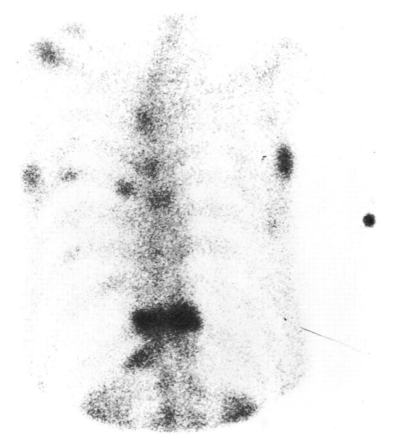

Fig. 9.43 Metastases: posterior scan of thorax showing several areas of increased uptake due to metastases from carcinoma of the prostate.

typical feature is the prognathous jaw with an increase in the angle between the body and ramus of the mandible.

Since many of these signs are subjective, attempts at measurement have been made. The thickness of the heel pad is one in common use. This has not proved particularly helpful for initial diagnosis as it only becomes unequivocal when the diagnosis is clinically obvious. The measurements are useful in following the response to treatment.

BONE SCANNING

If a ^{99m}Tc-labelled phosphate complex is injected intra-venously it is taken up selectively by the bones (Fig. 9.42). Increased uptake is seen in conditions when there is an increased blood supply, increased bone formation and high bone turnover including benign and malignant tumours, infection, infarction, Paget's disease and fractures; therefore correlation with the x-ray films is essential.

A bone scan is particularly valuable in demonstrating an abnormality even when the x-ray films are normal as may occur with bone metastases and early osteomyelitis. Bone scanning is the best survey procedure for detecting metastases (Fig. 9.43) since these may be more easily shown and at an earlier stage than with a radiographic skeletal survey.

10

Joints

Synovial joints have articular surfaces covered by hyaline cartilage. Cartilage is of the same radiodensity as the soft tissues and, therefore, is not visualised as such; only the space between the adjacent articular cortices can be appreciated (Fig. 10.1). Hence the commonly used expression 'joint space' is in fact a misnomer as it is not a space at all but consists of articular cartilage and synovial fluid. The synovium, synovial fluid and capsule also have the same radiodensity as the surrounding soft tissues and unless outlined by a plane of fat cannot be identified as discrete structures. The articular cortex forms a thin, well-defined line which merges smoothly with the remainder of the cortex of the bone.

SIGNS OF JOINT DISEASE

Signs indicating the presence of arthritis (but not the type)

Joint space narrowing

This is due to destruction of articular cartilage. It occurs in practically all forms of joint disease except avascular necrosis.

Soft tissue swelling

Swelling of the soft tissues around a joint may be seen in any arthritis accompanied by a joint effusion and whenever periarticular inflammation is present. It is, therefore, a particular feature of inflammatory and infective arthritis. The thickened synovium in synovial tumours and haemophilia may give rise to striking soft tissue swelling. Discrete soft tissue swelling around the joints can be seen in gout and rheumatoid arthritis.

Osteoporosis

Osteoporosis in the bones around joints occurs in many painful conditions. Underuse of the bones seems to be an important factor in the osteoporosis. Osteoporosis is particularly severe in rheumatoid and tuberculous arthritis.

Signs that point to the cause of the arthritis

Bone erosions

An erosion is an area of destruction of the articular cortex and the adjacent trabecular bone (Fig. 10.2). Bone erosion is usually accompanied by destruction of the articular cartilage. Erosions are easily recognised when seen in profile but when viewed *en face* the appearances can be confused with a cyst. Oblique views designed to show erosions in profile are often taken.

There are several mechanisms causing erosions:

1. Inflammatory overgrowth of the synovium (pannus) which occurs in:
    (a) Rheumatoid arthritis, which is by far the commonest cause of an erosive arthropathy
    (b) Juvenile chronic polyarthritis (Still's disease)
    (c) Psoriasis
    (d) Reiter's disease
    (e) Ankylosing spondylitis
    (f) Tuberculosis.
2. Response to the deposition of urate crystals in gout.

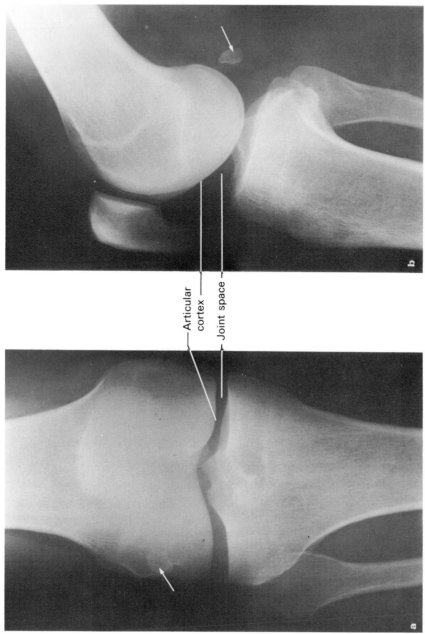

Articular cortex

Joint space

Fig. 10.1 Normal knee joint. Note the fabella (arrow). This is a sesamoid bone in the gastrocnemius.

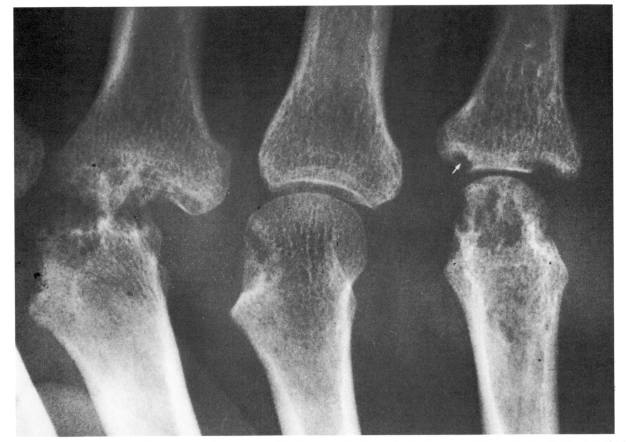

Fig. 10.2 Erosions. Areas of bone destruction are seen affecting the articular cortex in this case of rheumatoid arthritis. A typical erosion is arrowed. The joint spaces are also narrowed.

3. Destruction due to infection:
   (a) Pyogenic arthritis
   (b) Tuberculosis.
4. Synovial overgrowth due to repeated haemorrhage in haemophilia and related bleeding disorders.
5. Destruction due to neoplastic overgrowth of synovium, e.g. synoviosarcoma.

Osteophytes, subchondral sclerosis and cysts

These are the features of osteoarthritis and neuropathic joints. They are discussed further on p. 273. A characteristic increase in the density of subchondral bone is seen in avascular necrosis (see p. 275).

Alteration in the shape of the joint

There are a number of conditions that lead to characteristic alterations in the shape or relationship of the bone ends, e.g. slipped epiphysis, congenital dislocation of the hip, osteochondritis dissecans and avascular necrosis in its later stages.

When dealing with an arthritis it is important to know the following information:

1. Is more than one joint involved? Certain diseases typically involve several joints, e.g. rheumatoid arthritis, while others rarely do, e.g. infections and synovial tumours. This can be helpful since, for example, rheumatoid arthritis and infective arthritis may produce very similar changes in individual joints.

2. Which joints are involved? Many arthropathies have a predilection for certain joints and spare others. The distribution of joint disease can be very helpful when considering those diseases in which the changes in individual joints appear similar, e.g. various causes of erosion of the articular cortex. The following list highlights certain useful diagnostic features of the distribution of arthritis:

Rheumatoid arthritis virtually always involves the hands and feet, principally the metacarpo- and metatarsophalangeal and proximal interphalangeal joints, and also the wrist joints. Psoriatic arthritis usually affects the terminal interphalangeal joints.

Gout, characteristically, involves the metatarsophalangeal joint of the big toe.

When osteoarthritis is seen in the hands it almost always involves the terminal interphalangeal joints and often affects the carpometacarpal joint of the thumb. In the feet it is almost always the first metatarsophalangeal joint that is affected. In the large joints osteoarthritis is common in the hips and knees but relatively rare in the ankle, shoulders and elbows unless there is some underlying deformity or disease.

The distribution of neuropathic arthritis depends on the neurological deficit, e.g. diabetes affects the ankles and feet, tabes dorsalis affects the knees and syringomyelia and shoulders, elbows and hands.

3. Is a known disease present? Sometimes an arthritis is part of a known generalised disease, e.g. haemophilia.

Rheumatoid arthritis

Rheumatoid arthritis is a polyarthritis due to inflammatory overgrowth of the synovium often accompanied by a general systemic disturbance. Any synovial joint may be involved but particularly the wrists and small joints of the hands and feet with relative sparing of the terminal interphalangeal joints.

The earliest change in soft tissue swelling and osteoporosis around the joints. This periarticular osteoporosis is believed to be in part due to a combination of synovial hyperaemia and disuse due to pain. Destruction of the articular cartilage by pannus leads to joint space narrowing. Further destruction leads to small erosions which occur, initially, at the joint margins (Fig. 10.3). These are often seen first around the metatarso- or metacarpophalangeal joints, proximal interphalangeal joints and on the styloid process of the ulna. Later extensive erosions may disrupt the joint surfaces and ulnar deviation is usually present at this stage. With very severe destruction the condition is referred to as arthritis mutilans (Fig. 10.4).

Similar changes are seen in the large joints (Fig. 10.5). In these instances osteoarthritis may be superimposed on the changes of rheumatoid arthritis and may dominate the picture.

With severe disease there may be subluxation at the atlantoaxial joint due to laxity of the transverse ligament which holds the odontoid peg against the anterior arch of the atlas, so that the distance between this and the front of the odontoid peg becomes greater than 2 mm. Atlantoaxial subluxation may only be demonstrable in a film taken with the neck flexed (Fig. 10.6). It may be asymptomatic but there is always the possibility that it may give rise to neurological symptoms due to compression of the spinal cord by the odontoid peg.

The role of radiology in rheumatoid arthritis

Radiology assists in the diagnosis of doubtful cases. To this end the detection of erosions is extremely helpful. A widespread erosive arthropathy is almost diagnostic of rheumatoid arthritis. Radiology is also useful in assessing the extent of the disease and in observing the response to treatment.

Other erosive arthropathies

Juvenile chronic polyarthritis (Still's disease) shows many features similar to rheumatoid arthritis but erosions are rarer. The knee, ankle and wrist are the joints most commonly affected. Hyperaemia due to joint inflammation causes epiphyseal enlargement and premature fusion. In *psoriasis* there is an erosive arthropathy with predominant involvement of the terminal interpha-

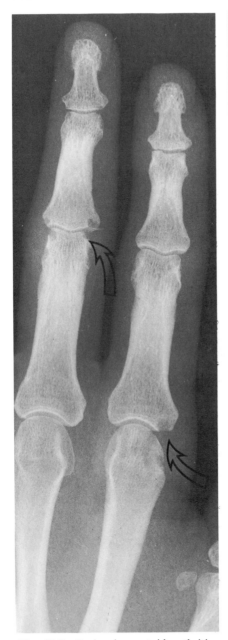

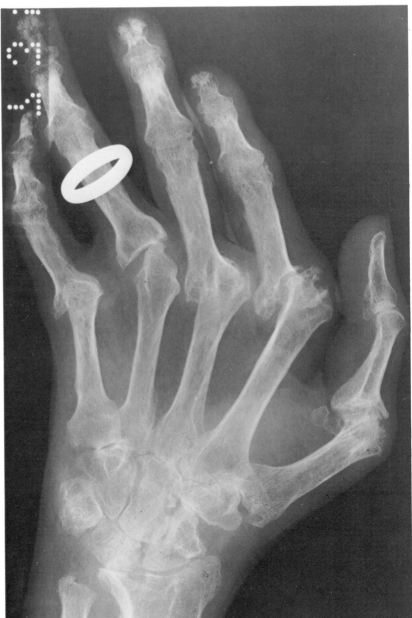

Fig. 10.3 Early rheumatoid arthritis. Small erosions are present in the articular cortex (arrows) and there is a soft tissue swelling around the proximal interphalangeal joints.

Fig. 10.4 Advanced rheumatoid arthritis (arthritis mutilans). There is extensive destruction of the articular cortex of the metacarpophalangeal joints with ulnar deviation of the fingers. Erosions are also present in the carpus, on the lower end of the radius and the styloid process of the ulna.

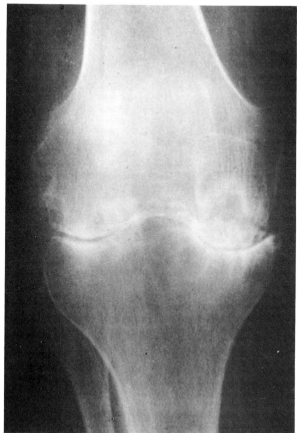

Fig. 10.5 Rheumatoid arthritis. There is uniform joint space narrowing and small erosions around the knee. Sclerosis is also present due to associated osteoarthritis.

langeal joints (Fig. 10.7). In *Reiter's disease* few joints are involved and there may be calcaneal erosions with spur formation. *Ankylosing spondylitis* causes an erosive arthritis which always affects the sacroiliac joints.

Gout

In gout, the deposition of urate crystals in the joint and in the adjacent bone gives rise to an arthritis which most commonly affects the metatarsophalangeal joint of the big toe.

The earliest change is soft tissue swelling. At a later stage erosions occur which, unlike rheumatoid arthritis, may be at a distance from the articular cortex. These erosions have a well-defined, often sclerotic, edge and frequently have overhanging edges (Fig. 10.8a). They are due to urate deposits in the bone. These deposits may be very large causing extensive bony destruction. There is usually no osteoporosis.

Localised soft tissue lumps due to collections of sodium urate, known as tophi, may occur in the periarticular tissues (Fig. 10.8b). These swellings may be large and may show calcification occasionally.

Joint infections

Many different bacteria can cause an infective arthritis but a feature common to them all is that they usually affect only one joint. Infective arthritis can be divided into:

1. Pyogenic arthritis—*Staphylococcus aureus* being the commonest organism.
2. Tuberculous arthritis.

Synovial biopsy or examination of the joint fluid are necessary in order to identify the infecting organism.

Pyogenic arthritis

In a pyogenic arthritis there is rapid destruction of the articular cartilage followed by destruction of the subchondral bone. A pyogenic arthritis may, occasionally, be due to spread of osteomyelitis from the metaphysis into the joint. A soft tissue swelling is seen around the joint (Fig. 10.9). At a late stage bony ankylosis may occur with complete obliteration of the joint.

Tuberculous arthritis

An early pathological change is the formation of pannus which explains why a tuberculous arthritis may be radiologically indistinguishable from rheumatoid arthritis. The hip and the knee are the most commonly affected peripheral joints. The features to look for are joint space narrowing and erosions which may lead to extensive destruction of the articular cortex. A very

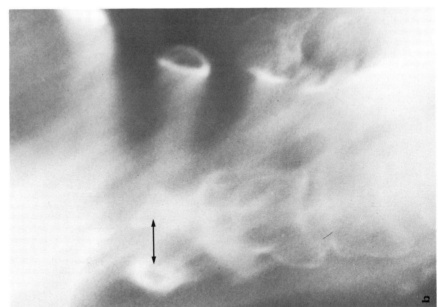

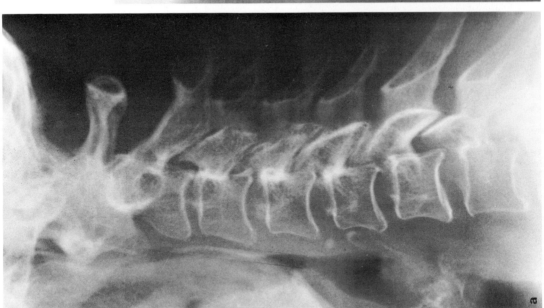

Fig. 10.6 Rheumatoid arthritis—atlantoaxial subluxation.
(a) Film in flexion shows C1 is displaced anteriorly upon C2
but the odontoid peg cannot be identified.

(b) Tomogram shows erosions of the odontoid peg. The distance between the arch of the atlas and the odontoid peg
(arrow) is increased from the normal value of 2 mm to 8 mm.

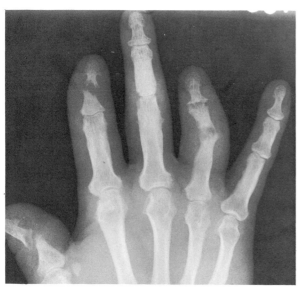

Fig. 10.7 Psoriatic arthropathy. There are extensive erosive changes affecting the interphalangeal joints but sparing the metacarpophalangeal joints.

Fig. 10.8 (*below*) Gout. (*left*) Erosion: there is a typical well-defined erosion with an overhanging edge (arrow) at the metatarsophalangeal joint of the big toe. (*right*) tophi: these are the large soft tissue swellings. A good example is seen around the proximal interphalangeal joint of the index finger. Several erosions are present (one of these is arrowed).

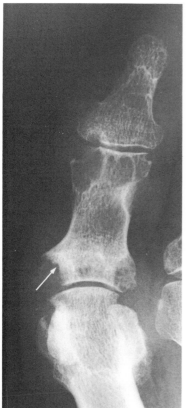
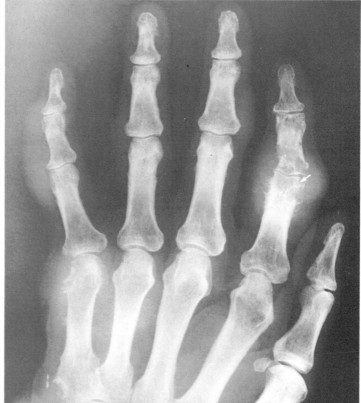

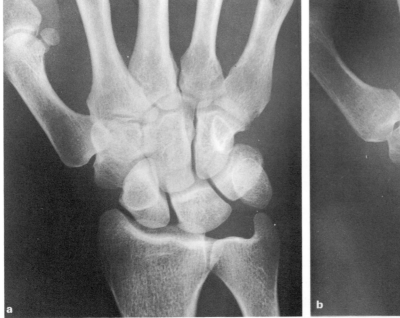
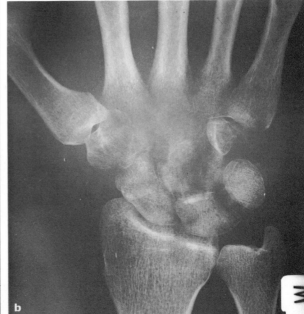

Fig. 10.9 (*above*) Pyogenic arthritis.
(a) Initial film of the wrist was normal;
(b) film taken 3 weeks later shows
destruction of the carpal bones and
bases of the metacarpals.

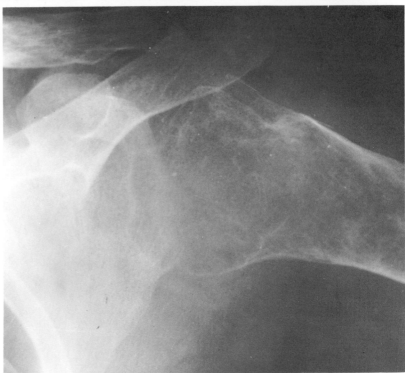

Fig. 10.10 Tuberculous arthritis of
the shoulder. Note the striking
osteoporosis and erosion of the
humeral head.

important sign is a striking osteoporosis which may be seen before any destructive changes are visible (Fig. 10.10).

At a late stage there may be gross disorganisation of the joint and calcified debris may be visible near the joint.

Haemophilia and bleeding disorders

In haemophilia and Christmas disease repeated haemorrhages into the joints result in soft tissue swelling, erosions and cysts in the subchondral bone. The epiphyses may enlarge and fuse prematurely (Fig. 10.11).

Osteoarthritis

This is the commonest form of arthritis. It is due to

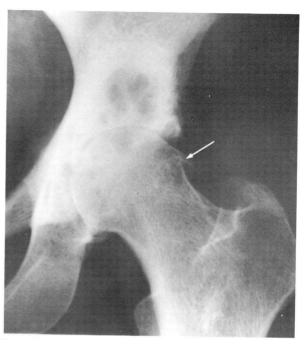

Fig. 10.12 Osteoarthritis of the hip. Note the narrowed superior part of the joint space, subchondral sclerosis and cyst formation and osteophytes (arrow).

degenerative changes resulting from wear and tear of the articular cartilage. The weight-bearing joints such as the hip and the knee are frequently involved but the ankle is infrequently affected.

In osteoarthritis a number of features can usually be seen (Fig. 10.12):

1. Joint space narrowing. The loss of joint space is maximal in the weight-bearing portion of the joint, e.g. in the hip it is often maximal in the superior part of the joint. Even when the joint space is very narrow it is usually possible to trace out the articular cortex.

2. Osteophytes are bony spurs, which may be quite large, occurring at the articular margins on both sides of the joint.

3. Subchondral sclerosis usually occurs on both sides of the joint.

4. Subchondral cysts may be seen beneath the articular cortex often in association with subchondral sclerosis. Normally, the cysts are easily distinguished from an

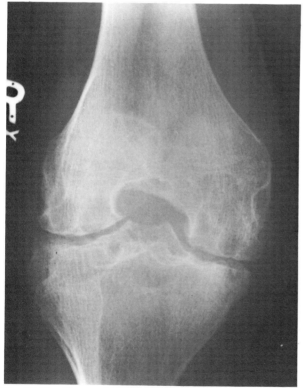

Fig. 10.11 Haemophilia. Subchondral cysts have formed due to repeated haemorrhages into the joint. Note the soft tissue swelling around the joint and the deep intercondylar notch; a characteristic feature of haemophilia.

Table 10.1   Comparison of osteoarthritis and rheumatoid arthritis

Osteoarthritis	*Rheumatoid arthritis*
Joint space narrowing maximal at weight-bearing site	Joint space narrowing uniform
Erosions do not occur but crumbling of the joint surfaces may mimic erosions	Erosions a characteristic feature
Subchondral sclerosis and cysts may be seen	Not a feature but erosions *en face* may mimic cysts
Sclerosis is a prominent feature	Sclerosis not a feature unless there is secondary osteoarthritis
No osteoporosis	Osteoporosis often present

erosion as they are beneath the intact cortex and have a sclerotic rim but occasionally, if there is crumbling of the joint surfaces, the differentiation becomes more difficult.

5. *Loose bodies* are discrete pieces of calcified cartilage or bone lying free within the joint, most frequently seen in the knee. It is important not to call the fabella, a sesamoid bone in the gastrocnemius, a loose body in the knee joint (Fig. 10.1).

Osteoarthritis and rheumatoid arthritis are the two types of arthritis most commonly encountered. They show many distinguishing features which are listed in Table 10.1.

Neuropathic (Charcot) joint

In a neuropathic joint there is an extreme form of degenerative change resulting from the loss of pain sensation. Typical examples are the shoulder and elbow in syringomyelia and the knee in tabes dorsalis. The joint is completely disorganised with much sclerosis of the surrounding bone. The joint is often subluxed with bone fragments or calcified debris around it (Fig. 10.13).

A different picture of a neuropathic joint is seen in the feet of diabetics with peripheral neuropathy. In these cases the predominant feature is absorption of the bone ends. There may also be bone destruction due to infection (Fig. 10.14). Calcification of the arteries in the foot is often present.

Chondrocalcinosis

Chondrocalcinosis is a descriptive term for calcification occurring in articular cartilage. In the knee, which is the

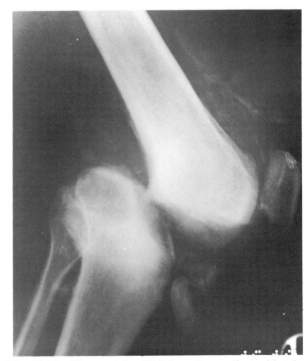

Fig. 10.13   Charcot joint. This knee in a patient with tabes dorsalis is grossly disorganised with sclerosis, calcified debris and bone fragments around the joint.

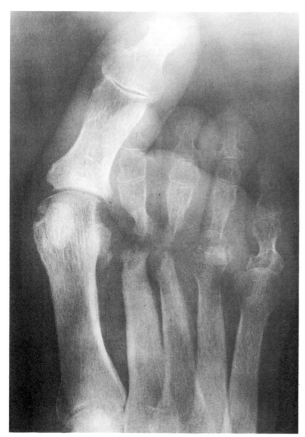

Fig. 10.14 Diabetic foot. There is resorption of the heads of the second and third metatarsals and bases of the proximal phalanges causing disorganisation of the metatarsophalangeal joints. The patient had a peripheral neuropathy with an anaesthetic foot.

most frequently affected joint, calcification may occur in the fibrocartilage of the menisci as well as the articular cartilage (Fig. 10.15). This is usually due to crystal synovitis (pseudogout) in which crystals of calcium pyrophosphate are deposited in the joint. A severe degenerative-like arthritis may follow. Chondrocalcinosis, occasionally, may be due to hyperparathyroidism or haemochromatosis.

Synovial sarcoma (synovioma)

This appears as a soft tissue mass that may contain calcium. Bone destruction on one or both sides of the joint occurs at a later stage.

Osteochondromatosis

This is a benign condition in which small, calcified bodies arise from the synovium. They become calcified and are often detached and lie in the joint (Fig. 10.16).

Avascular necrosis

Avascular or aseptic necrosis occurs most commonly in the intra-articular portions of bones. It is associated with numerous underlying conditions including:

—steroid therapy
—collagen diseases
—radiation therapy
—sickle-cell anaemia
—exposure to high pressure environments, e.g. tunnel workers and deep-sea divers (caisson disease)
—fractures

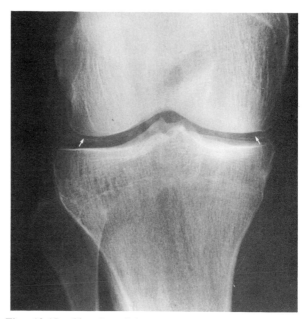

Fig. 10.15 Chondrocalcinosis. Calcification seen in the menisci in the knee (arrows).

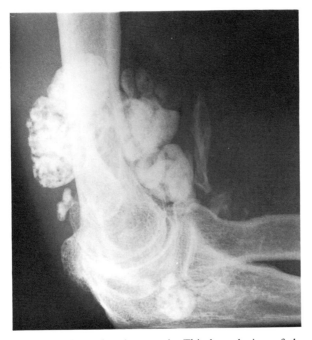

Fig. 10.16 Osteochondromatosis. This lateral view of the elbow shows many loose bodies in the joint.

There is also a group of conditions, some of which are called osteochondritis, in which no associated cause can be found. They are now regarded as being due to impaired blood supply associated with repeated trauma. Perthe's disease, an avascular necrosis of the femoral head in children, is the most important example.

The radiological features of avascular necrosis are increased density of the subchondral bone with irregularity of the articular contour or even fragmentation of the bone (Fig. 10.17).

Perthe's disease

The earliest change is increase in density and flattening of the femoral epiphysis which later may progress to collapse and fragmentation. The epiphysis becomes sclerotic due to the combined effect of deposition of new bone on the scaffold of the dead bone (Fig. 10.18). The epiphysis may widen and in consequence the femoral neck is widened and may contain small cysts. The joint space is widened and the acetabulum unaffected—these are features which distinguish Perthe's disease from most other joint disorders.

With healing the femoral head reforms but may remain permanently flattened which may, therefore, be responsible for osteoarthritis in later life.

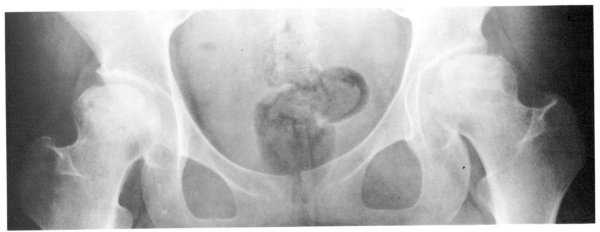

Fig. 10.17 Avascular necrosis. There is fragmentation with some sclerosis of both femoral heads.

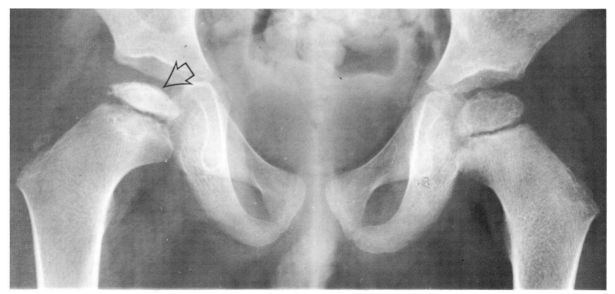

Fig. 10.18 Perthe's disease. The right femoral epiphysis (arrow) in this child is sclerotic and flattened. Compare it with the normal left side.

Other similar forms of avascular necrosis are:

1. Freiberg's disease affects the metatarsal heads, usually the second, which may be flattened, fragmented and sclerotic (Fig. 10.19).

2. Kohler's disease is avascular necrosis of the navicular bone of the foot resulting in fragmentation, flattening and sclerosis.

3. Osgood–Schlatter disease involves the tibial tuberosity which is fragmented with soft tissue swelling over it.

4. Kienböck's disease there is fragmentation and flattening of the lunate bone in the wrist (Fig. 10.20)

Posttraumatic avascular necrosis

After a fracture the blood supply may become interrupted and avascular necrosis supervene. This occurs, particularly, in subcapital fractures of the femoral neck and fractures through the waist of the scaphoid. The femoral head and proximal part of the scaphoid develop an impaired blood supply and become fragmented and dense (Fig. 10.21).

Osteochondritis dissecans

This is thought to be a localised form of avascular necrosis. A small fragment of bone becomes separated from the articular surface of the joint leaving a defect and the bony fragment can often be detected lying free within the joint (Fig. 10.22). It occurs most frequently in the knee.

Slipped femoral epiphysis

This occurs between the ages of 9 and 17 years and may present with pain in the hip or pain referred to the knee. The femoral epiphysis slips posteriorly from its normal position; this will be best appreciated on a lateral film of the hip (Fig. 10.23). With a greater degree of slip the condition can be recognised on the frontal view as a downward displacement of the epiphysis.

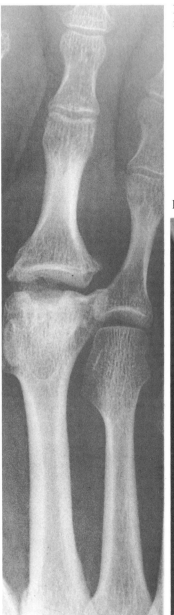

Fig. 10.19　Freiberg's disease. The head of the second metatarsal is flattened, irregular and sclerotic.

Fig. 10.20　Kienböck's disease. The lunate is flattened and sclerotic (arrow).

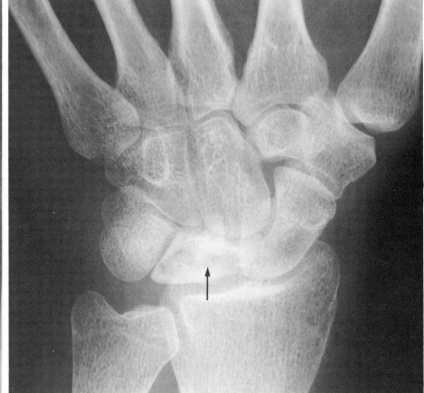

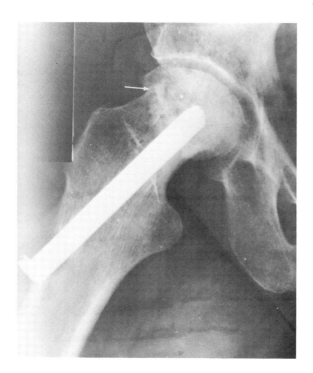

Fig. 10.21 Posttraumatic avascular necrosis. A pin has been inserted because of a subcapital fracture of the femoral neck (arrow) which occurred 10 months before this film was taken. Avascular necrosis has occurred in the head of the femur which has become sclerotic.

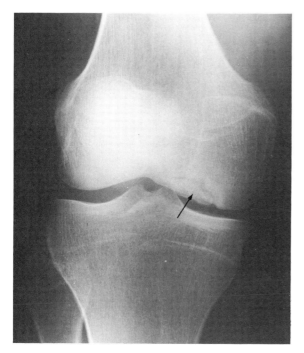

Fig. 10.22 Osteochondritis dissecans. A fragment (arrow) has become separated from the articular cortex of the medial femoral condyle.

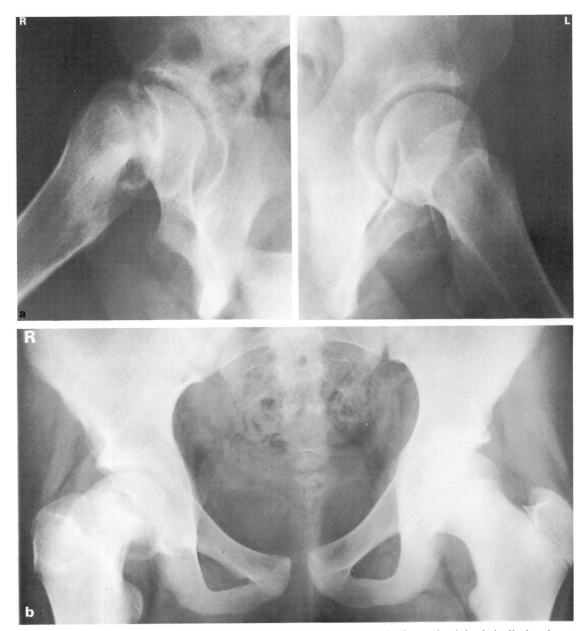

Fig. 10.23 Slipped femoral epiphysis. (a) Lateral view of the hips shows the right femoral epiphysis is displaced posteriorly (compare with the normal left side); (b) frontal view of same patient showing the right femoral epiphysis displaced downwards.

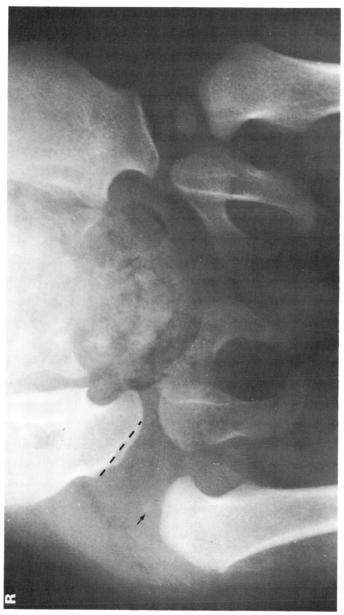

Fig. 10.24 Congenital dislocation of right hip. The right femoral epiphysis (arrow) is smaller than on the normal left side and it does not lie within the acetabulum. Note the sloping roof of the right acetabulum (dotted lines).

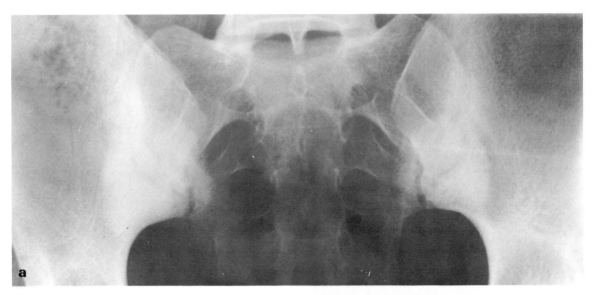

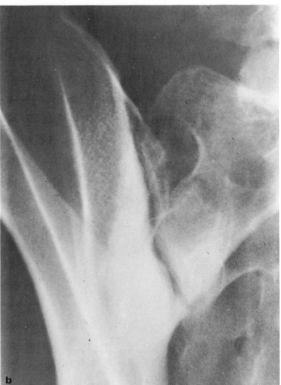

Fig. 10.25 Osteitis condensans ilii. (a) AP view: sclerosis is seen in both iliac bones just adjacent to the sacroiliac joints. The joints themselves, however, are normal. This is clearly shown on the oblique view (b). The patient was a young woman who had borne children.

Congenital dislocation of the hip

Although the diagnosis of congenital dislocation of the hips can be made by x-ray examination, this is not a reliable method of excluding dislocation in the neonatal period.

The features to look for are:

1. Lateral and upward displacement of the head of the femur (Fig. 10.24). The position of the femoral head may be difficult to assess in the newborn period as the femoral capital epiphysis does not appear until 3 to 6 months after birth. The position of the upper end of the shaft of the femur then has to be used to predict the position of the femoral head. The head of the femur develops more slowly on the affected side and may appear smaller compared with the normal side.

2. Increased slope to the acetabular roof.

Osteitis condensans ilii

Osteitis condensans ilii occurs almost exclusively in women who have borne children. The condition is thought to be a stress phenomenon associated with child bearing and is usually asymptomatic. There is a zone of sclerosis on the iliac side of the sacroiliac joints, but the sacroiliac joints themselves are normal (Fig. 10.25).

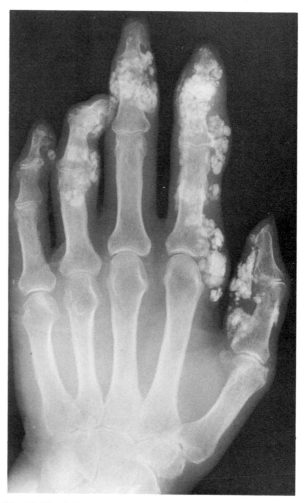

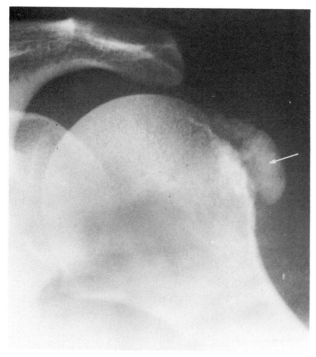

Fig. 10.26 Supraspinatus tendinitis. Calcification is present in the supraspinatus tendon (arrow).

Fig. 10.27 Scleroderma. Extensive soft tissue calcification is present as well as atrophy of soft tissues at the ends of the fingers.

Periarticular disease

Inflammation may occur in the soft tissues such as tendons and bursae around a joint especially the shoulder. The supraspinatus tendon is particularly affected and on the radiograph there may be amorphous calcification lying directly above the greater tuberosity of the humerus (Fig. 10.26).

Scleroderma

In the hands there may be calcification and atrophy of the soft tissues with loss of the tips of the terminal phalanges (Fig. 10.27).

11

The Spine

The detailed structure of the vertebrae differs in the cervical, thoracic and lumbar regions. The anatomy of the lumbar vertebrae is illustrated in Fig. 11.1. Note that the vertebral bodies are approximately rectangular in shape. There may be shallow indentations on the upper and lower surfaces of the vertebral bodies due to protrusion of disc material into the vertebral end plates. These indentations are known as Schmorl's nodes and are of no clinical significance (Fig. 11.2).

The bones of the spine may show manifestations of almost all the diseases that occur elsewhere in the skeleton; the emphasis in this chapter is on those features which are particular to the spine.

Signs of abnormality

1. Collapse of the vertebral body

A collapsed vertebra is one which has lost height (Fig. 11.3). Loss of height is best appreciated on the lateral view. The normal vertebral body is approximately rectangular in this view. If any collapse is present it is essential to check if any part of either pedicle is destroyed and to look at the adjacent disc to see if it is narrowed. The commoner causes of vertebral collapse are listed below together with a synopsis of the signs of importance in differential diagnosis:

(a) *Metastases and myeloma*. The pedicles may be destroyed. The disc spaces are usually normal.
(b) *Infection*. The disc space is usually narrow or obliterated. The pedicles are usually intact.
(c) *Osteoporosis and osteomalacia*. There is generalised reduction in bone density. The pedicles are intact. The disc spaces are normal or even increased in height.
(d) *Trauma*. A compression fracture is commonly due to forward flexion of the spine causing the vertebral body to become wedge shaped. The superior surface is usually concave (Fig. 11.4). The discs are normal.
(e) *Eosinophil granuloma*. Complete collapse of one or more vertebral bodies may occur in children or young adults with eosinophil granuloma. The vertebral body is flattened and is often referred to as a vertebra plana (Fig. 11.5). The adjacent discs are normal and the pedicles are usually preserved.

2. Disc space narrowing (Fig. 11.6)

The intervertebral discs are radiolucent as they are composed of fibrous tissue and cartilage. Normally, the disc spaces are the same height at all levels in the cervical and thoracic spine. In the lumbar spine the disc spaces increase slightly in height going down the spine, except for the disc space at the lumbosacral junction which is usually narrower than the one above it. Disc space narrowing occurs with degenerative disease and with disc space infection.

3. The pedicles

The pedicles are best assessed on the frontal view, except in the cervical spine where oblique views are necessary. Destruction of the pedicles causes them to disappear (Fig. 11.7). Loss of one or more of the pedicles is a fairly reliable sign of spinal metastases or myeloma.

Flattening and widening of the distance between the pedicles—the interpedicular distance—occurs in

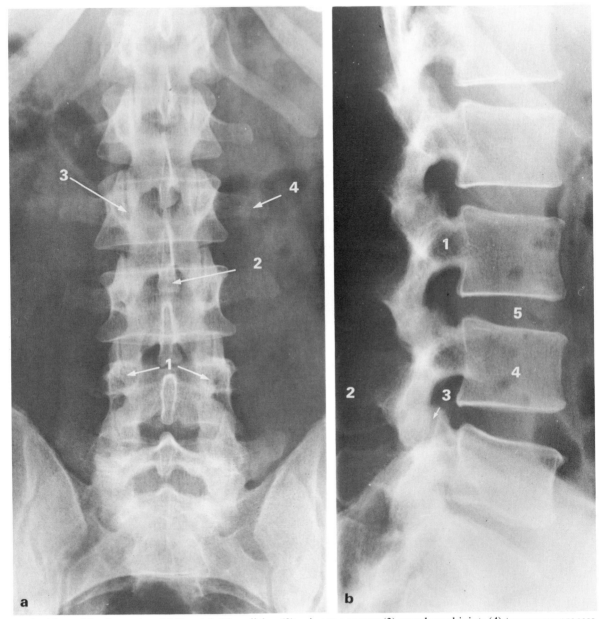

Fig. 11.1   Normal lumbar spine. (a) Frontal: (1) pedicles; (2) spinous process; (3) apophyseal joint; (4) transverse process. (b) Lateral: (1) pedicles; (2) spinous process; (3) apophyseal joint; (4) vertebral body; (5) disc space. Note how the height of the disc spaces increases from L1–L5 with the exception of the L5/S1 disc space which is normally narrower than the one above.

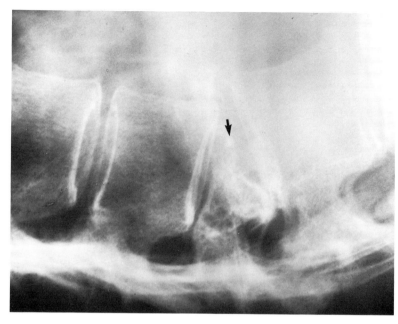

Fig. 11.3 Collapsed vertebra due to a metastasis (arrow). This has caused complete collapse of the vertebral body. The adjacent vertebral discs are unaffected.

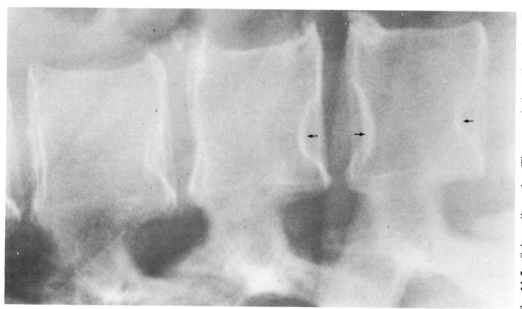

Fig. 11.2 Schmorl's nodes. These are indentations into the end plates of the vertebral bodies (arrows) and are without significance.

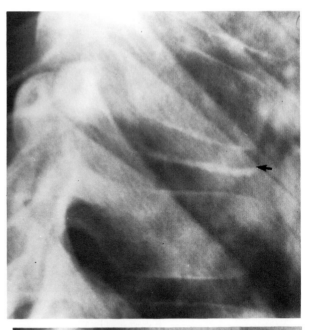

Fig. 11.4 (*below*) Traumatic collapse. Note the concave superior surface of the collapsed vertebral body. Some fragments have been extruded anteriorly (arrow).

Fig. 11.5 (*right*) Eosinophil granuloma. In this child the vertebral body is so collapsed it resembles a thin disc (arrow).

Fig. 11.6 (*below right*) Disc space narrowing. In this case it is due to disc degenerative changes between L3 and L4. Note the osteophytes (arrows) and sclerosis of the adjoining surfaces of the vertebral bodies.

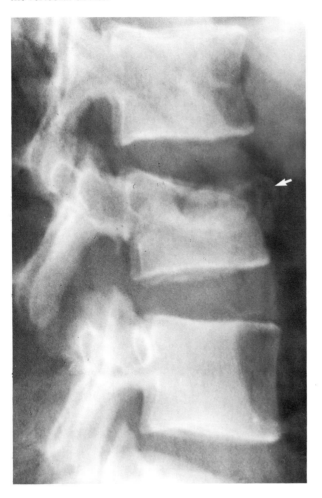

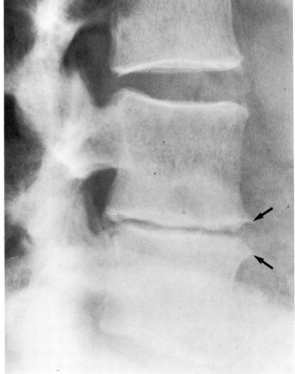

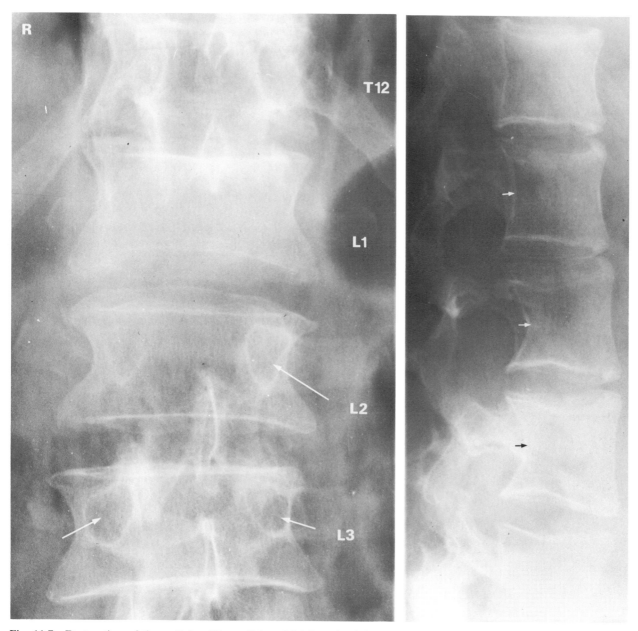

Fig. 11.7 Destruction of the pedicles. The pedicles of L1 have both been destroyed and the right pedicle of L2 is only partially visible. The normal left pedicle of L2 and the pedicles of L3 are arrowed. The patient had a renal cell carcinoma.

Fig. 11.8 Scalloping of the posterior surfaces of the bodies of the lumbar spine (arrows) due to a neurofibroma.

tumours arising within the spinal canal, e.g. neurofibroma or meningioma and there may be scalloping of the posterior border of the vertebral body (Fig. 11.8). Although neurofibromas may be completely intradural they may have a dumb-bell shape with a portion lying outside the spine. In these instances the intervertebral foramen will be enlarged (Fig. 11.9).

4. The dense vertebra

Sclerosis may be part of a generalised process involving many bones.

The common causes of a dense vertebra are:

(a) *Metastases*, particularly from primary tumours of the prostate or breast (Fig. 11.10).
(b) *Malignant lymphoma.*

(c) *Paget's disease*, which may be difficult to distinguish from malignant disease. An important diagnostic feature is increase in size of the vertebra. There may be a coarse trabecular pattern typical of Paget's disease but this is not invariably present in the spine (Fig. 11.11).
(d) *Haemangioma.* Characteristic vertical striations are seen in a normal-sized vertebra (Fig. 11.12).

5. Paravertebral shadow

A paravertebral shadow may first draw attention to an abnormality in the spine. It appears as a soft tissue swelling adjacent to the spine and can be distinguished from other shadows by its fusiform shape. The easiest place to recognise paravertebral swelling is in the thoracic region (Fig. 11.13). Swellings in the lumbar region

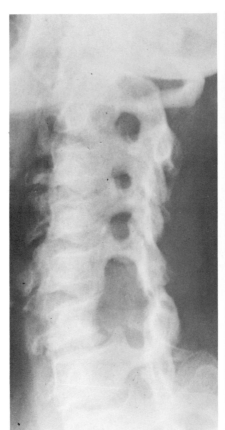
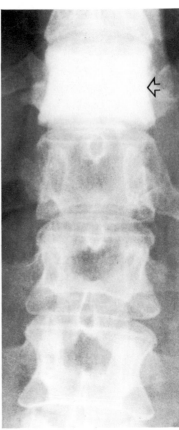

Fig. 11.9 (*left*) A dumb-bell neurofibroma has enlarged the intervertebral foramina in the cervical spine to cause a large bony defect.

Fig. 11.10 (*right*) Dense vertebra (arrow) due to metastases from carcinoma of the breast.

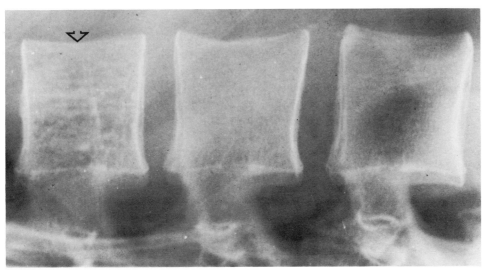

Fig. 11.12 Haemangioma. Vertical striations are present in this normal-sized vertebra (arrow).

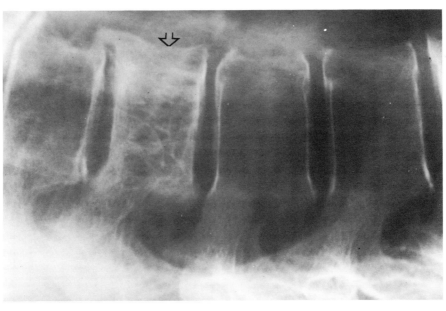

Fig. 11.11 Paget's disease. Note the increased density and coarse trabeculae in this thoracic vertebral body (arrow) which is slightly wider than the adjoining ones.

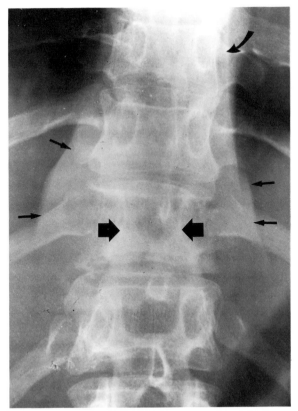

Fig. 11.13 Paravertebral shadow (small arrows) around a thoracic vertebra partially destroyed by Hodgkin's disease (large arrows). A paravertebral shadow has a fusiform shape and must be distinguished from normal shadows such as the edge of the descending aorta (curved arrow).

have to be very large to displace the psoas outline. Anterior swelling in the cervical region can also be readily recognised by the forward displacement of the pharyngeal air shadow.

A paravertebral shadow occurs with infection but may also be seen with malignant neoplasms and following trauma.

Metastases/myeloma

As with the remainder of the skeleton the important signs of metastases are areas of lysis or sclerosis or a mixture of both (Fig. 11.14). Multiple myeloma almost

always gives rise to lytic lesions which are frequently indistinguishable from lytic metastases. When several vertebrae are involved the diagnosis of malignant disease is virtually certain. Metastases in the spine frequently involve the pedicles.

Collapse of one or more vertebral bodies may occur with metastases and is a particular feature of myeloma. The collapse may mask the areas of bone destruction in the vertebral body.

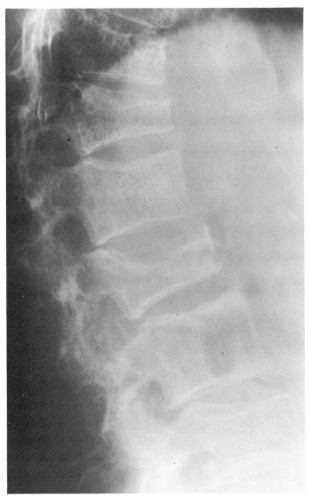

Fig. 11.14 Metastases. Lateral view of the upper lumbar spine. Note the abnormal bony architecture and varying degrees of collapse of several of the vertebral bodies.

True destruction of the disc space does not occur with metastases or myeloma.

Infection

The hallmark of infection is destruction of the intervertebral disc and adjacent vertebral bodies. Early in the course of the disease there is narrowing of the disc space with erosion of the adjoining surface of the vertebral body. Later, bone destruction may lead to collapse of the vertebral body resulting in a sharp angulation known as a gibbus (Fig. 11.15). A paravertebral abscess is usually present. The common infections are due to tuberculosis and *Staphylococcus aureus*. There are dif-

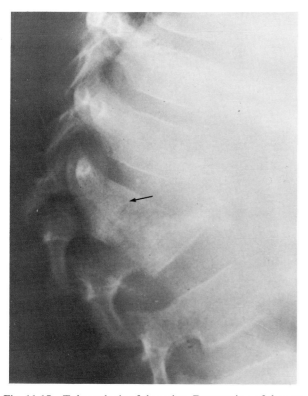

Fig. 11.15 Tuberculosis of the spine. Destruction of the vertebral bodies and the intervening discs has occurred with the formation of a sharp angulation (gibbus). One vertebral body is almost completely destroyed (arrow) and there is destruction of the upper part of the one below it.

ferences in the signs produced by these two infections but there is considerable overlap. The lesion in tuberculosis is usually purely lytic, whereas sclerosis of the edges of the destroyed area is more usual in pyogenic infection and there may be evidence of new bone formation. Paravertebral abscesses tend to be larger in tuberculosis.

Bony fusion of the vertebral bodies with obliteration of the disc spaces occurs with healing. Eventually, tuberculous paravertebral abscesses may calcify.

Spondylosis

Spondylosis results from degeneration of the intervertebral disc. The degenerate discs may herniate into the surrounding tissues and if the herniated disc presses on the spinal cord or spinal nerves pain and/or neurological deficit may result. The degenerate discs often stimulate the formation of osteophytes, which together with the thickened soft tissues may press upon the spinal cord or the nerves in the spinal canal and exit foramina. Osteoarthritic changes in the apophyseal joints may exacerbate this problem. Spondylosis occurs maximally in the lower cervical and lower lumbar regions.

In most cases in which spondylosis is seen on x-ray films, there are no neurological symptoms or signs. In patients with spondylosis, x-rays of the spine have limited clinical value as there is little correlation between the signs, symptoms and radiological changes. Even when there is disc protrusion producing neurological signs plain films of the spine may be normal. The major purpose of requesting x-rays of the spine is usually to exclude other disease that may be present.

The radiological signs of spondylosis are:

1. Disc space narrowing.
2. Osteophyte formation and sclerosis which frequently occur on the adjoining surfaces of the vertebral bodies. Osteophytes on the posterior surface of the bodies of the cervical vertebrae narrow the spinal canal and may also encroach on the exit foramina for the spinal nerves. In the cervical spine this is best shown on oblique views (Fig. 11.16).

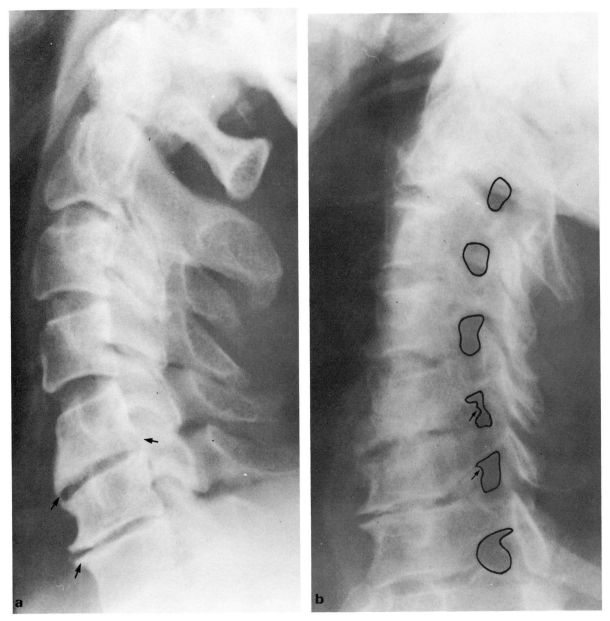

Fig. 11.16 Cervical spondylosis. (a) Lateral view: the C5/C6 and C6/C7 disc spaces are narrowed and there are osteophytes on the anterior and posterior aspects of these vertebral bodies (arrows); (b) oblique view: the intervertebral foramina have been drawn in. The osteophytes (arrows) are narrowing the foramina.

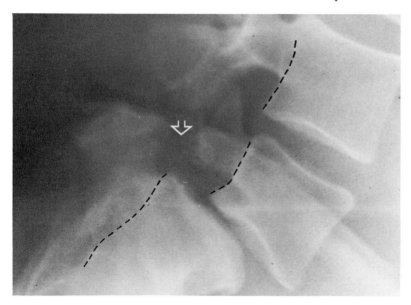

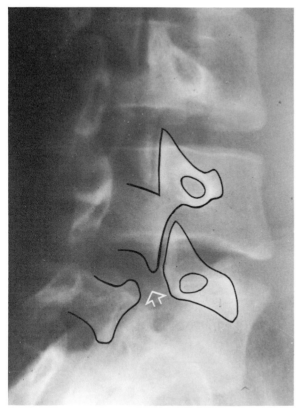

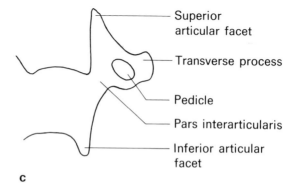

Fig. 11.17 Spondylolisthesis. (a) Lateral view: there is forward slip of L5 upon S1. The dotted lines which mark the posterior aspects of the vertebral bodies should form a smooth curve. The defect in the pars interarticularis is arrowed; (b) oblique view showing the defect in the pars interarticularis (arrow). On the oblique view a shape resembling the front end of a 'scottie dog' can be recognised; (c) the 'scottie dog': a defect in the pars interarticularis is seen as a break in the dog's neck.

Superior articular facet

Transverse process

Pedicle

Pars interarticularis

Inferior articular facet

c

Spondylolisthesis

This refers to forward slip of one vertebral body on the one below it, occurring most frequently at the lumbosacral junction and between the fourth and fifth lumbar vertebrae. It is usually the result of a defect between the superior and inferior articular facets (the pars interarticularis). The defect in the pars interarticularis is thought to be a stress fracture. It can sometimes be identified on the lateral projection but may be better seen on oblique films (Fig. 11.17). Spondylolysis is the term given to a defect in the pars interarticularis without a forward slip of one vertebral body on the other. Minor degrees of slip can also occur without a break in the pars interarticularis if there is degenerative disc disease with osteoarthritis in the apophyseal joints.

Spina bifida

Spina bifida is a result of incomplete closure of the vertebral canal, usually in the lumbosacral region,

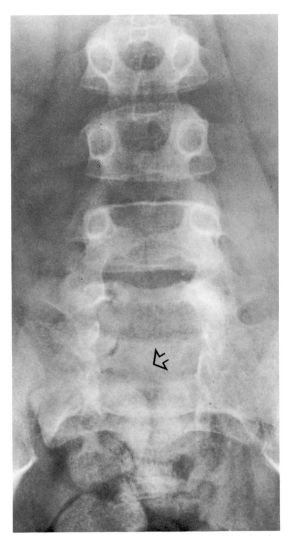

Fig. 11.18 Spina bifida. This child was born with a meningomyelocele. The laminae of L5 (arrow) and of the sacrum are absent.

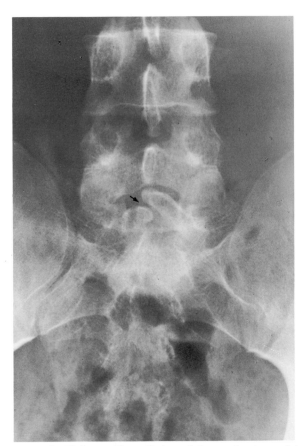

Fig. 11.19 Failure of bony fusion of the two laminae of the first part of the sacrum (arrow) is a common finding without significance.

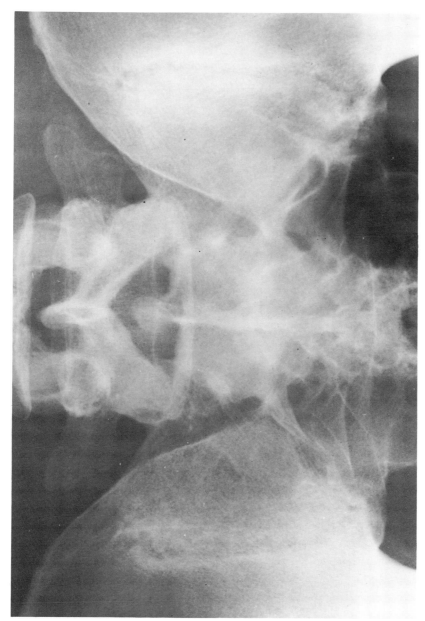

Fig. 11.20 Ankylosing spondylitis. The sacroiliac joints have an irregular fuzzy outline.

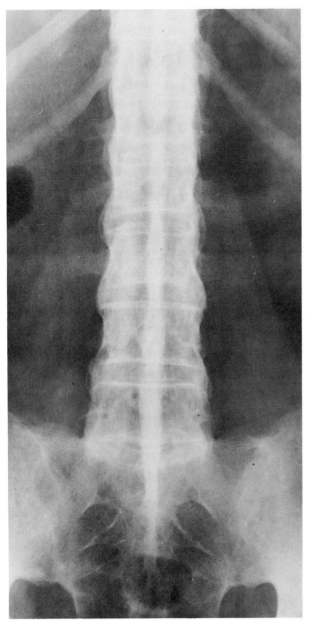

Fig. 11.21 Ankylosing spondylitis. With advanced disease the whole spine becomes fused (bamboo spine). Note the sacroiliac joints are also fused.

which may be associated with an abnormality affecting the spinal cord. In severe cases presenting at birth there may be protrusion of the spinal cord (meningo-myelo-cele) or its membranes (meningocele) from the spinal canal (Fig. 11.18). In these cases the laminae of several vertebrae will be absent and the distance between the pedicles will be increased. Complex malformations of the vertebral bodies may also be present.

Frequently, the patient has no visible abnormality and no neurological defect, but failure of bony fusion of the two laminae is visible radiologically. This may occur at any level but is common in the lumbosacral region and in these cases it is without significance (Fig. 11.19).

Ankylosing spondylitis

This condition affects principally the sacroiliac joints and the spine although occasionally other joints may be involved as well. Both sacroiliac joints are invariably affected by the time spinal involvement has occurred. The earlist radiological change is the fuzziness of the joint margins followed by frank erosions (Fig. 11.20). Eventually, the process leads to obliteration of the joint space.

In the spine the spinal ligaments ossify forming verti-cally oriented bony bridges between the vertebral bodies. The posterior apophyseal and the costotrans-verse joints become fused. In advanced cases the whole spine is rigidly fused and becomes a solid block of bone. From its radiographic appearance this is known as a 'bamboo spine' (Fig. 11.21).

Myelography and radiculography

The spinal cord and spinal nerves are invisible on plain films of the spine but they may be demonstrated if contrast is introduced into the subarachnoid space by lumbar puncture. The main indications are spinal cord on lumbar nerve root compression. Either water-soluble or oily contrast media (Myodil*) is used. Water-soluble contrast is rapidly absorbed from the subarachnoid space and it shows the nerve roots to advantage. Myodil

* Known as Pantopaque in the USA.

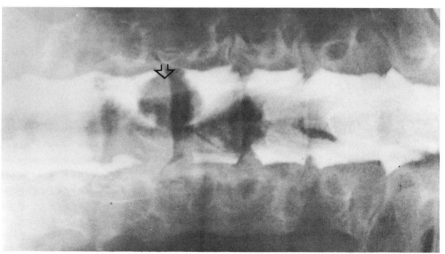

Fig. 11.24 Intradural neurofibroma. The myelogram shows a filling defect in the Myodil column (arrow) in the cervical region.

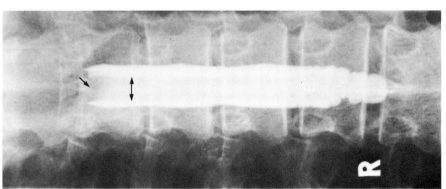

Fig. 11.23 Spinal cord compression. Myodil has been injected into the subarachnoid space. This film is taken with the patient lying head downwards and the flow of Myodil is arrested in the mid thoracic region due to an extradural block (arrow) caused by metastases. Note the normal half shadow in the Myodil column due to the spinal cord (arrows).

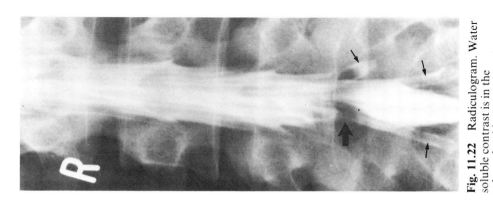

Fig. 11.22 Radiculogram. Water soluble contrast is in the subarachnoid space below the termination of the spinal cord. Note the filling of the nerve root sheaths (arrows). There is a large disc protrusion (large arrow).

is only very slowly absorbed and arachnoiditis is a serious though rare complication. When the spinal cord is examined, the examination is referred to as a myelogram. Examination of the lumbar nerve roots using a water soluble contrast is known as a radiculogram. This is normally carried out to demonstrate compression of nerve roots and to determine the level of a lumbar disc protrusion (Fig. 11.22).

Spinal cord compression

The causes of space-occupying lesions that cause compression of the spinal cord are usually divided according to the site of origin:

1. *Extradural*, e.g. cervical spondylosis, metastases (Fig. 11.23) and spinal tuberculosis.
2. *Intradural but extramedullary*, i.e. within the dura but not within the spinal cord, e.g. neurofibroma (Fig. 11.24) and meningioma.
3. *Intramedullary*, i.e. within the spinal cord, e.g. tumour (ependymoma).

A myelogram is performed in almost all cases to establish the diagnosis, but examining the plain films is most important as in some instances the cause will be visible. Although the level of the compression will be suspected on clinical examination a myelogram will demonstrate the exact site and often the nature and extent of the abnormality as well. This information is important should surgery be necessary.

12
Bone Trauma

Radiology in bone trauma is invaluable to diagnose the following:

1. Diagnosing the presence of a fracture or dislocation.
2. Determining whether the underlying bone is normal or whether the fracture has occurred through abnormal bone (pathological fracture).
3. Showing the position of bone ends before and after treatment of a fracture.
4. Assessing healing and complications of fractures.

The latter two subjects and details of individual fractures will not be discussed as they are best dealt with in textbooks of orthopaedics. Head injury is discussed on p. 323.

In any case of trauma it is essential to take at least two views, preferably at right-angles to one another. Sometimes a fracture or dislocation will be seen on only one view and so may be missed unless two views are taken. Similarly, the position of a fracture should never be assessed from a single film (Fig. 12.1).

Fracture

Frequently, a fracture is very obvious but in some cases the changes are more subtle. Fractures may be recognised or suspected by the following signs:

1. Fracture line. The fracture usually appears as a lucent line. This may be very thin and easily overlooked (Fig. 12.2). Occasionally, the fracture appears as a dense line due to overlap of the fragments (Fig. 12.3).
2. A step in the cortex may be the only evidence of a fracture (Fig. 12.4).
3. Interruption of bony trabeculae is of use in impacted

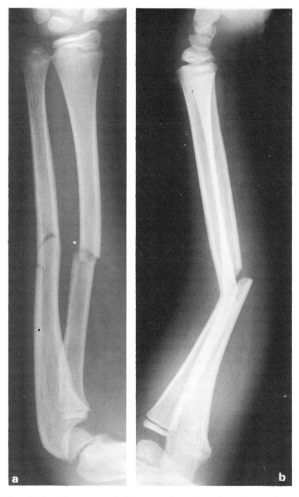

Fig. 12.1 Fracture of forearm showing the value of two views. (a) The fractures of the radius and ulna show little displacement on the frontal projection; (b) The lateral view, however, shows a marked angulation.

301

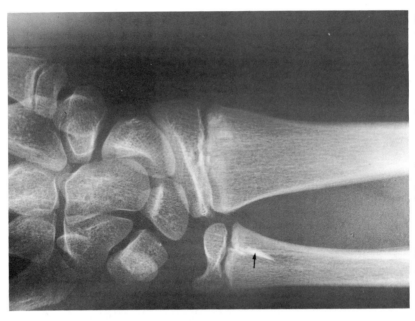

Fig. 12.3 Fracture of the ulna appearing as a sclerotic line (arrow).

Fig. 12.2 Fracture of the scaphoid appearing as a lucent line (arrow).

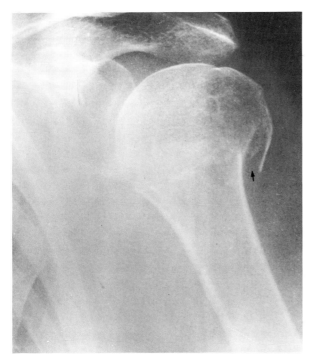

Fig. 12.4 Fracture of neck humerus appearing as a step in the cortex (arrow).

Further views

Injuries may sometimes be invisible even with two views taken at right-angles to each other. If the radiographic findings are equivocal or if there is clinical suspicion of bony injury with normal radiographs then further films such as the following should be taken:

1. Different projections, e.g. oblique views (Fig. 12.8).
2. Stress films. A film taken of a joint under stress may show that it is unstable due to ligamentous damage. Stress films are helpful in ankle injuries when forced inversion and eversion may show movement of the talus.

fractures where there is no visible lucent line. This is, however, a difficult sign to evaluate (Fig. 12.5).
4. Bulging or buckling of the cortex is a particularly important sign in children, where fractures are frequently of the greenstick type (Fig. 12.6).
5. Soft tissue swelling may be a valuable guide to the presence of an underlying fracture.
6. A joint effusion may become visible following trauma. In the elbow, where an effusion often indicates a fracture, fat pads lie adjacent to the joint capsule and if there is an effusion they will be shown on the lateral view to be displaced away from the shaft of the humerus (Fig. 12.7).

Dislocation

The joint surfaces no longer maintain their normal relationship to each other. Look very carefully for an associated fracture.

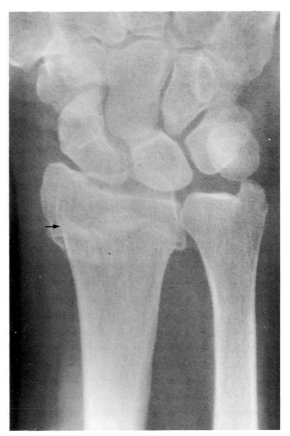

Fig. 12.5 Colles' fracture showing the bony trabeculae are interrupted across the fracture site (arrow). There is also a step in the cortex.

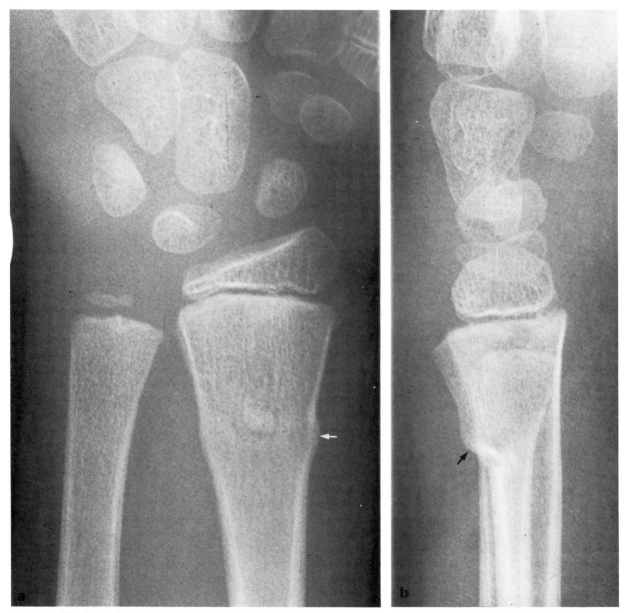

Fig. 12.6 Greenstick fracture of lower end of radius in a child. There is buckling of the cortex (arrows).

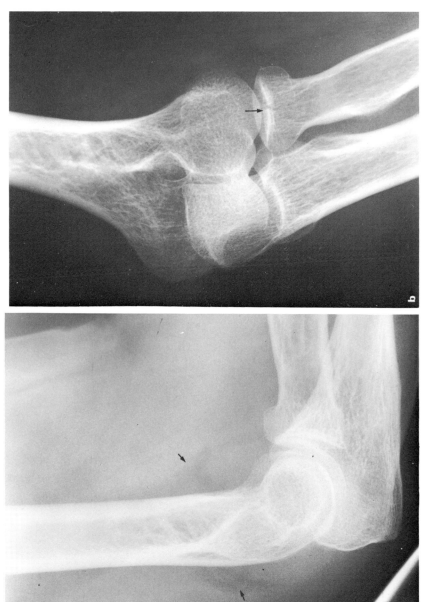

Fig. 12.7 Elbow effusion with fracture of radial head. (a) The anterior and posterior fat pads (arrows) are displaced away from the humerus which almost invariably means a fracture is present; (b) oblique view in this patient shows the fracture of the radial head (arrow) which was only demonstrated on the oblique view.

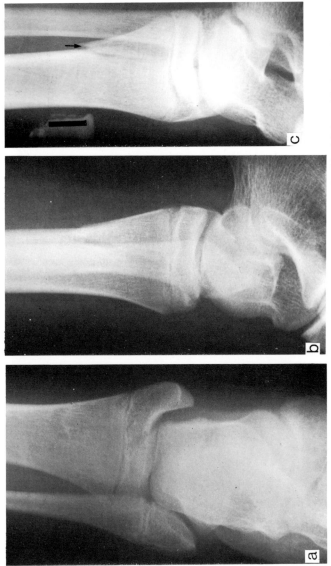

Fig. 12.8   Oblique view to demonstrate a fracture. (a) Frontal AP and lateral views in this child's ankle do not show an obvious fracture; (b) oblique view clearly demonstrates the fracture (arrow).

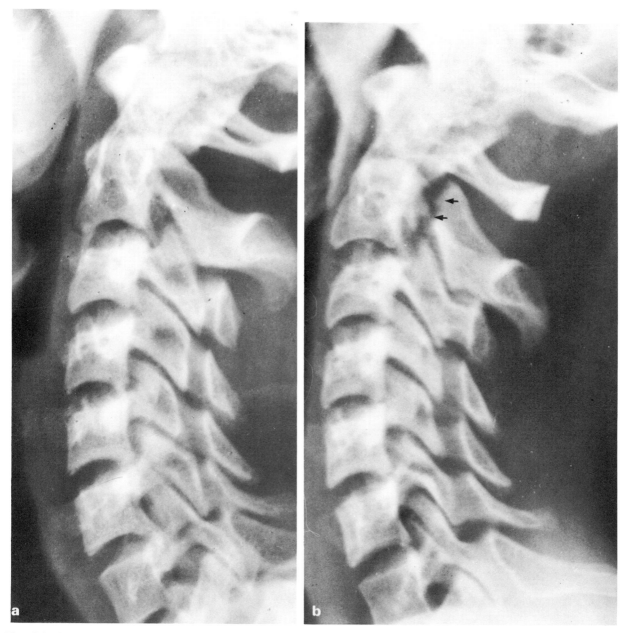

Fig. 12.9 Flexion and extension views to demonstrate a fracture. (a) Extension view of cervical spine does not reveal a fracture; (b) flexion view clearly shows the fracture of the arch of C2 (arrows).

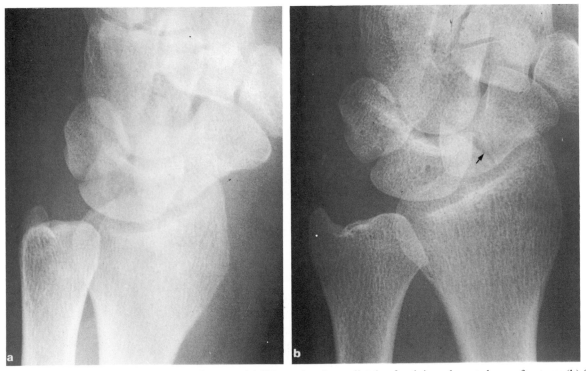

Fig. 12.10   Delayed films to demonstrate a fracture. (a) Films taken immediately after injury do not show a fracture; (b) films taken 2 weeks after injury show a fracture through the scaphoid (arrow).

3. Flexion and extension views. In the cervical spine, injury may cause alteration in the alignment of the posterior borders of the vertebral bodies. This is usually much more obvious on a film taken with the neck flexed than with it extended (Fig. 12.9). In the conscious patient pain prevents damage to the spinal cord from instability of the cervical spine, providing the movement is carried out by the patient himself. Flexion and extension views should not be carried out on the unconscious patient.

4. X-ray other side. Comparing with the normal side can be useful in the problem case particularly if expert help is not available. This applies largely in children where epiphyseal lines and unusual patterns of ossification may simulate a fracture.

5. Delayed films. If films are taken about 2 weeks after injury, resorption of the bone at the fracture site may then reveal the fracture line. This is often particularly useful in detecting scaphoid fractures which may be invisible immediately after the injury (Fig. 12.10).

Stress fracture

These are fractures due to repeated, often minor, trauma. A good example is the so-called march fracture occurring in the shafts of the metatarsals. Initially, although there is pain, a radiograph will show no evidence of a fracture in the metatarsals but if a further film is taken after 10–14 days a periosteal reaction may draw attention to the fracture site where a thin crack may be visible (Fig. 12.11).

Pathological fracture

A pathological fracture is one that occurs through abnormal bone. Pathological fractures may be the pre-

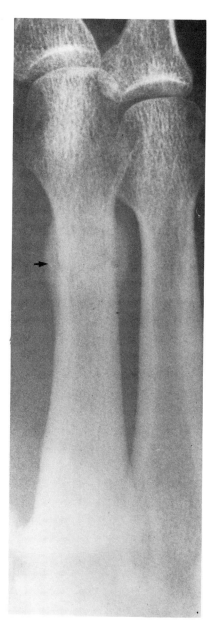

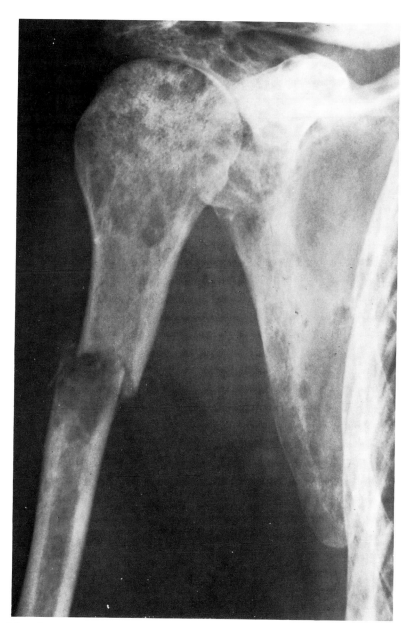

Fig. 12.11 March fracture. This film was taken 2 weeks after the onset of pain. It shows a periosteal reaction (arrow) around the metatarsal shaft although a fracture cannot be seen.

Fig. 12.12 Pathological fracture of the humerus. There are widespread lytic metastases from a carcinoma of the breast.

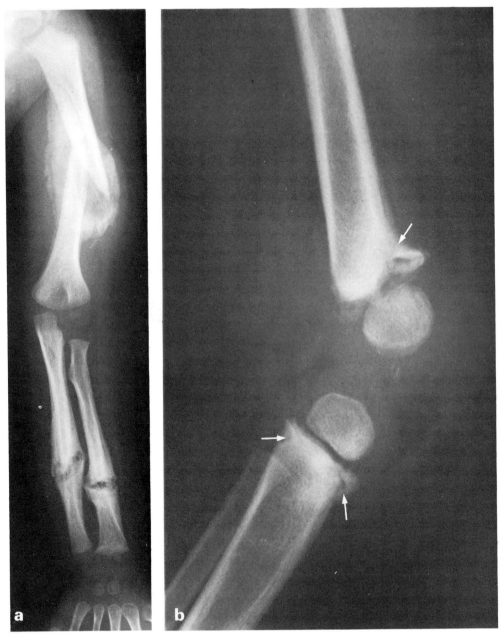

Fig. 12.13 Battered-baby syndrome. (a) Multiplicity of fractures. There is a recent fracture of the humerus with florid callus formation. The fractures of the radius and ulna are of longer duration and show healing with organised callus; (b) metaphyseal fractures (arrows) and sclerosis around the knee.

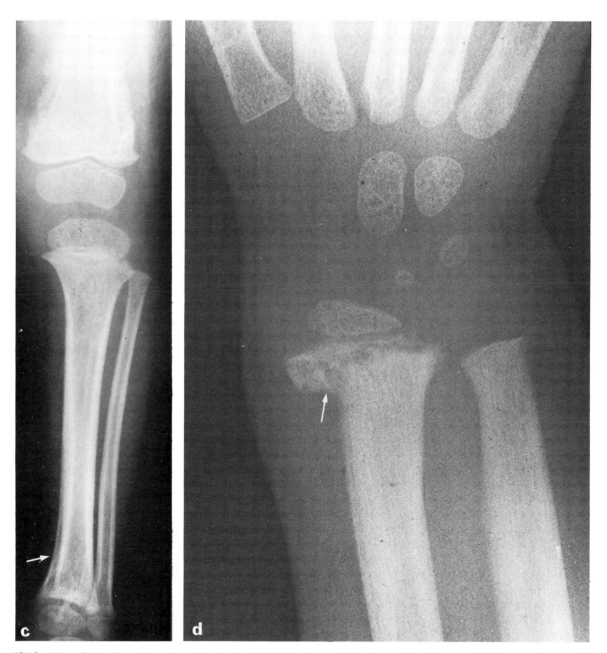

Fig. 12.13 (c) periosteal reaction along the shaft of the tibia (arrow) due to previous trauma with haemorrhage under the periosteum. There has been recent trauma to the lower end of the femur with a marked periosteal reaction; (d) metaphyseal fracture (arrow) of lower end of radius (same patient as in Fig. 12.13c). This fracture is very recent as there are not yet signs of healing.

senting feature of both primary and secondary bone tumours (Fig. 12.12). Often the causative lesion is obvious. However, metastases may be ill defined and difficult to see and yet be the site of a fracture. In such cases the diagnosis rests on recognising the irregularity of the margins of the fracture and the underlying bone destruction. This is a problem that often arises in the ribs. It is helpful to look elsewhere in the skeleton for other metastases if there is doubt about the diagnosis.

Pathological fractures may also occur in conditions causing demineralisation of bone such as osteoporosis and osteomalacia, but the underlying condition may be difficult to diagnose.

A transverse fracture often occurs through abnormal bone and this type of fracture is seen particularly in Paget's disease. In these instances the Paget's disease is always obvious (Fig. 9.33b).

Battered-baby syndrome (non-accidental injury)

It is essential that everyone looking at radiographs should be fully aware of this condition as the radiological findings may suggest the diagnosis in otherwise un-suspected cases. If child abuse is suspected the whole skeleton including the skull should be x-rayed, as clinically unsuspected injuries may be detected.

Certain patterns suggest that the injury is not accidental (Fig. 12.13):

1. Multiplicity of fractures is an important sign, particularly if the fractures are of different ages, because the injuries often take place on separate occasions. Some fractures appear recent, while others show signs of healing when a periosteal reaction is often an obvious feature (see also Fig. 9.2a).

2. Metaphyseal fractures frequently appear as small chips from the metaphyses of the long bones. They most probably result from twisting and pulling the limbs of a struggling baby.

3. Metaphyseal sclerosis probably due to repeated injury and repair.

4. Epiphyseal separation is frequently associated with a metaphyseal fracture.

5. Periosteal reactions. Haemorrhage under the periosteum occurs easily in children. The elevated periosteum lays down new bone which may be so extensive that it envelopes the shaft.

13

The Skull and Sinuses

Standard views of the skull are shown in Fig. 13.1

The vault

The bones of the normal vault have an inner and outer table of compact bone with spongy bone (diploë) between them. Blood vessels cause impressions on the bones of the skull vault resulting in linear or star-shaped translucencies (Fig. 13.2). These should not be confused with fractures (see p. 323). If the vascular markings appear large on the lateral view, check on the frontal view that they are equally prominent on both sides; in which case they are likely to be normal.

The sutures between the individual bones of the vault remain visible even when fused. A normal but inconstant suture—a metopic suture—is sometimes seen dividing the frontal bone; it must not be mistaken for a fracture.

Small lucencies are often seen in the inner table near the vertex due to normal arachnoid granulations. They

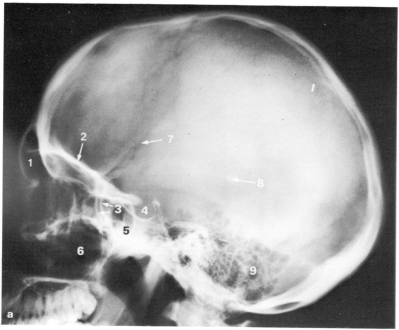

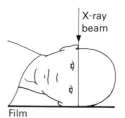

Fig. 13.1   Standard view of the skull (diagrams show the position in which the films are taken). (a) Lateral view: (1) frontal sinus; (2) roof of right and left orbits superimposed; (3) anterior border of middle cranial fossa; (4) pituitary fossa; (5) sphenoid sinus; (6) maxillary antrum; (7) vascular groove; (8) pineal; (9) mastoid air cells.

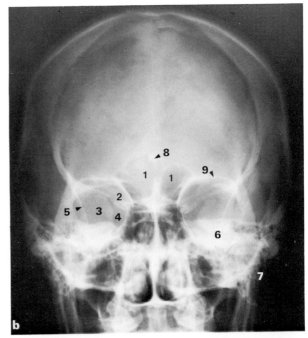

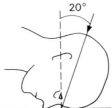

Fig. 13.1(b). Posteroanterior view: (1) frontal sinuses; (2) lesser wing of sphenoid; (3) greater wing of sphenoid; (4) superior orbital fissure; (5) wall of middle cranial fossa; (6) petrous bone; (7) mastoid air cells; (8) pineal; (9) superior orbital margin.

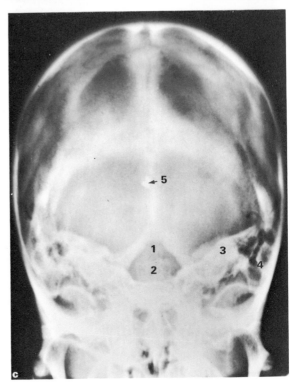

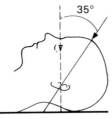

Fig. 13.1(c) Towne's view: (1) foramen magnum; (2) dorsum sellae of pituitary fossa; (3) petrous bone; (4) mastoid air cells; (5) pineal.

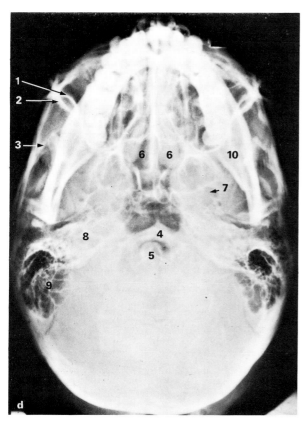

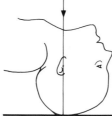

Fig. 13.1(d) Base or submentovertical view: (1) lateral border of maxillary sinus; (2) lateral border of orbit; (3) anterior border of middle cranial fossa; (4) anterior arch of atlas; (5) odontoid peg; (6) sphenoid sinus; (7) foramen ovale; (8) petrous bone; (9) mastoid air cells; (10) mandible.

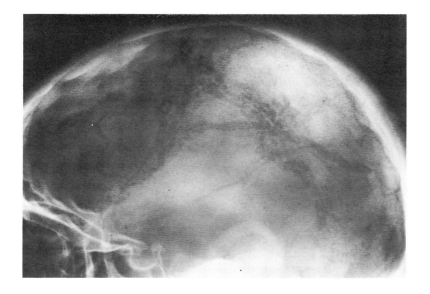

Fig. 13.2 Vascular markings. In this normal skull the vascular markings are very prominent. Note how they form a star-shaped translucency in the parietal region.

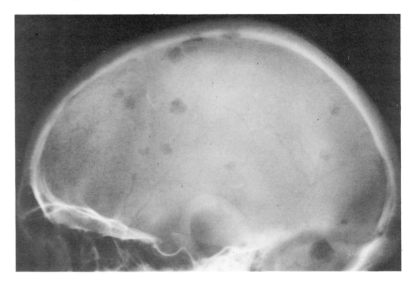

Fig. 13.3 Myeloma. Several well-defined lytic lesions of various sizes are seen in all areas of the skull vault.

may be difficult to distinguish from small lytic lesions but when lucencies are seen near the base or when large lytic areas are present they usually indicate metastases or myeloma (Fig. 13.3). Large areas of destruction are seen in histiocytosis X giving the appearance known as a

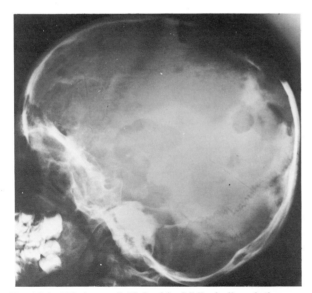

Fig. 13.4 Histiocytosis X. In this child's skull vault there are large irregularly shaped lytic areas. This appearance is known as a geographical skull.

geographical skull (Fig. 13.4), in metastases and in osteoporosis circumscripta which is a form of Paget's disease.

A common cause of sclerosis without significance is hyperostosis frontalis interna where there is irregular thickening of the inner table of the skull in the frontal regions sparing the midline (Fig. 13.5).

Localised sclerosis may be caused by a meningioma (Fig. 13.6), metastases and fibrous dysplasia but each of these lesions may also cause a localised area of mixed sclerosis and lysis. A meningioma may in addition show calcification and widened vascular channels together with changes of raised intracranial pressure in the pituitary fossa. An osteoma causes a well-defined density which may arise from the outer table of the skull. They occur particularly in the frontal sinuses.

The pineal

The position of a calcified pineal gland is the only method of identifying the midline on plain films. The incidence of calcification increases with age. It is rarely calcified in children but is seen in about 60% of adult skulls. A density in the occipital bone on the Towne's view can easily be mistaken for the pineal, so never diagnose a calcified pineal unless it can be seen on the lateral view.

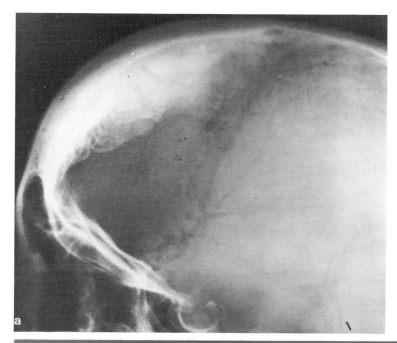

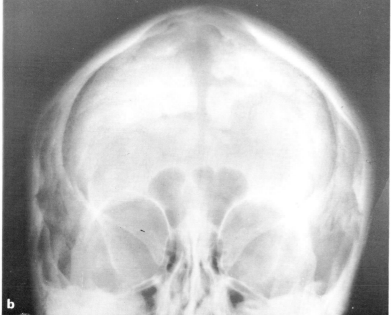

Fig. 13.5 Hyperostosis frontalis interna. (a) Lateral view showing sclerosis in the frontal region near the vertex; (b) the frontal view shows the characteristic sparing of the midline.

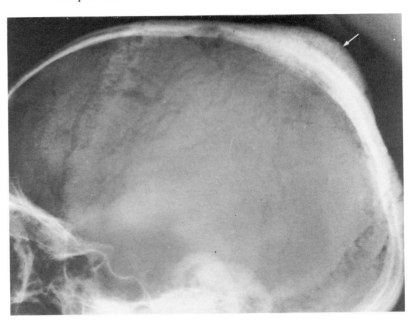

Fig. 13.6 Meningioma. A localised hyperostosis (arrow) is seen on the parietal bone.

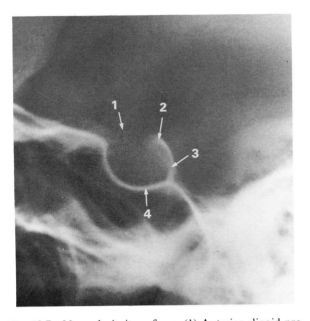

Fig. 13.7 Normal pituitary fossa. (1) Anterior clinoid process; (2) posterior clinoid process; (3) dorsum sellae; (4) floor. The white line forming the floor and the dorsum sellae is known as the lamina dura.

The pituitary fossa

The anatomy is shown in Fig. 13.7. Look at the pituitary fossa for two important features—for enlargement from a pituitary tumour and for erosion of the lamina dura of the dorsum sellae by raised intracranial pressure.

Enlargement of the pituitary fossa

Tumours of the pituitary are the commonest cause of enlargement or 'ballooning' of the pituitary fossa (Fig. 13.8). With chromophobe adenomas the rest of the skull will be normal but with an eosinophil adenoma there may be changes of acromegaly (p. 262). Basophil adenomas with Cushing's disease are usually too small to enlarge the pituitary fossa.

An early sign of a pituitary tumour is asymmetrical enlargement of the fossa which may then show a double line forming the floor on the lateral view (Fig. 13.8b), and a sloping floor can sometimes be identified on the frontal view.

Suprasellar extension of a pituitary tumour, which may press on the optic chiasm, cannot be detected on

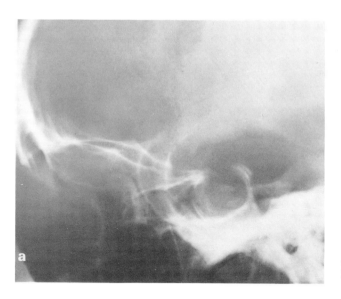

Fig. 13.8 Pituitary tumour. (a) Generalised enlargement or ballooning of the pituitary fossa.

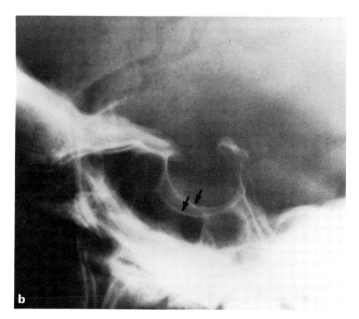

Fig. 13.8(b) Asymmetrical enlargement of the pituitary fossa giving a sloping floor which appears as a double line on the lateral view (arrows).

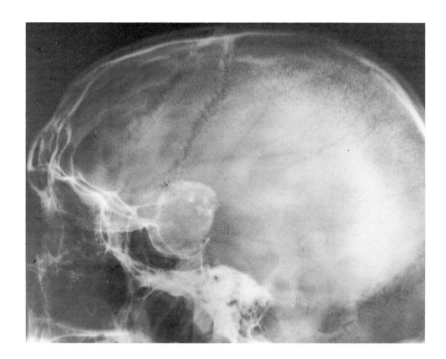

Fig. 13.9 Craniopharyngioma. Calcification is seen in the tumour which has enlarged and flattened the pituitary fossa.

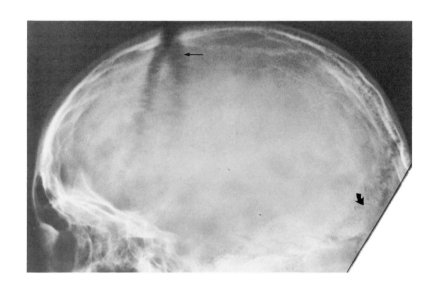

Fig. 13.10 Raised intracranial pressure in a child. The sutures become widened affecting the coronal suture first (straight arrow). Compare this widened suture with the normal lambdoid suture (curved arrow).

plain films. In order to assess this an encephalogram or computerised tomography (CT) scan may be needed.

A craniopharyngioma is the common suprasellar tumour in children. It may flatten or enlarge the pituitary fossa causing erosion of the dorsum sellae and it commonly calcifies (Fig. 13.9).

Raised intracranial pressure

The changes due to raised intracranial pressure depend upon the age of the patient. In children the sutures are still able to separate but in older children and adults the sutures are fused and the signs of raised intracranial pressure are then seen in the pituitary fossa.

In children under the age of 8 years the diagnosis of raised intracranial pressure depends upon recognising the widening of the sutures (Fig. 13.10). There may also be increased convolutional markings in the skull vault giving the 'copper-beaten appearance' due to pressure from growth of brain, but this is a difficult sign to evaluate because normal children sometimes show similar markings.

In adults the cardinal sign is erosion of the lamina dura of the dorsum sellae, i.e. it becomes ill defined (Fig. 13.11). This may be very hard to evaluate particularly in older people. The pituitary fossa does not enlarge. Raised intracranial pressure has to be present for at least 6 weeks before radiological changes are visible and papilloedema is usually present.

Intracranial calcification

Most intracranial calcification is normal or of no significance to the patient but it is important to recognise those few occasions where it is pathological.

Innocent calcification can usually be readily identified from its site (Fig. 13.12).

The common forms are calcification of the following:

1. Pineal.
2. Choroid plexus within the lateral ventricles. The calcification may be asymmetrical in position and only one choroid plexus may calcify.
3. Interclinoid and petroclinoid ligaments which appear as linear streaks around the pituitary fossa.
4. Falx resulting in midline calcification often seen best on the frontal projections.
5. Atheromatous calcification in arteries at the base of the brain may appear as curvilinear calcification projected over the pituitary fossa.

Pathological calcification is uncommon. At least two views, and often more, are required to determine the shape and anatomical site of any abnormal calcification, and this information helps to limit the differential diagnosis.

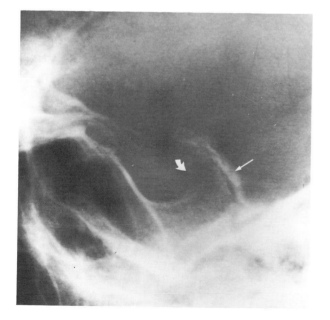

Fig. 13.11 Raised intracranial pressure in an adult. The dorsum sellae of the pituitary fossa (curved arrow) has lost its sharp outline and appears indistinct. The petroclinoid ligament is calcified (arrow).

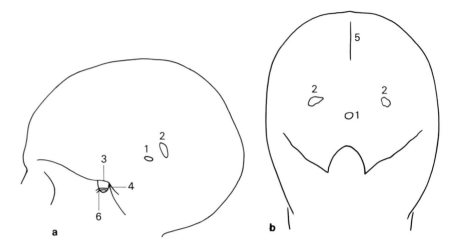

Fig. 13.12 Innocent calcification. (a) Lateral view; (b) Towne's view: (1) pineal; (2) choroid plexus; (3) interclinoid ligament; (4) petroclinoid ligament; (5) falx; (6) carotid artery.

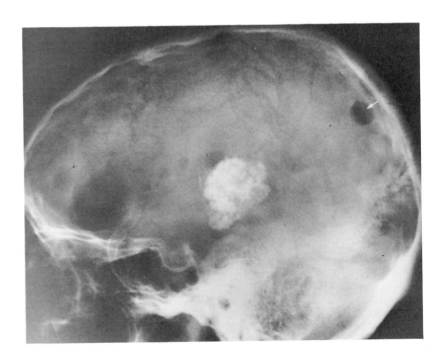

Fig. 13.13 Calcification in a glioma. Note the burr hole (arrow).

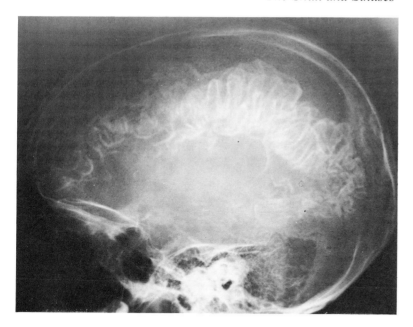

Fig. 13.14 Sturge–Weber syndrome. The calcification, which has a characteristic appearance, is in the superficial layers of the cerebral cortex.

Localised pathological calcification

Such calcification is seen with:

1. Primary tumours.
  (a) Gliomas particularly those that are slow growing (Fig. 13.13).
  (b) Meningioma.
  (c) Craniopharyngioma (Fig. 13.9).
  Calcification is not seen in metastases.
2. Vascular abnormalities—arteriovenous malformation and Sturge–Weber syndrome cause curvilinear calcification (Fig. 13.14).
3. Abscesses.

Multiple intracranial calcification

This occurs in toxoplasmosis when acquired *in utero*, and in tuberose sclerosis.

Paget's disease

Being the commonest cause of generalised sclerosis, Paget's disease may take the form of multiple patchy areas of increased density. A striking feature is the thickening of the skull vault. Recognition of this thickening is important in distinguishing Paget's disease from sclerotic metastases. Basilar invagination may also be seen. In basilar invagination the odontoid peg encroaches on the foramen magnum and may press on the brain stem giving rise to clinical signs and symptoms (Fig. 13.15).

Craniostenosis

In craniostenosis there is a premature fusion of one or more sutures. The actual shape of the skull depends on which sutures are fused, as there is overgrowth of the other sutures to allow for growth of the brain. Craniostenosis is associated with increased convolutional markings of the skull vault due to pressure from the growing brain (Fig. 3.16).

Head injury

Fractures appear as linear translucencies. The only exception is a depressed fracture which may appear as a curvilinear density due to the overlapping fragments (Fig. 13.17). A tangential view will show a fragment depressed inwards (Fig. 13.17b).

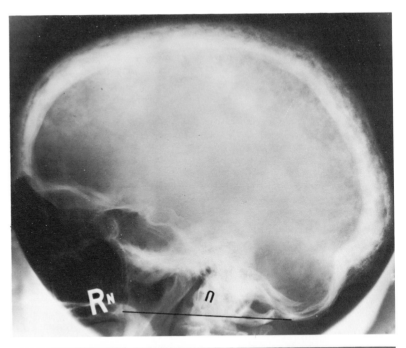

Fig. 13.15 Paget's disease. There is generalised sclerosis with marked thickening of the bone. Basilar invagination is present which can be detected by drawing a line from the back of the hard palate to lower most part of the occiput. The tip of the odontoid peg should not be more than 6 mm. above this line. In this case it is 18 mm.

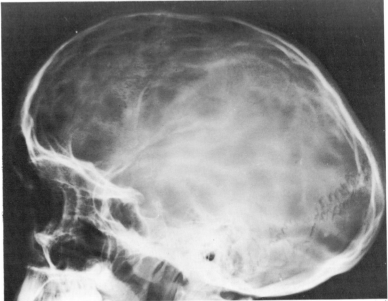

Fig. 13.16 Craniostenosis. The skull of this child is elongated from front to back and there are increased convolutional markings. The frontal view showed fusion of the sagittal suture.

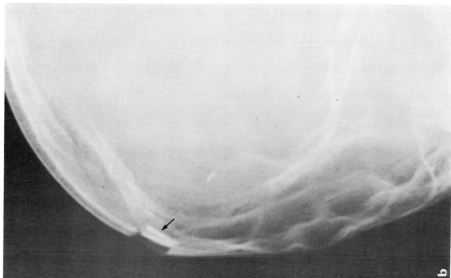

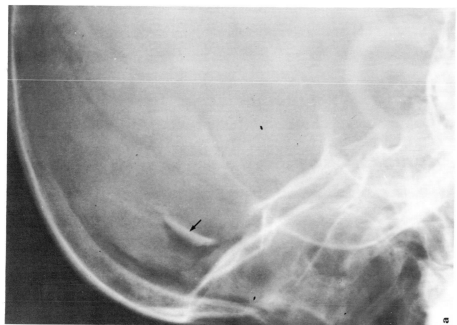

Fig. 13.17 Depressed fracture. (a) The fracture is seen as a curvilinear density (arrow) due to the overlapping fragments; (b) in a tangential view (another patient) the fragment can be seen to be depressed inwards (arrow).

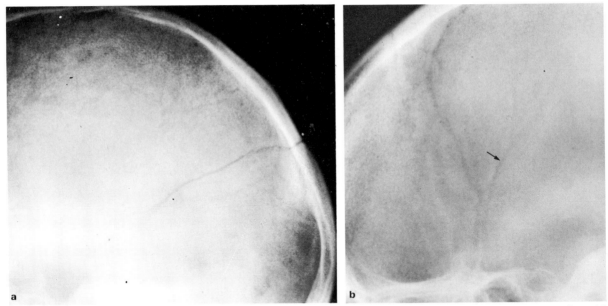

Fig. 13.18 Linear markings in the skull vault. (a) Fracture line with straight edges (arrow); (b) arterial groove for middle meningeal artery (arrow)—line with straight edges occupying a recognised site.

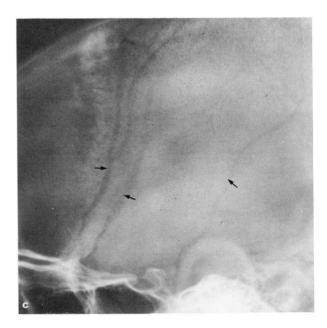

Fig. 13.18(c) Venous channels: wider, more undulating grooves (arrows)—the more posterior groove ends in a venous star.

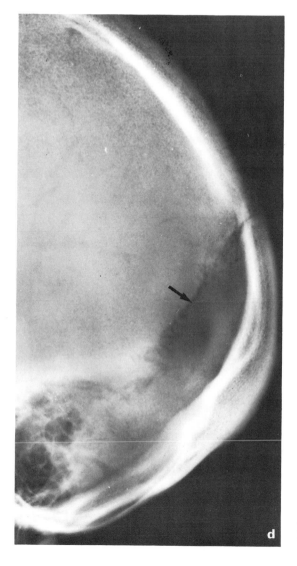

Fig. 13.18(d) Suture: these show regular interdigitations (arrow).

Sometimes it is difficult to distinguish a fracture from vascular markings and sutures. The following analysis may be helpful (Fig. 13.18):

1. Fractures may appear more translucent than vascular markings since they traverse the full thickness of the bone.

2. Fractures have straight or jagged edges. It is usually possible to see how these edges fit together.

3. Fractures may branch abruptly.

4. Venous channels have undulating irregular edges which cannot be fitted together.

5. Arterial grooves have parallel sides and are, therefore, more easily mistaken for fractures, but they occur in known anatomical sites.

6. Sutures also occur in recognised anatomical positions and show definite regular interdigitations. Widening of one or more sutures has the same significance as a fracture.

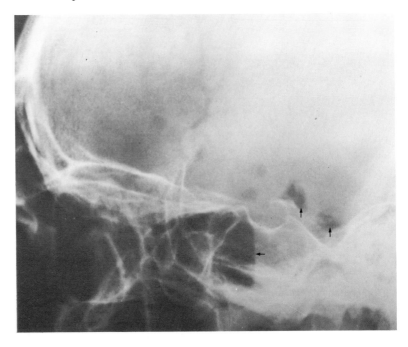

Fig. 13.19 Head injury. Horizontal ray lateral film showing a fluid level in the sphenoid sinus (horizontal arrow) and air in the subarachnoid space (arrows). Although a fracture cannot be identified these findings indicate that there has been severe skull trauma with a fracture of the skull base and tear of the dura.

The value of plain films following head injury

Although a fracture is a valuable sign of a head injury, it often bears little correlation to the underlying brain damage, and severe brain damage and subdural haematomas can occur in patients with normal skull films. A subdural haematoma may occur following only slight trauma, particularly in the elderly, and the onset of symptoms may be delayed.

Important plain film signs to be detected which affect management are:

1. Shift of the pineal from the midline is an important sign as it often indicates an extradural or subdural haemorrhage. Swelling of one hemisphere due to contusion of the brain can also cause pineal displacement.

2. Depressed fracture—since surgery may be necessary to elevate the depressed fragment.

3. A fracture crossing the groove of the middle meningeal artery—since an arterial laceration may be present resulting in an extradural haematoma.

4. Fluid levels in the sinuses or air in the subarachnoid space or ventricles (Fig. 13.19) indicate a fracture with a tear of the dura. Antibiotic treatment is often advocated with this type of injury to prevent meningitis. As with the detection of all fluid levels, it is necessary to take the lateral film with a horizontal ray.

NEURORADIOLOGICAL INVESTIGATIONS

In many neurological disorders the plain films are normal or the changes insufficient for a precise diagnosis to be made. Frequently it is necessary to know the exact anatomical extent of an abnormality, therefore complex radiological investigations may be necessary.

These can be divided into:

1. Radioisotope scanning.
2. Computerised axial tomography.
3. Contrast studies—arteriography and pneumography—which are uncomfortable procedures often performed under a general anaesthetic.

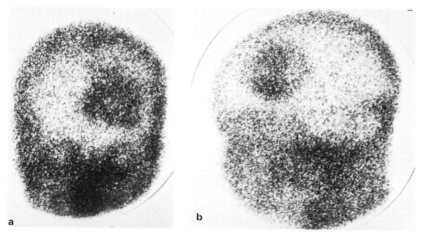

Fig. 13.20 ^{99m}Tc Brain scan.
(a) Anterior scan; (b) left lateral scan.;
There is a large area of increased
uptake in the left frontal region which
was subsequently proved to be a
glioma.

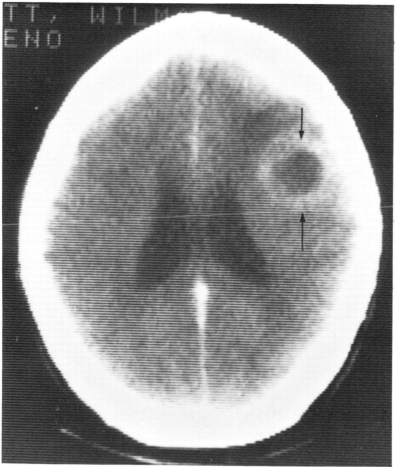

Fig. 13.21 Computerised
tomography: glioma. CSF can be
clearly seen in the ventricular system.
This film was taken after an
intravenous injection of contrast. A
round lesion with a contrast-enhanced
rim and a low density centre is seen in
the right frontal lobe (arrows). There is
minimal distortion of the adjacent
lateral ventricles.

Radionuclide brain scanning

Radioactive technetium (^{99m}Tc) is injected intravenously. This is excluded from normal brain tissue by
the blood brain barrier but when there is intracranial
pathology this barrier is broken down allowing the isotope to accumulate in the lesion which then appears on
the scan as an area of increased uptake. Blood flow can
also be assessed by obtaining rapid images shortly after
injection of the isotope.

Because of its simple non-invasive nature a brain scan
is often carried out as a screening test on the clinical
suspicion of intracranial disease. Tumours, abscesses,
cerebral infarction, intra- and extracerebral haemorrhage all give positive scans (Fig. 13.20). In cerebral
infarction and haemorrhage the scan becomes positive
in the first week after the incident, remains positive for
several weeks and then returns to normal so that serial
brain scanning can help to differentiate tumours from
vascular lesions.

Computerised tomography

The advent of computerised tomography has revolu-
tionised neuroradiology as it often enables a diagnosis
to be made with little or no discomfort to the patient. As
a result invasive examinations such as angiography and
particularly pneumography are now performed much
less frequently. Arteriography still retains an important
role in the evaluation of vascular lesions.

Computerised tomography has become established as
a primary diagnostic procedure. The technique has been
proven in the examination of all forms of intracranial
disease (Fig. 13.21) and it has an invaluable role to play
in the management of head injuries.

Contrast studies

Arteriography

Contrast is injected into the carotid or vertebral arteries
either following direct puncture in the neck or by passing a catheter into them from the groin via the aorta.

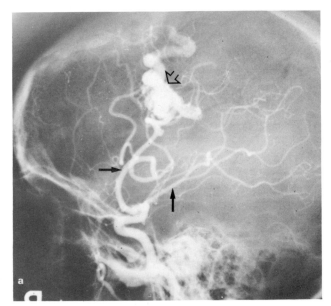

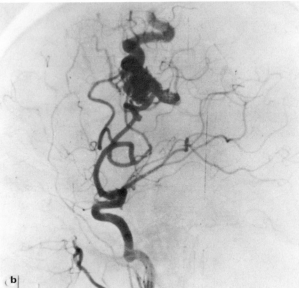

Fig. 13.22 Arteriovenous malformation. (a) Carotid angiogram showing a collection of large abnormal vessels (large
arrow) supplied by the middle cerebral artery (horizontal
arrow). On this injection the posterior cerebral artery (vertical
arrow) but not the anterior cerebral artery have filled; (b)
subtraction: with this technique (p. 5) the shadowing due to the
bones has almost been eliminated so that the contrast-filled
vessels stand out more clearly.

The commoner indications are to investigate the following:

1. Space-occupying lesions—tumours, abscesses, extradural and subdural haematomas.
2. Vascular lesions—intracranial aneurysms, arteriovenous malformations (Fig. 13.22) and atheroma of the extra- and intracranial vessels.

Pneumography

There are two main ways of performing pneumography; namely, pneumoencephalogram (air encephalogram) and ventriculogram:

1. Pneumoencephalogram. Air is injected into the subarachnoid space by means of a lumbar puncture with the patient sitting erect. The air rises and enters the ventricles through the foramina in the fourth ventricle.
2. Ventriculogram. Air is injected directly into the lateral ventricles through a burr hole in the skull vault.

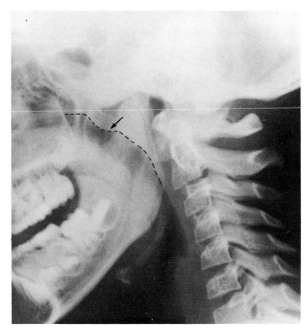

Fig. 13.23 Normal lateral view of postnasal space in a child. The posterior boundary of nasopharyngeal air passage has been marked with dotted lines and there is an impression into it (arrow) caused by the adenoids.

Many of the conditions previously investigated with pneumography are now carried out by computerised axial tomography. Pneumography is still sometimes used in the investigation of hydrocephalus and in pituitary tumours in order to evaluate any suprasellar extension, which will cause displacement of the anterior end of the third ventricle.

SINUSES AND NASOPHARYNX

The normal sinuses are translucent because they contain air. The normal mucosa lining the sinus is too thin to be visualised so there is an abrupt contrast between the air in the sinus and the bony wall giving the sinus a sharp outline. It so happens that on an occipitomental view the maxillary antra are of approximately the same lucency as the orbits. The frontal sinuses are often asymmetrical and part or all of them may be absent. The adenoids are normally seen in children on a lateral film as a bulge projecting into the nasopharyngeal air passage (Fig. 13.23).

Mucosal thickening

Thickened mucosa can be recognised provided there is some air in the sinus, by noting the soft tissue density between the air in the sinus and the bony wall. This is easiest along the medial and inferior walls of the antra. The mucosal thickening may be smooth in outline or it may be polypoid (Fig. 13.24). The polyps may be sufficiently large to extend into the nasopharynx.

Allergy and infection both cause mucosal thickening and it is impossible to say, radiologically, which condition is responsible.

Fluid level in a sinus

With the patient erect, a fluid level appears as a horizontal line across the sinus which remains horizontal even when the patient's head is tilted (Fig. 13.25). Fluid levels are seen with infection in the sinus and also with trauma, when a fracture allows blood or CSF to collect in the sinus.

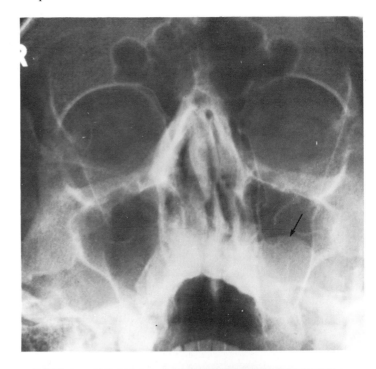

Fig. 13.24 Retention cyst in the floor of the left antrum (arrow). The right antrum and frontal sinuses are normal and have a sharp outline. Note the right antrum has the same translucency as the orbit.

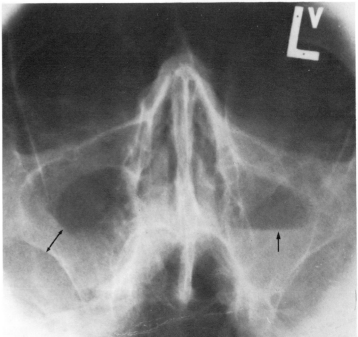

Fig. 13.25 Mucosal thickening and a fluid level. In the right antrum thickening of the mucosa (arrows) results in the sinus no longer having a thin outline. The horizontal line in the left antrum on this erect film (arrow) indicates a fluid level which remains horizontal even when the patient's head is tilted.

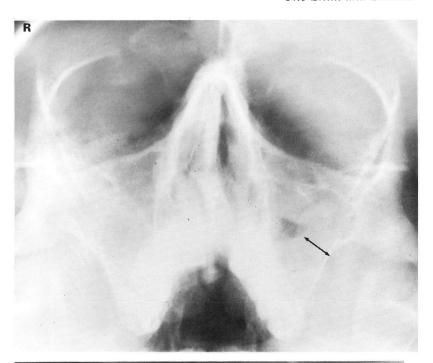

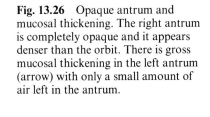

Fig. 13.26 Opaque antrum and mucosal thickening. The right antrum is completely opaque and it appears denser than the orbit. There is gross mucosal thickening in the left antrum (arrow) with only a small amount of air left in the antrum.

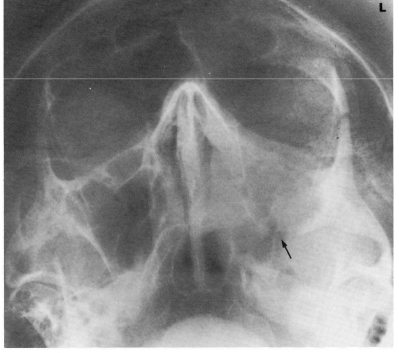

Fig. 13.27 Carcinoma of the antrum. The left antrum is opaque and there is extensive destruction of its walls (arrow). Compare with the normal right antrum.

The opaque sinus

The sinus becomes opaque when all the air is replaced and it then appears as dense or denser than the adjacent orbit (Fig. 13.26).

The causes of an opaque sinus are:

1. Infection or allergy. When the air in the sinus is replaced by fluid, or a grossly thickened mucosa, or a combination of the two.

2. Mucocele. Mucoceles are usually due to chronic infection with obstruction to drainage and accumulation of secretions. It occurs most frequently in the frontal sinus. The sinus becomes expanded and may erode the roof of the orbit and cause exopthalmus.

3. Carcinoma of the sinus or nasal cavity. In all opaque sinuses, particularly the antra, special attention should be paid to the bony margins, because if these are destroyed the diagnosis of carcinoma becomes almost certain (Fig. 13.27).

Index

(Page numbers in bold type refer to pages on which illustrations or tables appear)